Air Pollution Episodes

AIR POLLUTION REVIEWS

ISSN: 0219-9785

Series Editor: Robert L. Maynard
(Department of Health, Skipton House, London, UK)

Air Pollution Episodes

Editor

Peter Brimblecombe
City University of Hong Kong, Hong Kong

World Scientific

NEW JERSEY · LONDON · SINGAPORE · BEIJING · SHANGHAI · HONG KONG · TAIPEI · CHENNAI · TOKYO

Published by

World Scientific Publishing Europe Ltd.

57 Shelton Street, Covent Garden, London WC2H 9HE

Head office: 5 Toh Tuck Link, Singapore 596224

USA office: 27 Warren Street, Suite 401-402, Hackensack, NJ 07601

Library of Congress Cataloging-in-Publication Data
Names: Brimblecombe, Peter, 1949–   editor.
Title: Air pollution episodes / edited by Peter Brimblecombe
    (City University of Hong Kong, Hong Kong).
Description: [Hackensack] New Jersey : World Scientific, [2017] |
    Series: Air pollution reviews ; volume 6 | Includes bibliographical references.
Identifiers: LCCN 2017001102 | ISBN 9781786343406 (hc : alk. paper)
Subjects: LCSH: Air--Pollution--Case studies.
Classification: LCC TD883 .A478175 2017 | DDC 363.739/209--dc23
LC record available at https://lccn.loc.gov/2017001102

British Library Cataloguing-in-Publication Data
A catalogue record for this book is available from the British Library.

Desk Editors: Ram Mohan K/Mary Simpson

Typeset by Stallion Press
Email: enquiries@stallionpress.com

Printed in Singapore

About the Editor

Peter Brimblecombe is an atmospheric chemist whose research focuses on the effect of air pollutants on cultural materials and human health.

He was born in Australia, but went to university in Auckland, New Zealand where his PhD concerned atmospheric chemistry of sulphur dioxide, but more recent research includes the process of damage to cultural materials by air pollutants and climate. His current interest in air pollution relates to human exposure in the urban environment, especially Hong Kong. He is currently Associate Dean and Chair Professor in the School of Energy and Environment at City University of Hong Kong and an Emeritus Professor to the School of Environmental Sciences of the University of East Anglia.

Contents

Introduction

Peter Brimblecombe

*School of Energy and Environment, City University
of Hong Kong, Hong Kong*

Air pollution episodes have transformed our attitudes towards the environment. They have altered our perceptions, catalysed legislation and improved scientific understanding. It is therefore a little surprising that there are so few books about air pollution episodes. There are accounts of individual episodes, such as events that led to deaths through industrial accidents, for example Seveso, *The Poison that Fell from the Sky* (1977) by John Grant Fuller, or *Five Past Midnight in Bhopal* (2002) by Dominique Lapierre. Classic smog events such as those in Donora and London are found in *When Smoke Ran Like Water: Tales of Environmental Deception and the Battle Against Pollution* by Devra Davis (2002), or for London a whole suite of books (Brimblecombe, 2017a). The smoke released from burning oil wells at the end of the gulf war (Bell, 2017) are treated in *The Kuwaiti Oil Fires* (2005) by Kristine Hirschmann and *Kuwaiti Oil Fires: Regional Environmental Perspectives* (1995) by Tahir Husain and Husain, Tahir.

There are many books about environmental disasters, but these of course, cover issues that run well beyond air pollution. In many cases, they are written from the perspective of a journalist or an activist, so tend not to be focused on scholarly analysis. While it would be wonderful if this book was used in advocacy, it is not its primary intent. Rather, it wishes to show how air pollution episodes may be viewed from a wide range of perspectives.

Episodes, especially the tragic ones, don't merely lead to descriptive writing or technical writing in the scientific literature; they also have strong cultural impacts. The archetypical fogs of London and Los Angeles seem to have spawned fictional detectives such as Sherlock Holmes and Philip Marlowe (Brimblecombe, 1990). The Donora episode has had a range of literary outcomes, most recently the novel *After the Fog* (2012) by Kathleen Shoop, while the 1952 smog in London forms the backdrop of the children's books by Philippa Pearce, e.g. *A Dog So Small* (1962), and is a centrepiece to the alternative history in *Dominion* (2012) by C. J. Sansom. Naturally, the attacks on the World Trade Center attracted much attention, but it was easy for terrorism and heroism to dominate fictional works, such that Bob Minzesheimer (2007) of *USA TODAY* remained unimpressed by the emerging literature.

Even the recent events in Beijing (Ning *et al.*, 2017) have produced novels, perhaps the most direct outcome being the trilogy which includes *Smog is Coming* (2014) and *The Pains of Smog* (2015). The trilogy is written by Li Yuanchuan, a senior environmental protection official from the city of Langfang (SCMP, 2015) just fifty kilometres from downtown Beijing. Sometimes it is the lack of pollution or blue-sky days in Beijing that receive cultural prominence. Government actions to reduce industrial emission and vehicle numbers, during important events, such as the Asia-Pacific Economic Cooperation (APEC) meetings in the capital 10–12 November 2014 meant clean air. The colour of the skies over the capital was so intense at the time that it was referred to as *APEC Blue,* a term which has come to mean something wonderful, yet fleeting. Natural catastrophes, similarly inspire writers; as an example, the outgassing of Lake Nyos is the event that drives Vestal McIntyre's novel *Lake Overturn* (2009), a story about small town in Idaho. The ash and disruption to air traffic over the North Atlantic in 2010 were central to the 2013 French comedy film directed by Alexandre Coffre, *Eyjafjallajökull.*

This book is thus also concerned with air pollution episodes that derive from natural sources of air pollution, such as those that are produced by volcanoes (Baxter *et al.*, 2017; Kantha, 2017; Langmann, 2017). Forest fires in South East Asia created widespread pollution in historical times, and there were important South East Asian haze events in 1876–1878, 1972–1973 and 1982–1983. The more recent ones have become especially well known, with an extensive one in 1997/1998 when Indonesian forest fires spread smoke across the region, exacerbated by El Niño and possibly the economic crisis facing the country (Latif *et al.*, 2017). Subsequently, there were important episodes in 2005 (predominantly Malaysian), 2006 and then 2015. Elsewhere, a major forest fire in central Quebec occurred in the summer of 2002 and spread particulate material across the eastern United States (Begum, 2005). There were episodes (In and Kim, 2010) caused by the less well-known fires along the Russian border with Mongolia and China in May 2003. The Asia-Pacific Rim also experiences widespread incursions of dust from the deserts in the west, with the extensive East Asian dust storm of 2010 treated in this book (Lai *et al.*, 2017). Outbreaks of Saharan dust have been known since historical times, with a major Europe-wide episode in 1903 (Mill and Lempfert, 1904) and one as recently as 2014, although most of the particle load tended to be anthropogenic (Vieno *et al.*, 2016).

Some air pollution episodes are related to anthropogenic activities and are spread over large geographical areas, maintained by unfavourable synoptic meteorology. Among these, some of the best known were associated with events across the Eastern parts of the United States 27 November–5 December 1962, and later Thanksgiving of 1966 (Holzworth, 1969). In the East of England between 22 June and 12 July 1976, bright sunshine and a heatwave led to large areas of the country experiencing elevated ozone concentrations (Apling *et al.*, 1977). More recently, there was the European heatwave of 2003, where there were many deaths from the high temperatures, but elevated ozone episode during the period 4–13 August may have contributed between 423 and 769 excess deaths in England and Wales (Stedman, 2004). In December 2013, Eastern China experienced an episode that spread across many provinces.

No book can cover everything, and there are gaps both in the episodes covered and source types. In particular, none of the major releases of radioactive pollutants are dealt with in the chapters. Here, a number of key events could have been covered, including, for example: Windscale of 1957 (Arnold, 1992; Garland and Wakeford, 2007; Hernan, 2010), Three Mile Island of 1979 (Walker, 2004), Chernobyl of 1986 (Medvedev and Sakharov, 1991; River, 2014) and Fukushima in 2011 (Eisler, 2013; Lochbaum *et al.*, 2014). Fortunately, each of these has already been well treated in books and articles, so the interested reader can turn to them. These works range from historical and technical analysis all the way to works that argue against nuclear power. The approach can lead to very different interpretations of the data, especially in terms of the long-term impacts and the likely number of deaths. To take Windscale as an example, the technical account of Arnold (1992) concludes that "it seems unlikely that any health effects will ever be observed. . . . It was extremely fortunate that it did this with so little human damage, or perhaps none". Compare this with the alternate vision taken from the perspective of legal counsel: "It has been concluded that some 250 cases of thyroid cancer and between 120 and 300 deaths from other cancers occurred in the United Kingdom over a 40–50-year period as a result of Windscale" (Hernan, 2010). We see how authors can interpret available material in different ways, but it is interesting to observe the book titles: one calls Windscale an accident, while the other treats it as a disaster.

It is important not to neglect the more distant past, so there is a chapter on ancient air pollution episodes (Brimblecombe, 2017b), but the present book neglects some important historical studies of the South Wales copper industry (Newell and Watts, 2006), Selby in California (Holmes *et al.*, 1915), Trail, Canada (Bratspies and Miller, 2006) and smelters in Germany (Schramm, 1990). These accounts range from legal treatments to government assessments.

Industrial accidents are a frequent source of death and injury, and air pollution episodes that accompany many of these add to the mortality and morbidity. Important episodes include those at: Poza Rica in 1950 (McCabe and Clayton, 1952), Seveso in 1976 (Ballarin-Denti *et al.*, 1999) and Bhopal in 1984 (Mishra *et al.*, 2017). This book deals only with Bhopal from a medical and cancer perspective, and not some of the wider issues. These episodes have had wide-ranging effects on the regulatory approaches to industrial releases; in particular the *Seveso Directives* (82/501/EC, 96/82/EC and 2012/18/EU), which have altered the way that we mange safety and major-accident hazards involving dangerous substances.

Smog episodes in cities and industrial areas have also influenced the development of urban air quality regulation. Perhaps the earliest episode of the 20th century occurred as the St Louis Smog of 1939 (Tucker, 1941). This had a strong influence on Raymond Tucker, once the St Louis City Smoke Commissioner and later its mayor. He was notable as a key adviser on pollution problems in post-war Los Angeles (Brimblecombe, 2013). The Meuse Valley episode (1930) changed medical understanding and promoted the need to form working bodies to examine issues (Zimmer and Nemery, 2017). Donora (1948) remains a pivotal point in the development of clean air legislation in the US. However, the episode remains dogged with a sense that a conspiracy, cover-up and regulatory inaction have dominated (Brimblecombe, 2017c) and where fluoride has been often discussed as a controversial contaminant. Similarly, the history of the London Smog and the World Trade Center (Lippmann, 2017) still raise elements of controversy. However, these tragic episodes have typically catalysed change. The Great Smog of London (1952) triggered the Clean Air Act of 1956, and 10 years later the New York City smog of 1966 (Lioy and Georgopoulos, 2011) led to greater understanding of health impacts. Most recently the winter smog of January 2013, in Beijing led to a situation where the normal reticence among both Chinese politicians and the public seemed to be much reduced. It further raised awareness of the need for urgent action and a greater provision of information (Ning *et al.*, 2017).

The chapter on the eruption of Eyjafjallajökull and its effects on air traffic across the North Atlantic (Langmann, 2017) reminds us that air pollution episodes affect not only human health. Additionally, Bonazza (2017) treats the Kuwait oil fires not in terms of health effects, but rather the potential damage that the soot from fires may have imposed on the region's cultural heritage.

Air pollution episodes have had a huge influence on the science, regulation and perception of air pollution. Such episodes are often disasters, but they have coloured our perceptions, transformed our understanding and forged new legislation.

References

Apling, A. J., Sullivan, E. J., Williams, M. L., Ball, D. J., Bernard, R. E., Derwent, R. G., Eggleton, A. E. J., Hampton, L. and Waller, R. E. (1977). Ozone concentrations in South-East England during the summer of 1976, *Nature*, **269**, 569–573.

Arnold, L. (1992). *Windscale, 1957: Anatomy of a Nuclear Accident*, (New York: St. Martin's Press), p. 1992.

Ballarin-Denti, A., Bertazzi, P. A., Facchetti, S., Fanelli, R. and Mocarelli, P. (1999). *Chemistry, Man and Environment. The Seveso Accident 20 Years On: Monitoring, Epidemiology and Remediation* (Elsevier Science Publishers, Amsterdam).

Baxter, P. Bernstein, R. and Buist, S. (2017). Mount St Helens (1980): A severe air pollution episode from volcanic ash, in Brimblecombe, P. (ed.), *Air Pollution Episodes* (World Scientific, Singapore).

Begum, B. A., Kim, E., Jeong, C.-H., Lee, D.-W. and Hopke, P. K. (2005). Evaluation of the potential source contribution function using the 2002 Quebec forest fire episode, *Atmospheric Environment*, **39**(20), 3719–3724.

Bell, M. (2017). Kuwait oil fires (1991): A deliberate environmental disaster during wartime, in Brimblecombe, P. (ed.), *Air Pollution Episodes* (World Scientific, Singapore).

Bonazza, A. (2017). Kuwait oil fires (1991): Searching for long range deposits on heritage, in Brimblecombe, P. (ed.), *Air Pollution Episodes* (World Scientific, Singapore).

Bratspies, R. M. and Miller, R. A. (2006). *Transboundary Harm in International Law: Lessons from the Trail Smelter Arbitration* (Cambridge University Press, Cambridge).

Brimblecombe, P. (1990). Writing on smoke, in Bradby, H. (ed.), *Dirty Words* (Earthscan Publications, London), pp. 93–114.

Brimblecombe, P. (2013). Deciphering the chemistry of Los Angeles smog, 1945–95, in Fleming, J. (ed.), *Toxic Airs* (University of Pittsburgh Press, Pittsburgh, USA), pp. 95–108.

Brimblecombe, P. (2017a). London 1952: An enduring legacy, in Brimblecombe, P. (ed.), *Air Pollution Episodes* (World Scientific, Singapore).

Brimblecombe, P. (2017b). Early episodes, in Brimblecombe, P. (ed.), *Air Pollution Episodes* (World Scientific, Singapore).

Brimblecombe, P. (2017c). Donora (1948): Controversial contaminants, in Brimblecombe, P. (ed.), *Air Pollution Episodes* (World Scientific, Singapore).

Eisler, R. (2013). *The Fukushima 2011 Disaster* (CRC Press, Boca Raton).

Garland, J. A. and Wakeford, R. (2007). Atmospheric emissions from the Windscale accident of October 1957, *Atmospheric Environment*, **41**, 3904–3920.

Hernan, R. E. (2010). *This Borrowed Earth: Lessons from the Fifteen Worst Environmental Disasters around the World* (Palgrave Macmillan, New York).

Holmes, J. A., Franklin, E. C. and Gould, R. A. (1915). *Report on the Selby Smelter Commission.* United States.

Holzworth, G. C. (1969). Large-scale weather influences on community air pollution potential in the United States, *Journal of the Air Pollution Control Association*, **19**, 248–254.

In, H.-J., Kim, Y. P. (2010). Estimation of the aerosol optical thickness distribution in the Northeast Asian forest fire episode in May 2003: Possible missing emissions, *Atmospheric Research*, **98**, 261–273.

Kantha, L. H. (2017). Lake Nyos (1986): The geology of killer lakes, in Brimblecombe, P. (ed.), *Air Pollution Episodes* (World Scientific, Singapore).

Lai, I.-C., Brimblecombe, P. and Lee, C.-L. (2017). Asia dust storms 2010: Continent wide air pollution, in Brimblecombe, P. (ed.), *Air Pollution Episodes* (World Scientific, Singapore).

Langmann, B. (2017). North Atlantic volcanic ash (2010): Contemporary social vulnerability to a natural event, in Brimblecombe, P. (ed.), *Air Pollution Episodes* (World Scientific, Singapore).

Latif, M. T. Othman, M., Khan, M. F., Abdullah, A. M., Ahamad, F. and Juneng, L. (2017). Southeast Asian forest fires (1997/1998): El Niño as a driver of regional impacts, in Brimblecombe, P. (ed.), *Air Pollution Episodes* (World Scientific, Singapore).

Lioy, P. J. and Georgopoulos, P. G. (2011). New Jersey: A case study of the reduction in urban and suburban air pollution from the 1950s to 2010. *Environmental Health Perspectives,* **119**(10), 1351–1355.

Lippmann, M. (2017). World Trade Center (2001): Health effects of the dust, in Brimblecombe, P. (ed.), *Air Pollution Episodes* (World Scientific, Singapore).

Lochbaum, D., Lyman, E., Stranahan, S. Q. and The Union of Concerned Scientists (2014). *Fukushima: The Story of a Nuclear Disaster* (The New Press, New York).

McCabe, L. C. and Clayton, G. D. (1952). Air pollution by hydrogen sulfide in Poza Rica, Mexico. An evaluation of the incident of Nov. 24, 1950, *Archives Industrial Hygiene and Occupational Medicine,* **6**, 199–213.

Medvedev, G. and Andrei Sakharov, A. (1991). *The Truth About Chernobyl,* (Basic Books, New York).

Mill, H. R. and Lempfert, R. G. K. (1904). The great dust-fall of February 1903, and its origin, *Quarterly Journal of the Royal Meteorological Society,* **30**, 57–91.

Minzesheimer, B. (2007). Novels about 9/11 can't stack up to non-fiction, *USA TODAY,* http://usatoday30.usatoday.com/life/books/news/2007-09-10-911-novels_N.htm.

Mishra, P. K., Bunkar, N. Bhargaval, A., Jain, S. K., Khare, N. K. and Pathak, N. (2017). Bhopal (1984): Cancer risk among survivors and opportunities for translational environmental health research, Brimblecombe, P. (ed.), *Air Pollution Episodes* (World Scientific, Singapore).

Ning, Z., Zhang, J. and Gao J. Beijing (2017). A driver of change, in Brimblecombe, P. (ed.), *Air Pollution Episodes* (World Scientific, Singapore).

Newell, E. and Watts, S. (2006). The environmental impact of industrialisation in South Wales in the nineteenth century: 'Copper smoke' and the Llanelli Copper Company *Environment and History,* **2**(3), 309–336.

River, C. (2014). *The Chernobyl Disaster: The History and Legacy of the World's Worst Nuclear Meltdown* (Create Space Independent Publishing Platform, Seattle, US).

Schramm, E. (1990). Experts in the smelter smoke debate, in Brimblecombe, P. and Pfister, C. (eds.), *The Silent Countdown. Essays in European Environmental History* (Springer-Verlag, Berlin), 196–209.

SCMP (2015). Smog-hit Chinese city official adopts novel approach to raise awareness about need for environmental protection, *South China Morning Post*, http://www.scmp.com/news/china/society/article/ 1882340/smog-hit-chinese-city-official-adopts-novel-approach-raise.

Stedman, J. R. (2004). The predicted number of air pollution related deaths in the UK during the August 2003 heatwave, *Atmospheric Environment*, **38**, 1087–1090.

Tucker, R. R. (1941). Smoke prevention in St Louis, *Industrial and Engineering Chemistry*, **33**, 836–839.

Vieno, M., Heal, M. R., Twigg, M. M., MacKenzie, I. A., Braban, C. F., Lingard, J. J. N., Ritchie, S. Beck, R. C., Móring, A., Ots, R., Di Marco, C. F., Nemitz, E., Sutton, M. A. and Reis, S. (2016). The UK particulate matter air pollution episode of March–April 2014: More than Saharan dust, *Environmental Research Letters*, **11**, 044004.

Walker, S. J. (2004). *Three Mile Island: A Nuclear Crisis in Historical Perspective* (University of California Press, CA).

Zimmer, A. and Nemery, B. (2017). The Meuse Valley (1930): Just fog or industrial pollution?, in Brimblecombe, P. (ed.), *Air Pollution Episodes* (World Scientific, Singapore).

Chapter 1

Early Episodes

Peter Brimblecombe

*School of Energy and Environment, City University
of Hong Kong, Hong Kong*

1.1 Introduction

Air pollution has been experienced throughout human history. It is
hardly surprising that our ancestors would have been exposed to
dust, pollen and smoke from forest fires. Pollution has been regulated since ancient times as humans gathered in cities, and records
of environmental pollution and degradation begin to be more abundant from the classical period (Hughes, 1994). In the ancient world,
air pollution episodes that were lengthy and geographically extensive are more likely to have been natural in origin. Nevertheless, it
is still possible to trace some episodes of human origin from such
early times and follow them through much of our history. This
account will end at the beginning of the 20th century, because by
this point measurements and detailed scientific studies were available and change the nature of the evidence.

1.2 Ancient Egypt, Greece and Rome

Perhaps familiar, but sometimes seeming not to be part of our view
of urban air quality, is odour. It was rife in cities of the past where

small piles of dung and rubbish were widespread. There is an example of this type of air pollution from putrefaction in Egypt that was claimed to have serious outcomes. The Kushite King Piye (reigned ~741 to ~712 BC), founder of the 25th Dynasty, was responsible for expanding Nubian power into Lower Egypt and ultimately extended this control as far as the Nile Delta. He laid siege to a number of Egyptian towns including the siege of Hermopolis, which Lichtheim (2006) places at about 734 BC, and is described on *The Victory Stele of Piye*. The fate of the city is revealed: "Days passed and Ur [Hermopolis] was a stench to the nose for lack of air to breathe. Then Ur threw itself on its belly, to plead before the king." It is hard to interpret this, but one can only presume that the stench of rotting corpses and other organic matter made continued occupation within the city intolerable.

The problem of odour was important to regulate in ancient cities. In ancient Athens, there were:

> "… ten City Commissioners (*Astynomi*), of whom five hold office in Piraeus and five in the city. Their duty is to see that …. no collector of sewage shall shoot any of his sewage within ten stradia [possibly a stadion so this distance would be ~10*185 m] of the walls; they prevent people from blocking up the streets by building, or stretching barriers across them, or making drain-pipes in mid-air with a discharge into the street, or having doors which open outwards; they also remove the corpses of those who die in the streets"

Aristotle, *The Athenian Constitution*
Translated by Sir Frederic G. Kenyon

The Roman Governor of Britain, Sextus Julius Frontinus, was later appointed *Curator Aquarum* to Rome. He improved Rome's water supply, and argued in his book *De aquis urbis Romae*, presumably through a miasmatic connection, that he also reduced the effects of the city's "infamis aer … [or] gravioris caeli …" through proper administration of Rome's aqueducts. In the Middle East, there were related pollution problems in classical Judea where the Mishnah Laws demanded that polluting sources be more than 50 cubits

(~25 m) from a neighbour's residence. In the case of tanneries, the prevailing wind ought to be considered (Mamane, 1987) and a pre-occupation with organic contaminants continued in the Islamic world (Gari, 1987).

More regular forms of pollution, such as those from smoke, appear in later classical references. Relatively few solutions were available to solve industrial air pollution problems, but the idea of zoning was apparent, and in Rome glass making was moved to the suburbs to limit the impact of its smoke. Rural areas of Spain and the Pyrenees had a great deal of mining, and "furnaces for silver are constructed lofty, in order that the vapour, which is dense and pestilent, may be raised and carried off" (Hamilton and Falconer, 1854).

In Imperial Rome nuisance laws, which treated issues of neighbourly responsibility in cities (urban servitudes), dealt with smoke as though it was water (Brimblecombe, 1987a) arguing: you could no more let water drain across a house than smoke. In Jerusalem, kilns and furnaces were not allowed in the city, to avoid soiling of the walls and buildings (Mamane, 1987). We find a parallel concern in the writings of the Roman poet Horace, which mention the blackening of temples in the ancient city.

"Your fathers' guilt you still must pay,
Till, Roman, you restore each shrine,
Each temple, mouldering in decay,
And smoke-grimed statue, scarce divine"

Horace, *Odes and Carmen Saeculare*

Some of the best evidence of early exposure to air pollutants comes from human remains, but this is most likely due to indoor air pollution. Exposure to wind-blown sand and indoor smoke was widespread in the ancient past. Evidence of anthracosis from the deposition of smoke in the lung, and silicosis from fine sand particles is found in mummified tissue (Aufderheide, 2003). Capasso (2000) has argued that lesions found on the ribs of victims of the eruption of Vesuvius suggest that pleurisy was common in Roman times, and this may be evidence of indoor pollution from burning vegetable matter lamp oils. Air pollution seems to have contributed to the poor health of a child from Imperial Rome of the second century, with the mummified body showing evidence of severe anthracosis (Scheidel, 2009). There is skeletal evidence of increased incidence of sinusitis, due to indoor exposure to smoke, in populations from Britain (Wells, 1977). The is some evidence that women were affected more than men, perhaps because they spent more time at the fireplace (Panhuysen *et al.*, 1997), and rural communities seem less affected than those in urban areas (Roberts, 2007). Poor ventilation or the lack of chimneys may have increased the levels of indoor air pollution and sinusitis, as found in Viking communities (Sundman and Kjellström, 2013; Christensen and Ryhl-Svendsen, 2015).

1.3 Medieval Period

As with ancient Rome, wood-burning industrial activities could create a serious air pollution problem. The north German city of Lüneburg was renowned for its salt-making in the medieval and early modern periods. Solar evaporation is inefficient so far north, so brines extracted from beneath the city were heated in lead and ceramic pans to increase the concentration and allow salt to crystallise. Salt was a key driver of the city's wealth from the late 1200s. Across the active period of production between 1554 and 1614 the mean annual output was 21 ktonnes. During the 15th and 16th centuries, the salt-works consumed between 48,000 and 72,000 cm^3 of wood each year (Lamschus, 1993; Witthöft, 1989). The salt-works thus generated a large amount air pollution (Brimblecombe, 2011), as shown by the smoke cloud in an early woodcut of the town (Fig. 1.1).

The medieval period is especially interesting, because not only did industrial intensity grow, but also the use of coal became important, most notably in London. Coal was probably used in small quantities in Roman Britain, as it has been frequently found at archaeological sites (Dearne and Branigan, 1995; Smith, 1997). However, literary evidence for its use in classical times is limited and difficult to interpret (Dearne and Branigan, 1995). It is clear though by the 12th century it was used as a fuel in the north of Britain, and

Fig. 1.1. Smoke from the salt-works close to the now vanished church of St Lambert's at Lüneburg in the 16th century, from a woodcut of Sebastian Munster (1550) that appears in Braun and Hogenberg's *Civitates Orbis Terrarum* (1572).

in the following century imported into London, where shortages of wood as a fuel led to its increasing use, particularly in lime-burning and metallurgical activities. The lack of chimneys in most dwellings restricted its widespread use as a domestic fuel until the late 1500s.

The use of coal may have risen sharply in the 1280s and set the scene for some of the earliest air pollution episodes that warranted investigation by civic authorities. Wood prices increased in the late 13th century and pressured some industries to shift to coal (Brimblecombe, 1975, 1987b). The new fuel was probably not only cheap, but also convenient; especially for kilns used in making lime mortar from calcium carbonate, as these needed to operate at high temperatures. In rural areas, these kilns did not have tall chimneys as late as the 19th century (see Fig. 1.2). Thus, the pollutants, most notably smoke and sulphur dioxide, were released into the atmosphere almost at ground level. Often the quantities of coal used were

Fig. 1.2.   Smoke from a lime kiln at Dumbarton rock and castle on the banks of the River Clyde from Stoddart, John (1800), *Remarks on Local Scenery and Manners in Scotland.* William Miller: London. Note the low emission height.

large; deliveries in excess of 100 tonnes are known from the medieval period. In London, coal was delivered via the lost Fleet River (now Fleet Street) to the west of the city, and adjacent to nearby Old Seacoal Lane and Limburner Lane. These locations are upwind (i.e. the prevailing wind came from the west) of important places in medieval London, such as Smithfield Market and St Paul's Cathedral. Therefore, the smoke would have created great annoyance, especially as people believed that disease arose from strange smells or noxious airs (miasmatic theory). The coal smoke problem was serious enough to cause complaints in 1285 (and again in 1288). A relevant document reads:

> "Commission to Roger de Northwode, John de Cobbeham and Henry le Galeys to enquire touching certain lime-kilns (*rogis calcis*) constructed in the city and suburb of London and at Suthwerk,

of which it is complained that whereas formerly the lime used to be burnt with wood, it is now burnt with sea-coal, whereby the air is infected and corrupted to the peril of those frequenting and dwelling in those parts. In executing this commission they are to associate with themselves the mayor and sheriffs of London and the bailiffs of Suthwerk"

Calendar of Patent Rolls 13 Edward I — membrane 18d

This medieval record confirms the fuel and the type of industry, and asserts the reason for concern was health. A later document stresses the need to find remedies for the problem. Frustration was evident by the next century, when the use of coal was banned. Penalties for contravening air pollution regulations seem to have been fines or the removal of the offending kiln; a much-discussed case of an execution seems unlikely to have occurred (Brimblecombe, 1976). The ban on coal was apparently difficult to enforce in the long term, as by 1329 a Newcastle collier, Hugh de Hencham was making lime in London, presumably with coal as a fuel (Brimblecombe, 1975).

As in the classical Mediterranean, solutions to the air pollution problem were limited. Evidently, the initial response from officials, although not really a solution, was simply to ban the use of coal. Additionally, there is some evidence that chimney heights may have been specified in London as shown by a case brought before the London Assize of Nuisance in 1377. Here, it was argued that the offending chimney was 12 feet lower than was good practice. Furthermore, there are records indicating that some blacksmiths thought it would be wise to burn coal only during the day (Brimblecombe, 1976, 1987a), perhaps when the air was more turbulent and wind speeds greater.

The advent of the *Black Death* caused a decrease in the European population in the mid-14th century. This tended to cause a shortage of labour and an increase in wages. Such economic changes may have helped maintain wood as the preferred, but more expensive fuel. The wider incorporation of chimneys into houses made coal more acceptable as a domestic fuel in London by the late 1500s.

Thus, the 17th century saw it become the predominant fuel, with more than a million tonnes imported each year into London towards the end of the century (Brimblecombe, 1978; Hausman, 1980).

1.4 Early Scientific Interest

The 17th century also saw a growing interest in scientific discovery, and there were a number of academic papers on air pollution and its impact. In particular, early members of the Royal Society, such as Boyle, Digby, Evelyn, Graunt, Grew and Henshaw took an interest in the composition of air and the effects this might have on health (Brimblecombe, 1978). Sir Kenelm Digby speculated about air pollution, using atomistic ideas derived from his friend Margaret Cavendish (an early woman scientist and the subject of a recent novel, *Margaret the First*, by Danielle Dutton), and wrote of this in *A discourse on sympathetic powder* of 1658 (Brimblecombe, 1987b). The best-known treatise on air pollution from this period was John Evelyn's *Fumifugium* of 1661, which represents an excellent account of coal-burning London. Less well known from air pollution perspective is the work of John Graunt: *Natural and Political Observations Made upon the Bills of Mortality* of 1662. This pioneering work on demography partially attributes the high death rate in London to smoke from the use of coal.

The concentrations of sulphur and smoke were so high in London at this time that they encouraged the development of London fog, which characterised the city for some three centuries (also see Brimblecombe, 2017). These fogs were so thick, and visibility so low that it is said horses ran into each other in the streets. Following John Graunt's lead, data was extracted from the *Bills of Mortality* and daily weather observations from London diaries. This allows the number of deaths in London to days of fog to be compared for the last quarter of 1679, as shown in Fig. 1.3 (Brimblecombe, 1987b). The week with six days of fog (two of which were described as "great stinking fogs") show high death rates compared to the week that followed. Deaths were especially enhanced among the elderly, and some were attributed to *tisick*, a pulmonary disease. The event shown

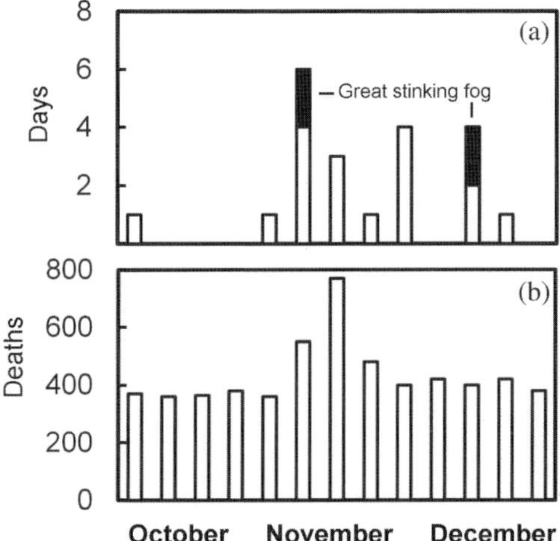

Fig. 1.3.   (a) Number of days with fogs each week for the last quarter of 1679, with intense days of fog in black. (b) The number of deaths in London each week for the last quarter of 1679.

in this figure probably represents one of the earliest air pollution episodes where we can attribute health affects to high pollutant concentrations. In the London smog of 1952, the sulphur dioxide and smoke concentrations were many hundreds of microgrammes per cubic metre. It would be possible for these to be as almost high in 17th century, as the emissions from coal were confined to a much smaller urban area (Brimblecombe, 1977).

1.5 Urban Industries and the Steam Engine

The previous sections have argued for the importance of a transition to coal in creating early air pollution episodes in London. This runs somewhat counter to assumptions that environmental pollution was a product of the development of the steam engine and the related issues of the industrial revolution. The industrial revolution began about 1760 in Britain, and some half a century later in Belgium along the Meuse, which was rich in coal (see the implications for air pollution in Zimmery and Nemery, 2017). The industrial revolution was soon causing change throughout Western Europe, and ultimately the entire world. However, as we have shown severe air pollution problems did not need to wait for the industrial revolution. The arrival of the steam engine may have had only a small impact at first, despite being dirty and dangerous, as many would have been associated with industrial activities (mining and smelting especially) in the countryside. The few steam engines in London would have used less coal than the numerous domestic fireplaces.

The American economic historian John Ulric Nef argued in *The Rise of the British Coal Industry* (1966) for the need to see industrialisation as having ramifications that extend back far earlier than the typical dates accepted for the beginning of the industrial revolution. This chapter shows a number of early air pollution events that precede the development of the steam engine. Mieck (1990) argues following French historians Le Roy Ladurie and Goubert, that anti-pollution decrees from the Middle Ages were essentially responses to single pollutant sources of what he terms *pollution artisanale.* This he distinguishes from the later *pollution industrielle,* which characterised the industrialising world. His typology goes on to include other forms of pollution. However, air pollution of the medieval and even classical period may have been extensive and had broader implications, and might be considered as industrial. Thus, pollution prior to the industrial revolution might not be restricted to *pollution artisanale.*

As examples, in medieval London or Lüneburg, industries could at times use very large quantities of fuel. The impact of these industrial emissions could be regional if not global in extent. Brännvall *et al.* (1999) show that there are clear signals of lead emissions from the classical period lake sediments from Northern Sweden. The Dark Ages by comparison show little contribution from lead, but after the ninth century "there was a conspicuous, permanent increase in atmospheric lead pollution fallout ... and peaks in atmospheric lead pollution at 1200 and 1530 AD comparable to present-day levels." They particularly note that lead deposition did not arise, as often assumed, primarily as a result of the industrial revolution. Rather the "atmospheric pollution climate in northern Europe was established in Medieval time[s]."

1.6 Victorian Episodes

Coal was so dominant as a fuel during the rapid urban growth of the 19th century it is hardly surprising that concern about the increasing levels of pollution grew in parallel. There is also evidence that air pollution episodes (often during bad fogs) were being recorded in terms of the number of excess deaths in cities such as London (see Table 1.1). This became part of the reports offices of the Registrar of Births Deaths and Marriages, so the information was increasingly used to press for legislation to reduce the emission of smoke. Regulations, which were widely demanded in the latter part of the 19th century, grew as part of the reform in approaches to urban government. Such regulations were often as sanitary legislation, and became necessary as population density urban grew with the increasing need for factory workers, usually drawn from the countryside.

Early regulations to abate smoke were not particularly effective, and needed both new approaches to the way this early abatement regulation was administered, but also improvements in the technology that could be applied (Bowler and Brimblecombe, 1990). It also set the stage for an increasing range of specialisation in central government as professional skills became necessary. They were increasingly needed by those who had to put the growing amount of legislation into practice through their roles as sanitary and smoke inspectors (Brimblecombe, 2003a, 2003b). The introduction of

Table 1.1. Major fog episodes in Victorian London with the estimated number of deaths.

Year	Month	Duration days	Deaths
1873	Dec	5	660
1880	Jan	4	1180
1882	Feb		610
1891	Dec	5	1630
1892	Dec	6	680

Source: Details from Brimblecombe (1987b), Anderson (1999).

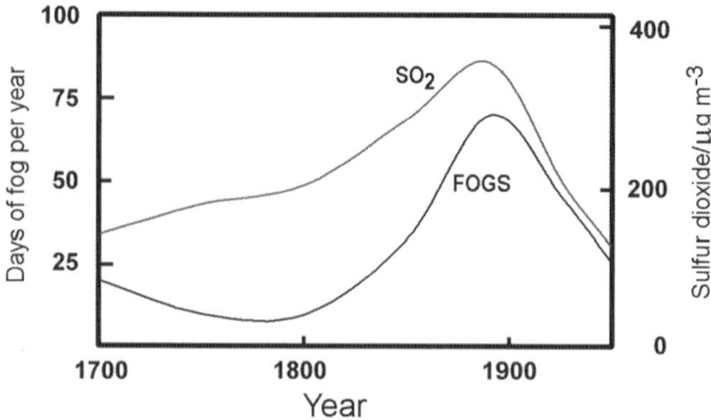

Fig. 1.4. The average concentrations of sulphur dioxide in London estimated from modelling compared with the number of days with fogs each year (see Brimblecombe, 1975, but also Brimblecombe, 1981, 1982).

automatic stoking, as a technology, was able to reduce the emission of smoke, but was not effective in reducing sulphur dioxide.

Fogs and air pollution came to be closely associated in the 19th and early 20th centuries. In London, simple modelling of the air pollutant concentrations (Fig. 1.4) derived from fuel use reinforce this link (Brimblecombe, 1975). The few air pollution measurements along with meteorological observations from the 19th century tend to support calculations that suggest high pollutant concentrations in Victorian London (Brimblecombe, 1981, 1982). The reasons for the rapid decline in sulphur dioxide shown by the model in the 20th century, arises from an expansion of the size of the city, which effectively reduced the areal emission strength. It was not so much the result of any improvements in regulation. The peak in primary pollutant concentrations about 1900 can be discussed in terms of an environmental Kuznets curve, or an inverted U, which is also typical of other cities, although the timing and the width of the curve varies greatly. The Kuznets view would see industrialisation initially occurring without great concern for the environment, but at some point society concludes that environmental quality can no longer be sacrificed for this growth (Brimblecombe and Grossi,

2009). Despite these improvements in the average concentrations of air pollution, episodes still occurred in London although perhaps not with the high frequency found in Victorian times (Bernstein, 1975). It was not until the disastrous London fog of 1952, that a modern approach to air pollution regulation began in the United Kingdom (Brimblecombe, 2017).

1.7 Conclusions

Large sources in a small area (i.e. a high source strength) could lead to elevated pollutant concentrations when meteorological conditions were stagnant. Smell provoked some of the earliest responses and legislation. In more recent episodes, although large amounts of pollutants are emitted, the effects of meteorology remain important. Some of the subsequent events had even a more profound effect on the development of environmental law. They promoted a realisation of the seriousness of the impacts, and fostered a growing awareness that began to separate wealth generation from air pollution

References

Anderson, H. R. (1999). Health effects of air pollution episodes, in S. T. Holgate, J. M. Samet, H. S. Koren and Maynard R. L. (eds.), *Air Pollution and Health* (Academic Press, San Diego, CA), pp. 461–484.

Bernstein, H. T. (1975). The mysterious disappearance of Edwardian London fog, *London Journal*, **1**, 189–206.

Aufderheide, A. C. (2003). *The Scientific Study of Mummies* (Cambridge University Press, Cambridge).

Bowler, C. and Brimblecombe, P. (1990). The difficulties of abating smoke in late Victorian York, *Atmospheric Environment*, **24B**, 49–55.

Brännvall, M. L., Bindler, R., Renberg, I., Emteryd, O., Bartnicki, J. and Billström, K. (1999). The Medieval metal industry was the cradle of modern large-scale atmospheric lead pollution in northern Europe, *Environmental Science & Technology*, **33**, 4391–4395.

Brimblecombe, P. (1975). Industrial air pollution in thirteenth century Britain, *Weather*, **30**, 388–396.

Brimblecombe, P. (1976). Attitudes and responses towards air pollution in medieval England, *Journal of the Air Pollution Control Association*, **26**, 941–945.

Brimblecombe, P. (1977). London air pollution, 1500–1900, *Atmospheric Environment*, **11**, 1157–1162.

Brimblecombe, P. (1978). Interest in air pollution among early members of the Royal Society, *Notes and Records of the Royal Society*, **32**, 123–129.

Brimblecombe, P. (1981). Long term trends in London fog, *The Science of the Total Environment*, **22**, 19–29.

Brimblecombe, P. (1982). Trends in the deposition of sulphate and total solids in London, *The Science of the Total Environment*, **22**, 97–103.

Brimblecombe, P. (1987a). The antiquity of smokeless zones, *Atmospheric Environment*, **21**, 2485–2485.

Brimblecombe, P. (1987b). *The Big Smoke* (Methuen, London).

Brimblecombe, P. (2003a). Historical perspectives on health: The emergence of the Sanitary Inspector in Victorian Britain, *The Journal of the Royal Society for the Promotion of Health*, **123**, 124–131.

Brimblecombe, P. (2003b). Origins of smoke inspection in Britain (*circa* 1900), *Applied Environmental Science & Public Health*, **1**, 55–62.

Brimblecombe, P. and Grossi, C. M. (2009). Millennium-long damage to building materials in London, *The Science of the Total Environment*, **407**, 1354–1361.

Brimblecombe, P. (2011). A history of the causes and consequences of air pollution, in M. Quante *et al.* (eds.), *Persistent Pollution: Past, Present and Future* (Springer-Verlag, Berlin Heidelberg).

Brimblecombe, P. (2017). London 1952: An enduring legacy, in Brimblecombe, P. (ed.), *Air Pollution Episodes* (World Scientific, Singapore).

Capasso, L. (2000). Indoor pollution and respiratory diseases in Ancient Rome, *Lancet*, **356**, 1774.

Christensen, J. M. and Ryhl-Svendsen, M. (2015). Household air pollution from wood burning in two reconstructed houses from the Danish Viking Age, *Indoor Air*, **25**, 329–340.

Dearne, M. J. and Branigan, K. (1995). The use of coal in Roman Britain, *The Antiquaries Journal*, **75**, 71–105.

Gari, L. (1987). Notes on air pollution in Islamic heritage, *Hamdard*, **30**, 40–48.

Hamilton, H. C. and Falconer, W. (1854). *The Geography of Strabo*, Vol. I (Henry G. Bohn, London).

Hausman, W. J. (1980). A model of the London coal trade in the eighteenth century, *The Quarterly Journal of Economics*, **94**, 1–14.

Hughes, J. D. (1994). *Pan's Travail: Environmental Problems of the Ancient Greeks and Romans* (Johns Hopkins University Press, Baltimore).

Lamschus, C. (1993). Die Holzversorgung der Lüneburger Saline in Mittelalter und früher Neuzeit, in S. Urbanski, C. Lamschus, and J. Ellermeyer (eds.), *Recht und Alltag im Hanseraum* (Förderkreis Industriedenkmal Saline, Lüneburg), pp. 321–333.

Lichtheim, M. (2006). *Ancient Egyptian Literature* (University of California Press Berkeley, US).

Mamane, Y. (1987). Air pollution control in Israel during the first and second century, *Atmospheric Environment*, **21**, 1861–1863.

Mieck, I. (1990). Reflections on a typology of historical pollution: complementary conceptions, in P. Brimblecombe and C. Pfister (eds.), *The Silent Countdown* (Springer-Verlag, Berlin), pp. 73–80.

Panhuysen, R. G. A. M., Coenen, V. and Bruintjes, T. D. (1997). Chronic Maxillary Sinusitis in Medieval Maastricht, the Netherlands, *International Journal of Osteoarchaeology*, **7**, 610–614.

Smith, A. H. V. (1997). Provenance of coals from Roman sites in England and Wales, *Britannia*, **28**, 297–324.

Roberts, C. A. A (2007). Bioarcheological study of maxillary sinusitis, *American Journal of Physical Anthropology*, **133**, 792–807.

Scheidel, W. (2009). Disease and death in the ancient city of Rome, Princeton/Stanford *Working Papers in Classics*. Available at SSRN: http://ssrn.com/abstract=1347510 or http://dx.doi.org/10.2139/ssrn.1347510.

Sundman, E. A. and Kjellström, A. (2013). Signs of sinusitis in times of urbanization in viking age-early Medieval Sweden, *Journal of Archaeological Science*, **40**, 4457–4465.

Wells, C. (1977). Diseases of the maxillary sinus in antiquity, *Medical and Biological Illustration*, **27**, 173–178.

Witthöft, H. (1989). *Das Maß der Arbeit an Sole und Salz*, in C. Lamschus (ed.), *Salz — Arbeit und Technik. Produktion und Distribution im Mittelalter und Früher Neuzeit* (Förderkreis Industriedenkmal Saline, Lüneburg).

Zimmer, A. and Nemery, B. (2017). The Meuse Valley (1930): Just fog or industrial pollution? in Brimblecombe, P. (ed.), *Air Pollution Episodes* (World Scientific, Singapore).

Chapter 2

Meuse Valley (1930): Just Fog or Industrial Pollution?

Alexis Zimmer* and Benoit Nemery[†]

*Université Paris 7- Diderot, Institut des humanités de Paris — Centre Georges Canguilhem, Paris, France
[†]Centre for Environment and Health, KU Leuven, Leuven, Belgium

2.1 Introduction

The Meuse Valley fog of 1930 may be considered as a landmark in the history of disasters because it arguably represents the first scientifically recognised instance of a catastrophe caused by an air pollution episode of industrial origin. The historical event and its main characteristics have been described in some detail in a *Lancet* publication in 2001 (Nemery *et al.*, 2001). Here, we will focus on the perception of the phenomenon and its causal attribution by different stakeholders, both during the events and in the wake of the disaster. These socio-historical aspects have been the object of a doctoral thesis [Zimmer] and other publications (Zimmer, 2016).

The many deaths that occurred in the Meuse Valley during the "killer" fog of December 1930 appeared in the newspapers as a complete surprise. Their supposed mysterious origin led initially to many conjectures about their cause, as well as controversy about who — if anybody — was to be held responsible for their

27

occurrence. It is important to realise that the Meuse Valley disaster occurred in an era when air pollution was not perceived as an issue of much concern by mainstream health experts or political authorities. In those days, levels of air pollutants were not monitored and the potential health effects of industrial emissions of pollutant gases and particulates into the ambient air were largely not subject of concern by doctors and scientists. The smoke emerging from industrial chimneys was in a sort of cliché viewed as a sign of progress, not as a health hazard. Urban air pollution, especially during winter fogs, was notorious in industrial cities such as London or Manchester, and its adverse consequences on human health had been advanced by some, but this was not recognised by the medical establishment or by environmental legislation. Also, no outbreaks of acute deaths had been shown to be caused by peaks of industrial air pollution.

2.2 Brief Description of the Events

During the first week of December 1930, a thick fog covered a large part of western Europe. The fog was particularly intense in the heavily industrialised portion of the narrow Meuse Valley between the towns of Liège and Huy, in Belgium. On the third day of the fog, i.e. on 3 December 1930, hundreds of people living in the villages along the Meuse started to suffer from stinging eyes, cough and phlegm production, and breathing difficulties with asthmatic characteristics. Animals were also affected. Over the course of the following 2 days, at least 60 people died in cardio-respiratory distress, which represented a 10-fold increase in mortality compared to the background. The symptoms — and the excess deaths — rapidly subsided after the fog had dissipated on 6 December (Fig. 2.1). The massive increase in mortality led to widespread commotion and coverage in the local, national and even international press. Experts were initially at a loss to explain what had happened. Numerous causes, including fanciful ones, were proposed to explain what was considered as a mystery in the Press. Some contended that the deaths had simply been caused by the "fog alone" (thus denying any human responsibility), while others incriminated the industrial emissions that had plagued the area for such a long time.

Fig. 2.1.   Funeral procession for smog victims in rue Wauters, Engis, in the Meuse Valley (http://www.hermalle-sous-huy.be/fr/Histoire-XXe.html#catastrophes).

In the end, the air pollution hypothesis was vindicated following a thorough investigation by a multidisciplinary judicial commission that had been appointed to "determine the mechanism of the accidents that had occurred in the valley" during the fog episode. The commission's findings and conclusions have been described in a remarkably detailed report (Firket, 1931). The report firmly concluded that people had become ill and had died as a result of their exposure to air pollutants that had originated mainly from the combustion of coal by industry and accessorily also domestic fires, and these pollutants (mainly sulphur dioxide and fine particulates) had accumulated to excessive levels because of a meteorological phenomenon known as thermal inversion. In its report, the commission concluded that such tragedies would occur again if the same circumstances were met, and it made the prophetic statement that the public health services (PHSs) of London could bear the responsibility of 3,200 deaths in 2 days if a similar phenomenon (*un phénomène de même envergure*) were to occur there (Brimblecombe, 2017b).

In national and international newspapers, the narrative of the simultaneous sudden deaths caused by the "killer" fog sounded like in a whodunit novel. Front pages and headings referred to the "homicidal fog", the "valley of death", "malignant forces" and "black death". Early speculation as to the cause of the catastrophe included a volcanic eruption, an industrial accident or the malicious release of toxic agents from a plant taking advantage of poor visibility during the fog. With the Belgian trenches having been the seat of chemical warfare less than 15 years before, a possible role of war gases was also evoked: gases were speculated to have been released from airplanes, from stocks of shells with chemical ammunition in an abandoned coalmine, or from a secret factory preparing an attack on communist Russia. Infectious causes were also proposed, such as a resurgence of Spanish flu (which had also killed thousands in a recent past), the emergence of an infectious agent from the ground as a result of recent floods, or the presence of unknown microbes blown with the wind from the Sahara. However, the prevailing tone of the earliest media reports was that "the mystery was as thick as the fog itself."

The first official explanation of the events was very simple. Dr. Lacombe, representative of the public health administration of the Province of Liège, and Dr. Timbal, representative of the public health administration of the ministry of interior adamantly stated that "the fog alone" and the sudden fall of temperatures were sufficient to explain the increase in deaths, thus exonerating industry from any responsibility. Their conclusions were based on interviews with physicians of the valley and on observations of victims in places which, allegedly, could not have been reached by industrial fumes or gases. They insisted that no gases, no industrial emissions could explain the deleterious effect of the fog. In other words, all deaths could be attributed to natural causes. Thus, one day after the dissipation of the fog, doctor Lacombe announced the end of the inquiry on the square of the town hall of Engis, the most affected town of the area: the case was closed.

In the early days, international experts did not really understand the situation either. In *The Lancet* of 13 December 1930 a piece titled "Fog panic on the Meuse" reported that "At Engis and several other villages on the river level a number of deaths occurred among men and animals which were so grouped and so rapid as to give rise to something of a panic. There was talk of poisonous fumes [...] and some deadly local epidemic [...]. When the fog lifted saner counsels prevailed." The text goes on in the same rather belittling tone to conclude, echoing the initial view of Belgian official, that "It seems that [a combination of frost, fog, and still air occurring for days together] united on the Meuse to produce so terrifying a situation." A famous French scientist Jules Amar, "father of exercise physiology", during a session of the French Academy of Sciences on 5 January 1931, defended the notion that the deaths had been due to progressive asphyxia brought about by the inhalation of water droplets, leading to a sort of drowning which he called "hydrodiffusion" or "poisoning by humidity".

However, "the fog alone" hypothesis was not tenable. Among others, former minister Pierre Nolf, President of the Red Cross and the king's physician, stated that "pure fog has never killed anybody". Moreover, a thick fog had been present over a large portion of western Europe, but severe symptoms of respiratory irritation and casualties had only occurred in the small portion of the Meuse Valley between Huy and Liège. This part of the Meuse Valley was less than 15 miles long, but it contained eight zinc plants, six coking plants, five steel works, two fertiliser plants, quarries, and many other factories. Inhabitants of the valley also pointed to damage to vegetation and they argued that it was impossible to cultivate anything near zinc smelter plants. The statistics of the province points out that more than 10,000 tonnes of coal per day were consumed during the week of the fog. One newspaper reader, among others, said in a letter to the daily paper *La Meuse*:

"All investigations would not convince the inhabitants of the valley, about the safety of the pestilential air they breathe. They live in those places, their smell and their lungs are protesting in their own fashion against all conclusions exculpating the factories".

Deadly fogs had previously occurred in the area (in 1897, 1902 and 1911) as reported by meteorologist Felix Bertyn. Thus, far from being a mystery, the role of pollutant emissions by industry appeared obvious from the beginning.

2.3 The Official Enquiries

Two official committees of inquiry were instituted on the same day as the public health administration gave its conclusions. One inquiry was to be conducted by the Mining administration, the body in charge of safety and surveillance of the chemical and the smelting plants of the area. The other inquiry was instigated by the public prosecutor of Liège who pressed charges against persons unknown and installed an investigating committee composed of scientific experts.

The investigation of the Mining Administration, commissioned by the Minister of Industry, Labour and Social Welfare and under the responsibility of its Director General Lebacqz, was headed by the Inspector General Vincent Firket. The investigation was conducted by three engineers, Massin, Guerin and Masson, who visited factories of the area, discussed with manufacturers and estimated the types and quantities of substances emitted in the atmosphere in order to evaluate "the possibility of intervention of fumes from the plants in effects caused by fog."[1] The judicial investigation committee was headed by Jean Firket, Professor of pathology and forensic medicine at the University of Liège, and composed of a toxicologist (M. Schoofs), a meteorologist (M. Dehalu), two chemists (J. Mage and G. Batta), a veterinarian (P. Rubay) and two physicians (J. Bovy and J. Firket). The explicit aim of the latter inquiry was to "determine the mechanism of the accidents, deadly or not, which occurred in the valley on the 4th and 5th of December."

---

[1] *Archives de l'État à Liège (AEL), Archives de l'administration des Mines, Division du bassin de Liège, nouveau fonds, no 161, Lettre du directeur général des Mines à l'inspecteur général, 6 décembre 1930.*

Medical data were obtained by questioning family physicians and patients and also by necropsies performed in the days immediately after the fog. More than 60 deaths that occurred on 4th and 5th December were attributed to the fog. The report did not take into account people who died after the fog, although it did mention, in a footnote, that the fog could have favoured the occurrence of infections in those who died later. The report does not contain demographic or clinical details of the victims, but from the names and ages of 48 of those listed in newspaper reports (e.g. Le Soir, 7 and 8 December 1930), one could calculate that their average age was about 62 years and ranged from 20 to 89 years. The 27 male victims were younger than the 21 female victims, with average ages of about 56 and 70 years, respectively. The hardest hit locality appeared to be the village of Engis with 14 deaths in 2 days against an expected 65 deaths per year for approximately 3,500 inhabitants.

Most people had laryngeal irritation, retrosternal pain, coughing fits and "dyspnoeic breathing characterised by paroxysms and slowed expiration, like asthma," which was rapidly relieved by adrenaline-based drugs. However, some patients had "dyspnoea which did not have an asthmatic rhythm, but which was accompanied by signs of pulmonary oedema such as cyanosis, rapid breathing, and even frothy sputum." Nausea and vomiting were common, but there were no clinical signs of systemic poisoning. Respiratory problems rapidly improved after the fog cleared on 5 December. The most affected, even when they had remained indoors, were people who were elderly, asthmatic, debilitated, or those with cardiac disease. Even people who were previously healthy exhibited wheezing. Children were described as being indisposed and a few were seriously affected, though none died. Ten necropsies that were done showed respiratory tract alterations only. The toxicological analyses of blood or viscera revealed arsenic, zinc and sulphuric anhydride, but in insufficient quantities to cause clinical poisoning. Experts considered that it is common to detect these agents in an industrial area.

Cattle were described as suffocating and were taken to areas above the fog. However, some animals had to be slaughtered with post-mortems showing the presence of pulmonary and subcutaneous emphysema. Living animals were reported to have dysfunction of various digestive functions — appetite, rumination, defecation, lactation and presence, during a few days, of subcutaneous emphysema.

Investigators concluded that the signs and symptoms were entirely and solely due to a local irritant action towards the mucosae directly exposed to the outside air. They ruled out the low temperature and the density of the fog since similar weather conditions had prevailed over large parts of Belgium. Respiratory infections were also excluded mainly on clinical grounds.

To find out which specific substances had caused the casualties, the investigators first tried to analyse samples of grass, whitewash and water. However, they had to abandon this method because it was impossible to distinguish the contribution of the usual background levels of pollutants from those that had prevailed during the few days of fog. Hence, the investigators tried to model the concentrations of more than 30 agents released during the fog (including CO_2, CO, H_2, H_2S, AsH_3, NO_2, NH_3, HCl, acetylene and other hydrocarbons, sulphites, chlorides, and fluorides and nitrates of zinc or iron). Experts indicated that approximately 71 tonnes of SO_2 and 210 tonnes of dust were normally emitted daily into the air of the valley. Why those ordinary quantities had suddenly killed people could be explained by the particular meteorological conditions that had prevailed during the first week of December: high atmospheric pressure, slow wind and a sudden temperature decrease. Such conditions had encouraged the production of fog, for which industrial dust particles provided the necessary condensation nuclei. A temperature inversion (i.e. an increase in temperature at a certain altitude) had allowed and the industrial emissions trapped in the fog to accumulate to toxic levels during 5 days.

Investigators attempted to determine which of the emitted agents were most likely to have caused the adverse health effects and deaths. Hydrofluoric acid (HF) was investigated first because it was released from a fertiliser plant in Engis, where it had previously affected plants and grazing animals, and damaged light bulbs and window panes. However, the estimated maximum concentration of

HF (0.4 mg m^{-3} for the whole valley) and, more importantly, the simultaneous occurrence of the first symptoms downwind and upwind of Engis meant that a single central source of pollution was unlikely. Nevertheless, HF could not be completely ruled out as a contributory cause in the vicinity of the plant.

Eventually, sulphur compounds proved the most abundant compounds emitted in the valley. The experts calculated that the 71,000 kg of SO_2 produced per day resulted in a SO_2 concentration of up to 100 mg m^{-3} after 4 days of fog. They considered that such SO_2 concentration exceeded its threshold of toxicity (20–30 mg m^{-3} for prolonged exposure) and could account for the victims' symptoms. Moreover, if only one-10th of the SO_2 had been oxidised, a toxic concentration of H_2SO_4 (then assumed to be 4 mg m^{-3}) would have been exceeded. All those conclusions led to the differential diagnosis which pointed out sulphuric compounds as the chemical substances responsible of the symptoms described earlier.

Eventually, experts concluded that: "After examining what were all substances that might explain the harmfulness of the fog, and have successively eliminated most of them, we are led to conclude that, above all, sulphuric compounds that come from burning coal have exercised their deleterious action either as anhydride or sulphurous acid either in the form of sulphuric acid whose production in sufficient levels was made possible by some extreme weather conditions at the beginning of December 1930."

2.4 Naturalisation and Contestation of the Conclusions

From the first conclusions of the public health representative Lacombe — which seems to be more a denial than the result of an inquiry — to these later conclusions, a certain kind of nature was differently called in order to explain the production of the catastrophe. First "the fog alone" and the medical pre-disposition of the dead and ill persons; then those predispositions combined with meteorological conditions. This insistence on the meteorological or weather conditions marks, with the other instances of medical pre-disposition, what we can call a *naturalisation* of the catastrophe: Nature was at last the main agent of production of the mortal fog. It is the fog that permits the accumulation of toxic gas in the atmosphere and their transformations in sulphuric acid. Saying that is to forget or minimise the function of industrial dusts as condensation nuclei, it means as fog producer or catalyst. It also minimises the industrial responsibilities and the consequences of the ordinary, daily pollution, yet was put forward by some mayors of the valley. In other words, this fog was essentially a tempor and geographically circumscribed problem of dilution: the peculiar weather conditions prevent the dilution in the atmosphere of the usual toxic gas produced by the industry in the valley.

The naturalisation was not just discursive. Logically, the investigation ended without any industrial responsibility being assigned (have one's case dismissed). But most of all the decisions taken by the Belgian authorities after the catastrophe reflects the approach to expertise. In October 1931, Timbal, Director of the Health Department, and Langelez, Inspector General at the Ministry of Interior and Labor, presented the measure envisaged by the authorities to 16 mayors of communes affected: that of establishing close relations between Brussels meteorological observatory, the mayors of the region and industry. This observatory is committed to preventing the latter from the occurrence of potentially catastrophic weather conditions, which will in turn inform populations and temporarily reduce their production. Dr. Timbal asks mayors, when they were informed by the observatory, "(1) to distribute gas masks; (2) bring the most vulnerable people (the elderly and the sick) to higher ground; (3) to engage residents whose health is shaky to pull in well-heated rooms."[2] Meteorology, topography and predispositions are within these recommendations. Additionally, since it is difficult to act on them, it must be possible to escape from the potential disastrous situation. The mayors were surprised by the

---

[2] *AEL, Archives de l'administration des Mines, Division du bassin de Liège, nouveau fonds, no 161, Procès-verbal de la réunion du 16 octobre 1931*, and *"comment lutter contre les gaz mortels?", Nation Belge, 17 octobre 1931.*

weakness of the proposed measures and rejected the exceptionality invoked by the expertise and relayed by Timbal. They emphasised the "permanent danger" that prevailed in the area and requested sustainable and effective measures to curb the "effects of smoke". According to them, the meteorological exceptionality demonstrated the persistence of the threat.

To prove their assertions, they mentioned the previous fog episodes of 1897, 1902 and 1911, during which many animals succumbed and deaths were reported. They also drew attention to the untimely degassing of the new plant of the *Société des engrais et des produits chimiques de la Meuse*, installed in Tilleur, which occurred many times in January 1931, just after the fog and during the investigation. The release of gas, occurring in the wake of the fog, was much publicised; the administration of Mines and the judicial experts independently went to the scene. The plant manager announced on 24 January that no further suspect release would take place. However, on 15 April 1931:

> "On a small farm [in Tilleur and Sclessin-Ougrée], the elderly mother of 60 years has had a dry throat after which she had nausea, vomit, then coughed up blood. A cow died of causes attributable to gas poisoning. [...] By examining furniture, tool material, copper headlights of a motor truck, we see that all these metal objects have lost their sheen, they are become dull and dirty by oxidation (or some action of nitrous sour gas).
>
> In a house nearby, a small child was obviously disturbed and since that day, presenting an acute conjunctivitis. All metal objects are also oxidised, and the oxidation reproduced daily despite repeated cleaning. [...] During our investigation, heavy rain arose, gutters, roofing zinc dropped on the floor a liquid impregnated by oxidised products (ZnO)."[3]

Judicial experts recognised these effects. They interpreted them, however, not as indicative of a danger reigning permanently in the

---

[3] AEL, *Archives de l'administration des Mines, Division du bassin de Liège, nouveau fonds, no 161, Rapport de l'inspecteur principal d'hygiène Locombe (sic) à Monsieur le Gouverneur de la Province de Liège, 16 avril 1931.*

valley, but as a proof of any dangerousness of certain concentrations of SO_2 in the atmosphere.

"Symptoms of mild irritation of laryngo-tracheal mucosa, with cough and voice alteration, had shown in people living near the home producer of SO_2; but no serious accident respiratory (asthma, oedema, or cardiac collapse) was identified; the air was dry, there was no fog and the wind soon had to sweep the harmful gas; probably very little H_2SO_4 and H_2SO_3 could they occur in the air, yet relatively rich in SO_2".

Judicial experts not only tended to minimise the danger of the usual emissions from industry, but also those less frequent nuisances. Three days after the meeting with the mayors of the valley, three representatives of the Administration of Mines, accompanied by Dr. Timbal, as well as Langelez, met with representatives of the electricity company of Liege, as well as industrial zinc manufacturers, to present them with the same conclusions and ask them if it would be "possible in case of a fog, similar to that of December, to restrict factories." They stated that "it is only for urgent measures, which only refer to exceptional cases, intense and long fogs only those appearing very rarely." In response, manufacturers indicated notably putting the plants on standby could damage their equipment. These reservations about the usefulness of such a measure and the dangers to industrial facilities were even more marked in correspondence which followed the meeting.[4] Engineers from zinc factories are discussing the relevance of the expertise, believed that "the evidence is far from certain" on the role of sulphur compounds in the resurgence of mortality; they raised the hypothesis of that "fog alone", also suggested too little importance in the conclusions of the experts was assigned to the influence of domestic households in calculating the emissions of sulphur compounds into the atmosphere. They argued that peremptorily it is simply impossible for them "to stop or to shelve the ovens for a period of brouillard."

---

[4] *Archives générales du royaume, deuxième inspection générale des Mines à Liège, no 133, Procès-verbal de la réunion du 19 octobre 1931.*

In summary, the measure presented by the authorities satisfied neither the mayors of the valley to which the latter denied the effects of ordinary pollution, nor industrialists that call into question the responsibility of industrial fumes and considered it technically impossible to implement change.

Despite those two types of opposition, on 15 December, 1931, a circular issued by the Interior Ministry was sent to the mayors of the towns of the industrial basin of the Meus. It specified the paths taken by the administration to avoid recurrence of such a disaster. Two modes of action are preferred. First, it was confirmed that the Belgian Royal Institute of meteorology would notify by telegram the local authorities and industry of the "possibility of the return of a similar atmospheric situation to that of last December." On receipt of this warning, the mayors must "immediately alert the population so that they could follow the recommendations laid out, in the meantime, to it by [their] care."[5] In such a case, politics are nothing but a contradiction of that described by experts. The interaction between the emissions from industry and the weather is thus reproduced here at an institutional level. Finally, and on their side, "the government is actively concerned with the issue of pollution of the atmosphere by industrial gas emissions and research preventative measures that can mitigate this nuisance."

2.5 Legislative Change and Repetitive Fogs

Two other commissions took over these initial investigations. The commission Berryer — the name of a former health minister who directs it — and a commission analysing the conclusions of the latter. They planned to focus particularly on legislative and technical development likely to offer a response to such a disaster. The initial programmes for these study groups were less ambitious. These included proceedings, for a whole year, (i) to test air at various locations in the

---

[5] *AGR, deuxième inspection générale des Mines à Liège, no 133, Circulaire du ministère de l'Intérieur, Inspection d'hygiène, 15 décembre 1931.*

valley, at different altitudes and under varying weather conditions, including fog; (ii) to experiment on the conditions for production of toxic fogs using artificially fogs produced in laboratories of the army and to verify the effects on animals in order to develop a gas harmfulness scale according to some species; (iii) to settle in various places of the valley witnesses gardens that can inform investigators specifically on the effects of smoke on a wide range of plants and thus lead to the creation of reliable criteria linking damage to fumes specific and (iv) to find technical ways to reduce the danger of fumes, mainly by the recovery of dust contained in the flue gases, etc.[6]

However, point by point, the program lost its original intentions. The cost and difficulty of organising this research explain its abandonment. As for the study of industrial smoke recovery techniques, the conclusions of the Commission indicated that "if it is important to avoid air pollution by sulphur dioxide released by factories, it must, on the other hand, do not lose sight that we can advocate measures that would kill the industry and would take away their livelihood to the people." The Committee then contented itself to recommending that manufacturers conduct research aimed at capturing dust and sulphur products present in the fumes.

Finally, a royal decree from 10 August 1933, modified previous legislation from 1923, which governed oversight of industrial plants. The legislation introduced the administration of hygiene for the review of the site authorisation application for new industrial establishments, it reinforced the authority of the Administration of Mines notably toward mayors and clearly stated that they have no police power against factories, even in the case of a proven infringement. Finally, little changed.

Fogs loaded with toxic gas regularly occurred. Ironically, during the week of 24 November 1934, when a new fog spread through the valley, lasting more than 48 h, it was the Director of the Cockerill

---

[6] *AGR, administration des Mines, troisième série, no 565, Commission administrative chargées d'examiner les conclusions déposées par la commission instituée pour l'étude des pollutions de l'atmosphère, Rapport de la sous-commission de fonctionnaire, 22 avril 1932.*

plant in Seraing who signalled to the Royal Institute Meteorology in Belgium, which had failed to notice anything unusual.[7]

Failing to succeed or to try to impose measures requiring manufacturers to drastically reduce their releases of gas and dust — fuel switching, reduced production, improving the effectiveness of certain industrial processes — the Ministry of Health undertook experiments aimed "to the adoption of a permanent means of collective protection against the possible return of the 1930 accident." It was Dr. Dautrebande, Professor at the Faculty of Medicine of Liège and Director of the College for protection against poison gas, who headed the investigation. He imagined a convection current created from bottom to top of the valley to prevent toxic accumulation in the latter. To implement this, Dautrebande proposed the use of "huge fires loaded with a substance for judging the height of the riser." To make this effective, it will require number and volume foci required per unit area proceed to be determined. On 5 March 1938, Dautrebande conducted such experiments.

2.6 Conclusion

While the deadly fog of the Meuse Valley was sufficient to create a new industrial policy, it became the subject of scientific studies which illustrate the relevance of a forensic examination to outcomes. In some cities, it mobilised a legitimacy and defense of necessary government actions against atmospheric pollution. After the end of World War II, it became, with the fogs of Donora in 1948 (Brimblecombe, 2017a) and the London one in 1952 (Brimblecombe, 2017b), a landmark in the history of acute episodes of air pollution. In the Meuse Valley, Jean Firket, the physician, who led the investigation of the criminal investigation, initiated questions about the increase in lung cancer. He noted as Director of the cancer centre of the province of Liège that in the Meuse Valley, cancers develop as

---

[7] *AEL, Archives de l'administration des Mines, Division du bassin de Liège, nouveau fonds, no 161, Lettre Secrétariat Cockerill à l'ingénieur en chef directeur de l'administration des mines, 23 novembre 1934.*

new outcomes from air pollution, a legacy of change since the early 19th century. Pollution of the atmosphere gradually changes its form: from localised concerns to be the subject of further investigation of its composition and multiple outcomes. At the same time, it continues to grow: its components multiply and it is spread more widely. But this is another story, or rather an extension of the one we have initiated here.

But to understand this long history of repetition, relocation (think toxic fogs in India or China), transformations of air pollution, we have to open the narrow framework of scientific expertise in order to renew the way problems of air pollution are defined. We have to understand how scientific practices and expertise have accompanied the 19th century employment of coal. We also have to understand how the medical predisposition of those suffering in industrial areas was not a natural condition, but rather the product of long-term relationships with a deteriorated and polluted environment. Finally, we have to understand the industrial transformations of the atmosphere and the production of new industrially mediated meteorology not as something scientists and people discovered in December 1930, but as a long-term and well-documented phenomena. In this way, the toxic fog of December 1930 appears not so much as the first scientific demonstration of the air pollution mortality but rather as the occasion where it was no longer possible to deny it.

References

Brimblecombe, P. (ed.) (2017a). Donora (1948): Controversial contaminants, *Air Pollution Episodes* (World Scientific, Singapore).

Brimblecombe, P. (ed.) (2017b). London (1952): An enduring legacy, *Air Pollution Episodes* (World Scientific, Singapore).

Firket, J. (1931). The cause of the symptoms found in the Meuse Valley during the fog of December, 1930, *Bull Acad R Med Belg*, **11**, 683–741.

Nemery, B., Hoet, P. H. and Nemmar, A. (2001). The Meuse Valley fog of 1930: An air pollution disaster, *The Lancet*, **357**, 704–708.

Zimmer, A. (2016). Brouillards toxiques. *Vallée de la Meuse, 1930, contre-enquête*, Bruxelles, Zones Sensibles.

Chapter 3

Donora (1948): Controversial Contaminants

Peter Brimblecombe

*School of Energy and Environment, City University
of Hong Kong, Hong Kong*

3.1 Introduction

Air pollution has long associations with metal refining and smelting. These go back to classical times (Hughes, 1996) and are mentioned in John Evelyn's writings from 17th century England, and later in reports from Germany, California and the Meuse Valley (Schramm, 1990; Holmes *et al.*, 1915; Zimmer and Nemery, 2017). Pittsburgh was famed for its steel as much as for its air pollution, which required those who needed to look tidy to carry an extra white shirt for the afternoon. Pittsburgh's problems were so serious in the years that followed the Second World War that it became important to clean up the city (Davidson, 1979). Ironically, as Davidson points out, that during this period of desire for improvement one of the most noted air pollution episodes in the US occurred in the nearby town of Donora, some 30 km to the south of Pittsburgh. There had been some serious air pollution episodes before in the US, the best known being the St Louis smog of 1939, the infamous Black Tuesday (Tucker, 1941) and the events in Los Angeles during the

Second World War (Brimblecombe, 2013). These had failed to really capture public attention, as they were not so readily associated with identifiable deaths. The episode in Donora was transformational because of the impact it had on the development of environmental legislation in the US, causing air pollution to be seen as an urgent issue. This is often a role played by episodes, notably London of 1952 (Brimblecombe, 2017) and more recently Beijing in 2013 (Ning *et al.*, 2017).

Controversy surrounds many air pollution episodes, and disagreements don't always fade over time. There are frequent concerns over government inaction and the idea that the "real culprits" escape, perhaps most recently in relation to the Environmental Protection Agency (EPA) report (2003-P-00012) on its *Response to the World Trade Center Collapse: Challenges, Successes, and Areas for Improvement* (see also Lippmann, 2017). Such controversy is also inherent in media visions of air pollution episodes; e.g. a production about smog from the long-running Channel 4 Secret History series. The documentary series aims to give new insights into historical events, with a particular interest in the possibility that our understanding of them is misinformed, or that there has been a cover-up. Thus, *The Great Fog* which was part of this series, in September 1999 took a conspiratorial view of the 1952 smog (Brimblecombe, 2017). There is a long tradition of secret histories since *The Secret History* of Procopius, so it is hardly surprising to find such a treatment of London's famous smog, as memories of the incident fade. The episode at Donora also has a Secret History (e.g. http://donora.fire-dept.net/1948smog.htm), in which US Steel is seen as conspiring with US public health service (PHS) officials to cover up the role fluoride played in the tragedy (http://donora.fire-dept.net/1948smog.htm) and documented by Bryson (2006). In this chapter, we look at the Donora episode in terms of the persistence of controversy, as an example of the potential for extreme air pollution events to become the subject of continued dispute.

3.2 The Donora Smog

Donora started to grow as an industrial town in the opening years of the 20th century (Fig. 3.1). This growth took advantage of nearby

Fig. 3.1.   The Wire Mill, Donora, PA, taken by Bruce Dresbach in 1910. Retrieved from the Library of Congress.

coal such that it rapidly became a steel and wire-making town. By 1910, more than 8,000 people lived there. The rapid growth of industry in the river valley meant that air pollution problems were frequent even in those early years, with much smoke and the destruction of local vegetation.

The serious episode of 1948 developed from 27 October, 1948, when pollutants from the local industry began to accumulate under a temperature inversion. The slack winds meant that the pollutants were not removed from the city, and during inversions the air does not disperse vertically. Fog formed on the particles and acids in the atmosphere, with the high humidity providing the water needed for droplet growth. The residents soon began to experience respiratory problems. The thick fog additionally prevented any sunlight which would typically raise ground temperatures thus break up the inversion. These conditions allowed the fog to persist and led to sulphur dioxide concentrations that were said to be as high as 5,500 μg m^{-3}. The story is beautifully told in Berton Roueche's account in *The New Yorker* (1950a). As people began to die there was a realisation that a tragedy was unfolding. The town council called officials of the State

Health Department and asked for emergency aid. In the period immediately after the episode residents were still unwell, and of course there was fear that this smog could return (Fig. 3.2, Townsend, 1950). The Health Department team arrived the following week to undertake a preliminary survey, aiming "to find out precisely what caused the deaths, and to establish ways and means of preventing future tragedies of this kind."

In the first interviews with townspeople after the incident, they complained frequently about industry representatives and spoke of the lawsuits they had filed against the companies. However, bitterness toward the industries was short-lived. Once Donora was no longer the subject of media attention, residents were aware that continued complaints might lead to loss of employment, as local industries were forced to close (Davidson, 1979). Despite weakening

Fig. 3.2.   Woman wearing a surgical mask, Donora, PA, 21 April, 1949, image of Corbis-Bettmann.

environmental concerns, the industries eventually closed, not so much as a result of environmental pressures as to the collapse of the global steel market during the 1970s recession, and the enormous impact this had on employment by the 1980s (Skrabec, 2015). Pittsburgh had seen a steady decline in manufacturing employment since the mid-1950s. The anger Donora residents felt over the gradual loss of industry from a once-vibrant town is eloquently described in an essay for the Senior Division Historical Paper, National History Day 2011 Competition: "Boarded windows and nearly empty streets have replaced thoroughfares crowded with shoppers sporting Donora Dragons' orange and ebony colours. The high school, services stations and cinema all closed years ago" (Schroeder, 2011).

Thus, there was universal desire for steps to be taken so the disaster would not be repeated, but only as long as jobs were secure. The Pennsylvania Bureau of Industrial Hygiene worried about the possible increase in the frequency of smog episodes due to increasing industrialisation, so stressed the importance of adopting strong smoke control measures, but in the end the anticipated economic growth failed to materialise, and industry in the area declined; indeed Schroeder's essay (2011) argues that "Eventually, however, lawmakers mentioned "Donora" as a warning to temper ecological zeal in order to avoid economic catastrophe."

The report on Donora as Public Health Serv. Bull. No. 306 (Schrenk *et al.*, 1949) came out late in the year following the episode. Townsend (1950), the Medical Director and Chief, Division of Industrial Hygiene of the US PHS, presented the report on Donora to the American Public Health Association's annual meeting in New York, in October of 1949. He admits that investigators were "under a searching public scrutiny and often in the midst of much controversy". The publication of the report failed to stem controversy. Dr. Clarence A. Mills of the University of Cincinnati found the reported work inadequate, particularly in its inability to identify the toxic agent. Although one must be a little sympathetic to his views, given the rambling and guarded nature of some of the commentary, the abstract to the report was very clear: "The report's authors conclude that, of all possible pollutants, SO_2 acting synergistically with particulates was the probable causative agent". Mills (1949) had long been critical about the way research had been carried out in Donora, and the lack of funds available, but it was this position that caused his views to be countered as he had frequently complained about issues related to funding (Mills, 1948b).

Mills (1950) was convinced that in the long-term residents of Donora would suffer for the rest of their life, and that those who had died had somehow escaped: "These 20 suffered only briefly, but many of the 6,000 made ill that night will face continuing difficulties in breathing for the remainder of their lives". However, a follow-up study did not find much evidence to support the notion of continued suffering of residents, although this might be seen as an attempt to maintain the position taken by the 1949 *Bulletin on Donora*. It argued that findings 10 years on demonstrated that there were differences in the subsequent health experienced by those who had become ill during the 1948 the episode and those that did not. The differences appeared to relate to prior health status (Ciocco and Thompson, 1961). There was not a convincing difference in mortality. It was seen as somewhat disturbing that it was hard to find clear evidence of these effects, which could have led to clearer policy outcomes.

Davis (2002) has often argued that the number of deaths estimated was too small, suggesting that 50 more died in Donora, but thousands remain uncounted, among them many of her relatives, whose deaths she attributes to the smog. Certainly, the ideas of missing deaths parallel those from the later episode, the London smog of 1952, where there are well developed arguments that the number of deaths had been underestimated immediately after the event (Anderson, 1999; Bell *et al.*, 2004). This can happen because initial studies can readily become too narrowly focused, both temporally and spatially.

When reconsidering the events and the way they unfolded some of the more recent writing has relied on investigative reporting and interviews with elderly participants who were connected to various aspects of the tragedy. An influential article redirecting attention to Donora relies on an interview with the industrial chemist Philip Sadtler almost 50 years after the event. This article appeared as

The Donora Fluoride Fog: A Secret History of America's Worst Air Pollution Disaster in the Fall 1998 issue of *Earth Island Journal* by Chris Bryson. This work eventually formed part of a book *The Fluoride Deception* (Bryson, 2006) and revisits the possibility that fluoride was the injurious component of the Donora smog. Such oral histories can be difficult to assess many years after the events, but even those undertaken closer to the event can appear as fictional, e.g. William Wise's account of the London Smog of 1952 (1968) is easily read as fiction. In my own discussions with people, who lived through the London fog of 1952, I was surprised to hear, a seemingly conflated account from so many people I met, that they were fortunate to be led home by a blind man. Even Devra Davis' (2002) in her clearly historical study of the smog in Donora and London suffers when Wikipedia (2016) mislabels her book a novel; it was certainly not meant as such. Conversely accounts regarded as fictional can be highly informative, and just as with the London smog of 1952 (Brimblecombe, 2017), the Donora smog forms the backdrop to non-technical writing, such as Kathleen Shoop's novel *After the Fog* (2012) or Berton Roueche's story A Pig from Jersey in *The New Yorker* (1950b). Even Roueche's earlier account (1950a) has what must be fictional conversations, but are nevertheless informative as descriptions of the smog. Other elements of the account remind us of social behaviour in a very different age: "When Miss Stack [a nurse] had finished, she lighted a cigarette and sat down at her desk to go through the mail. It struck her that the cigarette had a very peculiar taste."

3.3 Origin of Health Outcomes

Despite increasing knowledge about the health effects of air pollution, the cause of the deaths is still discussed and remains a source of disagreement. In particular, questions persist about the mechanisms that caused the damage. Of course, it is perfectly true that in London as in Donora the weather conditions had an important part to play. However, it is not simply the weather as the contaminants obviously required pollutant sources. These were often underplayed in Donora much as they were in the Meuse Valley, in an attempt to construct a "… a *naturalisation* of the catastrophe" (Zimmer and Nemery, 2017). It is always much harder to agree on the sources of pollution. In London, there appear endless arguments; was it the trains, the diesel buses (Hughes and Mason, 2014), or was it Battersea Power Station (Bowler and Brimblecombe, 1991; Davis, 2002). Others were more interested in the nature of the contaminants present in the air. Did the deaths derive from the effect of coal smoke particles (Schwartz and Dockery, 1992), or was it the sulphur dioxide? From an advocacy point of view, some are concerned about the specific industry that should have been held responsible.

Given that the war was barely over, it is not surprising to hear a Donora resident claim that the death arose from "poison dropped from airplanes flying over" (Townsend, 1950). Even among experts there was persistent disagreement, and the choices regarding the causative agent became shrouded in controversy. The great American toxicologist Mary Amdur worked hard and in the face of much opposition to establish the importance of the sulphur compounds, in particular sulphuric acid droplets (Amdur *et al.*, 1953). Despite resistance from industry (Davis, 2002), her views were accepted, certainly by the authors from the PHS (e.g. Heimann *et al.*, 1958). This seems to fit with the importance of particles, but also population sensitivity to the acidic droplets typical in these intense winter smogs.

Mills (1950) strongly criticised weaknesses in the PHS report, and believed that the effects of red oxides of nitrogen (presumably NO_2) were neglected. This criticism did not go unchallenged, with Silverman and Drinker (1950) arguing that modelling showed that the nitrogen oxides were not critical, and fell below levels likely to cause damage. The tone of both these letters leaves something wanting, and the accuracy such that relevance of their observations may easily be missed by the contemporary reader.

Fluoride has perhaps remained as the most enduring of the alternatives to smoke and sulphur dioxide as the causative agent. Its relevance was certainly considered in the health response from exposure during early episodes at Meuse, Donora and London (e.g. Roholm, 1937). However, it has become associated with bitter controversies about the fluoridation of water supplies and too easily linked to the broader conspiracy theory associated with this issue: "there is good reason to believe that fluoridated drinking water and toothpaste — and the development of the atom bomb — are closely related" (Bryson, 2006).

In the Meuse Valley, fluoride was ultimately considered relevant only very close to the source. In Donora, the argument has remained a continuing alternative to the view that the health effects derived from the sulphur-rich haze (Davis, 2002; Bryson, 2006). The role of fluorides at Donora was proposed by Sadtler (1948), who was present at the time. He reported evidence of the role of fluoride and believed that the chronic poisoning found in Donora was to be blamed for the deaths. He had worked actively on this topic for some time, and wrote of the importance of this pollutant and its origin in the zinc mills of the area. Bryson (2006) traces the repression of his work, but its neglect may ultimately derive more from the consensus that the key role was played by sulphur dioxide and smoke; this ultimately gained ascendency. The view that fluoride caused the disaster is supported with a detailed interview. However, the case for the fluoride position seems also to depend on records and analyses that appear to have vanished or that have been withheld. Positions based on the absence of historical evidence make it hard for one to be convinced the conclusions drawn are well founded. It is not really surprising that the consensus view regarding the role of sulphur dioxide and particles is so strong.

Fluoride remains an important pollutant, but it may be more from water pollution than air pollution. Nevertheless, coal burning produces much of the fluoride found in the air, and China uses a large amount of coal, including domestic use in villages, where it can be mixed with clay. It is possible for the clay to be the dominant source of fluoride contamination, especially indoors (Wu *et al.*, 2004; Liu *et al.*, 2007). Although it has been shown to contribute to the particulate fluoride in Beijing's air, the long-range transport to Japan shows that dust is also important (Yang *et al.*, 2009).

3.4 Conclusion

Donora remains a turning point, and for some histories the assignment of culprits, chemical or commercial has been central to the continuing debate. It probably gets no easier to tease out the truth over time and as memories blur. Despite a consensus view of the relevance of sulphur dioxide and smoke in the health impacts of early episodes, there was a persistent view that fluorides played in major role in events in the Meuse Valley and at Donora and London. Some still believe that the fluoride story has been covered up, but it has also become entangled with the contentious issue of fluoridation of water supplies, so it is hard to unravel. However, coal smoke has largely vanished as a key contributor to urban air pollution episodes, and it is fine particles, from transport or through reactions in the atmosphere that have dominated urban health concerns in recent decades.

References

Amdur, M. O., Melvin, W. W. and Drinker, P. (1953). Effects of inhalation of sulphur dioxide by man, *The Lancet*, **262**, 758–759.

Anderson, H. R. (1999). Health effects of air pollution episodes, in Holgate, S. T. Samet, J. M. Koren H. S. and Maynard R. L. (eds.), *Air Pollution and Health* (Academic Press, San Diego, California), pp. 461–484.

Bell, M., Davis, D. L. and Fletcher, T. (2004). A retrospective assessment of mortality from the London smog episode of 1952: The role of influenza and pollution, *Environmental Health Perspectives*, **112**, 6–8.

Bowler, C. and Brimblecombe, P. (1991). Battersea Power Station and environmental issues 1929–1989, *Atmospheric Environment. Part B. Urban Atmosphere*, **25**, 143–151.

Brimblecombe, P. (2013). Deciphering the chemistry of Los Angeles smog, 1945–1995, in Fleming, J. and Johnson, A. (eds.), *Toxic Airs: Body, Place, Planet in Historical Perspective*, Vol. 31, (University of Pittsburgh Press, Pittsburgh, USA), pp. 95–108.

Brimblecombe, P. (ed.) (2017). London 1952: An enduring legacy, *Air Pollution Episodes* (World Scientific, Singapore).

Bryson, C. (2006). *The Fluoride Deception* (Seven Stories Press; New York, NY).

Ciocco, A. and Thompson, D. J. (1961). A follow-up of Donora ten years after: Methodology and findings, *Journal of Public Health*, **51**, 155–164.

Davidson, C. I. (1979). Air pollution in Pittsburgh: A historical perspective, *Journal of the Air Pollution Control Association*, **29**, 1035–1041.

Davis, D. L. (2002). *When Smoke Ran Like Water* (Basic Books, New York).

Heimann, H., Otis E. L., Prindle, R. A. and Fisher, W. M. (1958). Progress in medical research on air pollution, *Public Health Reports*, **73**, 1055–1069.

Holmes, J. A., Franklin, E. C. and Gould, R. A. (1915). *Report on the Selby Smelter Commission.* United States.

Hughes, J. D. (1996). *Pan's Travail* (Johns Hopkins University Press, USA).

Hughes, P. and Mason, N. J. (2014). *Introduction to Environmental Physics: Planet Earth, Life and Climate* (CRC Press, Boca Raton, Florida).

Lippmann, M. (2017). World Trade Center (2001): Health effects of the dust, in Brimblecombe, P. (ed.), *Air Pollution Episodes* (World Scientific, Singapore).

Liu, G., Zheng, L., Qi, C. and Zhang, Y. (2007). Environmental geochemistry and health of fluorine in Chinese coals, *Environmental Geology*, **52**(7), 1307–1313.

Mills, C. A. (1948a). Present distribution of medical research funds by governmental agencies, *Science*, **107**, 127–130.

Mills, C. A. (1948b). Present distribution of medical research funds by governmental agencies, *Science*, **108**, 438.

Mills, C. A. (1949). The Donora disaster, *Journal of the American Medical Association*, **140**, 556–557.

Mills, C. A. (1950). The Donora episode, *Science*, **111**, 67–68.

Ning, Z., Zhang, J. and Gao J. Beijing (2017). A driver of change, in Brimblecombe, P. (ed.), *Air Pollution Episodes* (World Scientific, Singapore).

Roholm, K. (1937). The fog disaster in the Meuse Valley, 1930: A fluorine intoxication, *The Journal of Industrial Hygiene and Toxicology*, **19**, 126–137.

Roueche, B. (1950a). The fog, *The New Yorker*, **30**, 1950, 33–51.

Roueche, B. (1950b). A pig from Jersey, *The New Yorker*, **18**, 1950, 157.

Sadtler, P. (1948). Fluorine gases in atmosphere as industrial waste blamed for death and chronic poisoning of Donora and Webster, Pennsylvania inhabitants, *Chemical and Engineering News*, **26**, 3692.

Schramm, E. (1990). Experts in the smelter smoke debate, in Brimblecombe, P. and Pfister, C. (eds.), *The Silent Countdown. Essays in European Environmental History*, (Springer-Verlag, Berlin), pp. 196–209.

Schrenk, H. H., Heimann, H., Clayton, G. D., Gafafer, W. M. and Wexler, H. (1949). *Air Pollution in Donora, Pennsylvania: Epidemiology of the Unusual Smog Episode of October 1948.* Public Health Service, United States, Washington, DC.

Schroeder, G. (2011). "Just Plain Murder": Public debate and corporate diplomacy in Donora's fight for clean air, *The History Teacher*, **45**, 93–116.

Schwartz, J. and Dockery, D. W. (1992). Particulate air pollution and daily mortality in Steubenville, Ohio, *American Journal of Epidemiology*, **135**, 12–19.

Silverman, L. and Drinker, P. (1950). The Donora episode — A reply to Clarence A. Mills, *Science*, **112**, 92–93.

Skrabec, Q. R. (2015). *The 100 Most Important American Financial Crises: An Encyclopedia of the Lowest Points in American Economic History* (Santa Barbara, California, Greenwood).

Shoop, K. (2012). *After the Fog* (CreateSpace Independent Publishing Platform).

Townsend, J. G. (1950). Investigation of the Smog Incident in Donora, Pa., and vicinity, *American Journal of Public Health and the Nation's Health*, **40**, 183–189.

Tucker, R. R. (1941). Smoke prevention in St Louis, *Industrial and Engineering Chemistry*, **33**, 836–839.

Wikipedia (2016). https://en.wikipedia.org/wiki/1948_Donora_smog, accessed 4th December 2016.

Wise, W. (1968). *Killer Smog: The World's Worst Air Pollution Disaster* (Rand McNally, Illinois, USA).

Wu, D., Zheng, B., Tang, X., Li, S., Wang, B. and Wang, M. (2004). Fluorine in Chinese coals, *Fluoride*, **37**, 125–132.

Yang, X.-Y., Yamada, M., Ning Tang, Lin, J.-M., Wang, W., Kameda, T., Akira Toriba, A. and Hayakawa, K. (2009). Long-range transport of fluoride in East Asia monitored at Noto Peninsula, Japan, *Science of the Total Environment*, **407**, 4681–4686.

Zimmer, A. and Nemery, B. (2017). The Meuse Valley (1930): Just fog or industrial pollution?, in Brimblecombe, P. (ed.), *Air Pollution Episodes* (World Scientific, Singapore).

Chapter 4

London (1952): An Enduring Legacy

Peter Brimblecombe

*School of Energy and Environment, City University
of Hong Kong, Hong Kong*

4.1 Introduction

Air pollution has been a problem in London since medieval times, with some bad smog in Victorian times (Brimblecombe, 2017). This was the classical smog, a term coined by Dr. Henry Antoine Des Voeux, who formed the word by combining the smoke and fog (Anon, 1905). During the 1940s the term began to be applied to Los Angeles smog, even though there was much awareness that its chemistry was quite different (Brimblecombe, 2013). During the early part of the 20th century, it appeared that the fogs vanished from London (Brodie, 1905; Bernstein, 1975; Brimblecombe, 1981) and there were some improvements in average air quality as a number of changes such as automatic stoking were adopted (Bowler and Brimblecombe, 1990). Other issues, such as burning bings (coal spoil heaps) and the construction of large urban power-stations such as the one at Battersea became important (Sheail, 2005; Bowler and Brimblecombe, 1991). There were increasing shifts in domestic fuel use to gas and electricity, which were seen as clean and modern. The studies of the late 19th century provide ample evidence that winter weather with its attendant "frost, fog and still air for days

57

together" presented a health risk. However, wars and a depression intervened and it was the London episode of 1952 that presented just this combination and drew attention to the dangers of air pollution (Whittaker *et al.*, 2004).

In the years following the Second World War, Britain was desperate to rebuild an economy shattered by the conflict. Many items were rationed; among these were domestic fuels. However, a low-quality *nutty slack*, consisting of slack (coal dust) and small lumps of coal (nuts), remained unrationed and became an alternative choice for home fireplaces. It often supplemented other low-quality fuel supplies during cold weather, and was blamed for the creation of London's smog (HANSARD, 1953). Although the *Great Smog* of 1952 is London's known best episode, there was a mid-century precursor that started in November 1948 and lasted 6 days, contributing some 800 excess deaths (Anderson, 1999). The comparatively small number of deaths, compared to the recent war-time blitz of London where roughly 40,000 civilians died, might explain the lack of overt political consequences (Parker in Berridge and Taylor, 2005).

4.2 Conditions During the Episode

The conditions that contributed to the formation of dense and polluted smog in London were: cold weather, low winds and a shallow inversion. Early December 1952 had been very cold, with heavy snowfall. On 5 December 1952, the sky was clear, and winds were light and damp. Conditions were ideal for the formation of radiation fog beneath an inversion of some 50–150 m. The air became stagnant under the anticyclonic conditions. Pollutants, notably smoke and sulphur dioxide, accumulated in the air along the Thames Valley. Initially, the fog was not particularly dense, but by evening visibility was at times less than 10 m (Chandler, 1965). For 114 h visibility remained continuously below 500 m, and below 50 m for 48 h in the centre of London. Finally, on 10th December, westerly winds dispersed the fog so London's worst air pollution episode came to an end.

As Sir Donald Acheson (Berridge and Taylor, 2005) remarked "As for my personal recollection of the smog itself, at its worst it had the effect of completely disorientating me in a part of London I knew well, so that I lost my way on a minor errand from the Middlesex Hospital to Oxford Street, 400 yards away. To get my bearings and to discover where I was, I had to creep on the pavement along the walls of the buildings, to the next corner, to read the name of the street. I do not recall any smell, but I do remember an eerie silence as there was little or no traffic. Visibility was less than 3 m, and it was bitterly cold."

London of the 1950s had a range of air pollution monitoring equipment, and although primitive this had the potential to provide information about the concentrations of pollutants. Unsurprisingly, concentrations were extremely high, although it is not clear whether these might be underestimates as the equipment was not really designed to work at such extreme concentrations. The daily maximum smoke concentration recorded at County Hall, Lambeth, was 4,460 μg m^{-3} with the daily average for all London sites 1,600 μg m^{-3}, while the maximum sulphur dioxide concentration, which occurred at County Hall was 1,340 ppb (Wilkins, 1954).

During the smog, there was a lack of awareness of the rising health consequences and journalists were preoccupied with other issues such as the failure of transport and cancellation of sporting fixtures. Weekend football had to be abandoned, and travel was much restricted (Parker in Berridge and Taylor, 2005). There is a widely-held view (e.g. Heimann, 1961), that the health effects went unnoticed until much later, but witnesses, such as Acheson (Berridge and Taylor, 2005) recollect that: "morticians ran out of space in the mortuary, and in the chapel of rest, and we had to use the anatomy department's dissecting room in another building." Nevertheless, it is likely that public perceptions of the smog may not have focused on health in the days of the smog or those that followed. Despite the apparent neglect of the health impacts even by the news media, on 12 December the *Daily Express* revealed London's hospitals were crowded with people suffering from fog illness. Ian Macleod, the Minister of Health, was questioned in the House of Commons about the number of deaths only a week after the fog had cleared. His estimates were: 4,703 during the fog compared with 1852 for the same period for the previous year. This was widely known, and for example, we find a Birmingham school boy reflecting in his diary on the year that had just ended: ". . . December and the Great Fog — the worst London had ever known — it killed hundreds of people" (Williams, 2005).

4.3 Health Effects of the Episode

The health impact of the Great Smog, particularly the large number of deaths that were readily attributable to the episode meant that it was well studied at the time. Even quite recently further papers have appeared on the health outcomes, varying from epidemiological re-analysis (Bell *et al.*, 2004), reviews (Anderson, 1999), records (Black, 2003) and re-examination of pathological materials that are still available (Hunt *et al.*, 2002; Whittaker *et al.*, 2004). "Bearing in mind the extreme loss of visibility in the streets, I would expect that many people died at home without medical help" (Acheson in Berridge and Taylor, 2005; Black, 2003).

In 1952, it was not necessarily easy to analyse recent epidemiological data, and often the counts were returned only slowly. Nevertheless, the London County Council's Chief Medical Officer of Health made a report 6 weeks after the event, suggesting that excess deaths amounted to 445 for the week ending 18 December. A month later the Chief Statistician at the General Register Office published its estimate of 4,000; a figure that remains oft-quoted after 60 years (Parker in Berridge and Taylor, 2005).

The early data analyses showed elevated deaths in London, while other cities of England showed only modest changes, ruling out simply the effect of temperature (Anderson, 1999). Nevertheless, there continued to be higher than expected death rates in January and February 1953, which added to the overall toll for the winter and worried contemporary investigators (Committee on Air Pollution, CoAP, 1953). The enhanced mortality of early 1953 was often attributed to influenza, but continued effects of air pollution may have been incorrectly assigned to influenza, such that the number of deaths resulting from the smog could easily have been higher (Anderson, 1999; Bell *et al.*, 2004). The impact of the smog may have had enhanced because of the low prevalence of influenza in the year prior to the 1952 smog. It is argued that it had sensitised the population, and there was an upward trend in respiratory disease outbreaks (Anderson, 1999).

Some 80–90% of the deaths immediately following the episode, were from cardiorespiratory disease, with bronchitis deaths at nine times the expected level. Most of the deaths were among the elderly, but the increases were proportional and similar across the 45–64, 65–75 and 75+ age groups. There was also a considerable increase in the younger population. Autopsy reports suggested an increase in the inflammation of the airways derived from overall toxicity. The event was striking because there was little need for subtle statistical analysis, everything was obvious. Perhaps more importantly, it was hard to see a qualitative difference between the health impacts of the air pollution during episodes and that which occurred on a day to day basis (Anderson, 1999).

4.4 Development of the Clean Air Act 1956

It is often claimed that legislation came only slowly, and that well over 3 years had to pass before the Clean Air Act (CAA) received royal assent on 5 July 1956 (Ashby and Anderson, 1981). However, legislation often develops rather slowly. A particular problem for the government was that it had made the construction of 300,000 houses part of their election manifesto of 1951 to rejuvenate post-war Britain. Harold Macmillan (Minister of Housing and Local Government: October 1951–October 1954; Prime Minister: January 1957–October 1963) says his outstanding preoccupation was housing as so many houses had been destroyed during the war years, and none built. Consequently, there was overcrowding and the stock was extremely poor, not least with respect to heating and insulation (Parker in Berridge and Taylor, 2005). The issue of clean air was split between central government departments. It had only recently been passed to the Ministry of Housing and Local Government from the Ministry of Health. There was just one civil servant in this ministry responsible for all aspects of air pollution.

In early 1953, parliament pressed for the fog episode to be taken seriously. However, there was an argument that there were pre-existing smoke clauses of Public Health Act 1936, so smoke could be controlled under this earlier legislation. Although this was true, the resulting CAA of 1956 ultimately had more than 40 clauses, compared with only six that were to be found in the earlier Act. The new legislation adopted an environmental tone. Fortunately, pressures on the government (Thorsheim, 2004) were sufficient to force it to assemble the Beaver Committee, in the summer that followed the smog, to investigate the issue. London County Council also set up its own Committee of Inquiry on Air Pollution, which reported in November of 1953 (Rivers and Drainage Committee, 1953). The Beaver Committee produced an Interim Report for presentation in December 1953, and a year later submitted its final report, which was generally well received. This report recommended that there be a CAA, and most notably that this would move beyond industrial sources to cover domestic emissions, a very dramatic shift from previous legislation. It wanted to reduce the impact of future smog episodes by issuing warnings to the public, giving advice to vulnerable people and maintaining the availability and use of smokeless fuels (CoAP, 1953). It was thus more stringent than the recommendations envisioned by the London County Council, which had rightly observed that smokeless fuels would not have any effect on sulphur dioxide emission.

When the government appeared likely to shelve the *Report of the Beaver Committee on Air Pollution,* the Tory MP Gerald Nabarro brought it forward as a private Bill. He was not a politician known for modern thinking, so his efforts with the Bill (along with the ever-controversial Enoch Powell) may seem a little surprising. However, Nabarro had been thoroughly briefed by the National Smoke Abatement Society before the Bill was read, which meant he was able to give well-informed comment on the Beaver Committee's recommendations. The Bill appealed to such a wide range of elected members of parliament, that the government was forced to put forward its own bill. This was debated in November of 1955 and received royal assent on 5 July 1956 (Ashby and Anderson, 1981). There followed other legislation, with the CAA being extended in 1968. Nearly 40 years later, the CAA of 1956 was repealed and consolidated into the CAA of 1993. This newer legislation included pollutants such as particles, sulphur dioxide, polycyclic aromatic hydrocarbons, dioxins and furans, in addition to smoke.

4.5 Changing London Air Quality

In terms of the major pollutants, air quality was generally improving from the beginning of the 20th century. The reasons for this may arise because London was expanding in response to the development of better urban transport, so the population density was decreasing along with the source strength (Brimblecombe, 1977), as seen for sulphur dioxide in the inset to Fig. 4.1. These trends continued through the 1950s and reinforced the improvements

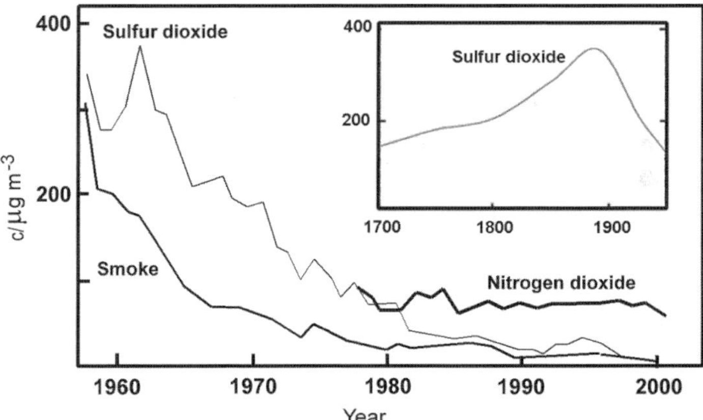

Fig. 4.1.   Changes in concentration of London's air pollution over 50 years. Data: Smoke and sulphur dioxide at County Hall until 1989 and then Westminster (Greater London Authority 2002) and nitrogen dioxide London Victoria until 1990 and then Bloomsbury. The inset shows the modelled change in sulphur dioxide since 1700 (Brimblecombe, 2006, 2017).

that had already occurred. These changes have often been attributed to the CAA (1956), but evidence of this relationship is not always convincing. The Act was not passed until 1956 and smokeless zones required by the act were not implemented in many areas until much later.

A few studies that look at long-term observations of air pollutants showed improvements to air quality that in some cases can go back to almost 1900 (Brimblecombe, 1977, 1987), so improvements may well have been in evidence prior to the CAA of 1956. Even in the early 20th century, there was a sense that things might be improving, which was partly attributed to better stoking. In the years after the CAA, there was evidence of a continued improvement in air quality. However, the legislation covered smoke, so the fact that sulphur dioxide was also in decline was a little surprising as this pollutant was not controlled by the 1956 Act. Recent appraisals of the mid-20th century improvements in air quality have been more thoughtful. The Quality of Urban Air Review Group of 1992 concluded that dramatic reductions in the concentrations of smoke and sulphur dioxide from the mid-20th century were "brought about as a result of burning cleaner fuels, especially the use of gas; tall stacks on power stations, and their relocation outside cities; and the decline of heavy industry." Giussani (1994) attributed the improvements arising from the CAA to activities in the industrial sector. However, the domestic sector also reflected huge social transitions in Post-war Britain; the all-electric house and gas fired central heating became desirable. Such lifestyle changes in mid-century Britain meant that the old coal-fired house was no longer attractive to modern urban dwellers (Brimblecombe, 1987).

Prior to the implementation of CAA 1956, there were worries that not enough smokeless fuel would be available, yet in reality such solid fuels had become less important by the time the act was being enforced. Areas declared black by the Beaver Committee were to implement the provisions of the CAA rapidly, but elsewhere things were slower and it was not until the 1990s that some locations finally had smokeless zones in place across the entire city. Eventually, the concerns arising in the original formulation of the CAA in that it failed to control sulphur were finally addressed in the CAA 1993, which enables local authorities to designate areas in which smoke may not be emitted, and only permit fuels if the sulphur content is less than 2% and which are effectively smokeless.

Despite the declining emissions of pollutants from coal, there are oft repeated stories of coal merchants selling unauthorised smoky fuels in their shops within smokeless zones; especially as solid fuels became more popular again in the 1990s, with an enthusiasm for solid fuel cookers (e.g. AGA, 2013). You can now purchase coal online, so it may be more difficult to trace, although some coal merchants are very clear on their websites that: "It is an offence to use bituminous coals as a fuel within a local authority designated smoke control area" and go on to emphasise that coal is not authorised for use in Smoke Controlled areas, so they cannot deliver this product to residential premises (http://www.abitofcoal.co.uk/winter.html).

Figure 4.1 shows the changes in pollutant concentrations for London across the years that followed the passage of the CAA. There has been a noticeable reduction in pollutants such as sulphur dioxide and black smoke, but less improvement in nitrogen dioxide. The difficulties in reducing nitrogen dioxide are widespread because it is produced from the oxidation of nitric oxide. Additionally, modern engines with pollution control devices tend to produce a higher ratio of NO_2/NO_x. Despite some reduction in emissions of nitric oxide emissions in the UK the NO_2 has been more difficult to reduce, and nitrogen oxides have been recognised as playing a role in winter smog during the late 20th century (Bower *et al.*, 1994).

More dramatic perhaps has been a shift to summer haze or photochemical smog. Here, change has come from the shift to the automobile as a key source of pollutants, with the volatile organic compounds as a precursor to photochemical smog. Liquid and gaseous fuels have also replaced solid coal, except in the largest plants such as power stations. However, even in the electricity industry, there has been a transition away from coal. There was a dash to gas after privatisation in the 1990s. At the time, high-interest rates and a need for simplicity, high efficiency and ease of construction favoured gas turbine power stations. Changing fuels meant larger amounts of reactive organic material in London's air, such that they serve to promote photochemical smog under sunlit conditions and mean higher ozone concentrations, a characteristic of London since the 1970s.

Additionally, a rising number of diesel vehicles can mean an increase in particulate material. This is a problem because the increasing amount of small particles PM-10 and PM-2.5 is likely to have an impact on health. London's former Mayor Ken Livingstone was uncompromising about the importance of the congestion charge, although this applies to only a small area within the city, so the effect on air quality is at best modest and always hard to detect (Atkinson *et al.*, 2009). A *Low Emission Zone* was subsequently proposed, aimed at reducing the emissions from commercial vehicles across much of Greater London. Nevertheless, the city still suffers from poor air quality typical of large modern cities, and continuing improvements are required by European Union regulations. Sulphur dioxide is now at reasonably low concentrations, although some concern remains that urban health may still be correlated with its concentrations.

4.6 Enduring Legacy of the Fog

Despite the fact that London's Great Smog occurred more than 60 years ago, it has an enduring influence. Naturally, memories fade and some argue that few know about the episode, and it has been covered up. This position might be sensed within Channel 4's documentary *The Great Fog* (28 September 1999) which treats the event of 1952. The programme is part of a long running series, Secret History that looks at issues that are concealed or manipulated so that the truth can become submerged beneath propaganda. Despite a view that the fog had faded from interest, its 50th anniversary was widely commemorated in 2002, with booklets and web pages produced by the BBC, the UK Met Office, etc.

Stories, true or untrue have grown up around the fog. As an example, it is frequently reported that cattle died in Smithfield Market, although Professor Patrick Lawther, Head of the Air

Pollution Unit at St Bartholomew's Hospital London, questioned this (Berridge and Taylor, 2005). *Secret History* television program seems to imply mass graves were used to bury the dead, although it is hard to establish this. However, it is clear that there were problems coping with mortuary space, much as in Paris during the heatwave of 2003. Sir Donald Acheson recalled the lack of mortuary space at Middlesex Hospital. Roy Parker (Berridge and Taylor, 2005) mentions that there was a delay in actually appreciating the scale of death, as the media focused on other issues, in particular, the disruption of transport. Additionally, as the smog occurred over a weekend it resulted in "the cancellation of almost every sporting fixture in the south-east of England".

Parker also notes that the media "concentrated on what was claimed to be the enormous increase in street crime", and suggested that the fog had increased housebreaking, which is hardly likely if home dwellers were trapped within (Berridge and Taylor, 2005). This association of crime and fog was long part of the Victorian detective novel that seems to have resulted from continuing fictional association of fog with crime since the days of Sherlock Holmes (Brimblecombe, 1990). In 1952, there was plenty of opportunities to link crime with fog. The mystery film by director Sam Newfield *Lady in the Fog* was released that year, along with the detective novel by Margery Allingham *Tiger in the Smoke*.

A literary legacy of the fog remains. *Neverwhere* by Neil Gaiman suggests: "The river mist had become a thick yellow fog.... the yellow-green fog became thicker and tasted of ash, and soot, and the grime of a thousand urban years." An imaginative novel which features the 1952 fog is *Dominion* by C. J. Sansom, which is set in London after Germany had won the war. This book took the Channel Four Secret History documentary on the Great Smog as a basis for its descriptions. Fascinating, as it seemingly links secret and alternate histories. The smog also has a presence in children's books such as Philippa Pearce's *A Dog So Small*.

There are enduring views that the smog was the outcome of London Power stations such as Battersea or a failure of its scrubber system (Pearson *et al.*, 1935) that should have cleaned the effluent, although detailed support for the claims tends to be absent. Others suggest that the replacement of electric trams by diesel busses caused the smog or that vehicles were responsible (Black, 2003). The modern episodes in Beijing (Ning *et al.*, 2017) are frequently compared to the London Smog of 1952, even though an even broader type of comparison is made when comparing the episode of January 2013 to those of London (e.g. Marshall *et al.*, 2013) or even Los Angeles (http://opinion.huanqiu.com/editorial/2013-02/3649030.html), hardly a city with winter smog from coal-smoke. Media reports in 2012, generated by the 60th anniversary of London's *Great Smog* of December 1952 encouraged comparisons with Beijing. Growing interest in China in London's historic air pollution led to the publication of a translation of *The Big Smoke* (Brimblecombe, 2015), a history of London's air pollution, into Chinese.

4.7 Enduring Legacy of the Clean Air Act

The Act itself has also left an impressive legacy. The previous section has argued that the CAA, rather than initiating an improvement in air quality, reinforced changes that were already underway. However, this should in no way be taken as lessening its importance, which should be seen as much greater than just changing air quality in London. The CAA (1956) was truly modern, and contained more than the few smoke abatement clauses of earlier acts. The new Act tried to foster practical approaches: a focus on smoke, exempting some practices and funding research. It is fascinating to see that much of the important technical background is embodied in memoranda associated with the legislation, such as those on chimney heights, smoke control areas and industrial premises. This practice, where the technical background is placed into larger ancillary documents and guidelines also is characteristic of more modern legislation. As an example, the formulation of the European Council Directive 96/62 on *Air Quality Monitoring and Management* saw the underlying science as embedded in its supporting position papers. The principles embodied in the directives are then detailed at the level of individual air pollutants in daughter directives or position papers. The approach to managing the CAA (1956) was also modern in terms of the desire for subsequent evaluation. In 1959, the government began the *National Survey of Air Pollution,* which sought evidence that could be used to test the effectiveness of the Act. The monitoring data gathered as part of the *Survey* were published as regional analyses of changes in air pollution over the years, 1961–1971 and have contributed to the notion that air pollution from coal use typically declined during these years.

The development of the CAA required a reduction in smoke from domestic sources. This was extremely innovative and is in some ways a major novelty of the Act, as earlier smoke abatement focused on industrial sources. Thus, the CAA challenged personal freedom by attempting to control what took place inside individual homes and so limited the choice of what could be burnt in home fires. A lingering problem of freedom would emerge with issues such as garden bonfires. While there are no specific laws on bonfires, if the problem was significant and persistent it could be an offense (under the *Environmental Protection Act,* 1990). In recent years, there have been a number of serious pollution episodes in the capital, of which the best known is the 1991 episode, but there was a more recent and widespread one in the early spring of 2014 (Elliot *et al.,* 2016). Fortunately, these have not been associated with the dramatic increase in death rate experienced in the smog of 1952. The lower impact has made it harder to be certain of the relationship between air pollution and mortality, but air pollution seems the most plausible explanation. This also meant it was not possible to establish the relative importance of nitrogen dioxide and particulates on health outcomes (Anderson *et al.,* 1995).

The CAA (1956) was seen both in the UK and abroad as an important example of environmental legislation, even though many correctly observed that it failed to address the problem of sulphur dioxide. Despite questions about exactly what the act achieved, and the reasons for improved air quality, it was a turning point in the regulation of air pollution.

References

AGA (2013). How the AGA Became an Icon., http://portal.agamarvel. com/dl/415/AGA-Cast-Iron-Ranges-How-AGA-Became-an-Icon.pdf.

Anderson, H. R. (1999). Health effects of air pollution episodes, in Holgate, S. T., Samet, J. M., Koren, H. S. and Maynard R. L. (eds.), *Air Pollution and Health* (Academic Press, San Diego, California), pp. 461–484.

Anderson, H. R., Limb, E .S., Bland, J. M., De Leon, A. P., Strachan, D. P. and Bower, J. S. (1995). Health effects of an air pollution episode in London, December 1991, *Thorax*, **50**, 1188–1193.

Anon. (1905). Smog, *Journal of the American Medical Association*, **45**, 637.

Ashby, E. and Anderson, M. (1981). *The Politics of Clean Air* (Clarendon, Oxford).

Atkinson, R. W., Barratt, B., Armstrong, B., Anderson, H. R., Beevers, S. D., Mudway, I. S., Green, D., Derwent, R. G., Wilkinson, P., Tonne, C. and Kelly, F. J. (2009). The impact of the congestion charging scheme on ambient air pollution concentrations in London, *Atmospheric Environment*, **43**(34), 5493–5500.

Bell, M., Davis, D. L. and Fletcher, T. (2004). A retrospective assessment of mortality from the London smog episode of 1952: The role of influenza and pollution, *Environmental Health Perspectives*, **112**, 6–8.

Berridge, V. and Taylor, S. (2005). The Big Smoke: Fifty years after the 1952 London Smog. Centre for History in Public Health, http://history. lshtm.ac.uk/files/2013/07/BigSmokeNS.pdf.

Black, J. (2003). Intussusception and the great smog of London, December 1952, *Archives of Disease in Childhood*, **88**, 1040–1042.

Bower, J. S., Broughton, G. F. J., Stedman, J. R. and Williams, M. L. (1994). A winter NO_2 smog episode in the U.K., *Atmospheric Environment*, **28**, 461–475.

Bowler, C. and Brimblecombe, P. (1990). The difficulties of abating smoke in late Victorian York, *Atmospheric Environment*, **24B**, 49–55.

Bowler, C. and Brimblecombe, P. (1991). Battersea Power Station and environmental issues 1929–1989, *Atmospheric Environment*, **25B**, 143–151.

Bernstein, H. T. (1975). The mysterious disappearance of Edwardian London fog, *The London Journal*, **1**, 189–206.

Brimblecombe, P. (1977). London air pollution, 1500–1900, *Atmospheric Environment*, **11**, 1157–1162.

Brimblecombe, P. (1981). Long term trends in London fog, *The Science of the Total Environment*, **22**, 19–29.

Brimblecombe, P. (1987). *The Big Smoke* (Methuen, London).

Brimblecombe, P. (1990). Writing on Smoke, in Bradby, H. (ed.), *Dirty Words* (Earthscan Publications, London), pp. 93–114.

Brimblecombe, P. (2006). The clean air act after 50 years, *Weather*, **61**, 311–314.

Brimblecombe, P. (2013). Deciphering the chemistry of Los Angeles smog, 1945–1995, in Fleming, J. R. and Johnson, A. (eds.), *Toxic Airs: Body, Place, Planet in Historical Perspective* (University of Pittsburgh Press, Pittsburgh), pp. 95–108.

Brimblecombe, P. (2015). *The Big Smoke* (Shanghai Academy of Social Science Press, Shanghai) (translated from Brimblecombe, P. (1987), Methuen, London).

Brimblecombe, P. (ed.) (2017). Early episodes, *Air Pollution Episodes* (World Scientific, Singapore).

Brodie, F. J. (1905). Decrease in London fog during recent years, *Quarterly Journal of the Royal Meteorological Society*, **31**, 15–28.

Chandler, T. (1965) *Climate of London* (W. Heffer & Sons, Cambridge, England).

CoAP (1953). Interim Report of the Committee on Air Pollution (Cmd. 9011). HMSO, London.

Elliot, A. J., Smith, S., Dobney, A., Thornes, J., Smith, G. E. and Vardoulakis, S. (2016). Monitoring the effect of air pollution episodes on health care consultations and ambulance call-outs in England during March/April 2014: A retrospective observational analysis, *Environmental Pollution*, **214**, 903–911.

Giussani, V. (1994). The UK Clean Air Act 1956: An Empirical Investigation. (Centre for Social and Economic Research on the Global Environment, University of East Anglia, Norwich).

HANSARD (1953). Nutty slack, HC Deb 02 February 1953 vol 510 cc1460-2.

Heimann, H. (1961). Effects of air pollution on human health, *Air Pollution*, 159.

Hunt, A., Judson, B., Berry, C. L. and Abraham, J. L. (2002). Histo-compartmental analysis of retained fine particles in the lungs of London residents who expired at the time of the great smog of 1952, *Annals of Occupational Hygiene*, **46**, 377–381.

Marshall, M., Coghlan, A. and Slezak, M. (2013). China's struggle to clear the air, *New Scientist,* **217**(2903), 8–9.

Ning, Z., Zhang, J. and Gao J. Beijing (2017). A driver of change, in Brimblecombe, P. (ed.), *Air Pollution Episodes* (World Scientific, Singapore).

Pearson, J. L., Nonhebel, G. and Ulander, P. H. N. (1935). The removal of smoke and acid constituents from flue gases by a non-effluent water process, *Journal of the Institution of Electrical Engineers,* **77**, 1.

Sheail, J. (2005). 'Burning bings': A study of pollution management in mid-twentieth century Britain, *Journal of Historical Geography,* **31**, 134–148.

Thorsheim, P. (2004). *Inventing Pollution: Coal, Smoke, and Culture in Britain Since 1800* (Ohio University Press, Athens, Ohio).

Whittaker, A., BeruBe, K., Jones, T., Maynard, R. and Richards, R. (2004). Killer smog of London, 50 years on: Particle properties and oxidative capacity, *Science of the Total Environment,* **334**, 435–445.

Wilkins, E. T. (1954). Air Pollution and the London Fog of December, 1952 *Journal of the Royal Sanitary Institute,* **74**, 1–15.

Williams, B. D. (2005). *The Diary of a Birmingham Schoolboy, 1952,* http://web.ukonline.co.uk/brian.david.williams/diary/1952.html.

Chapter 5

Mount St Helens Eruption (1980): A Severe Air Pollution Episode from Volcanic Ash

Peter Baxter*, Robert Bernstein[†] and A. Sonia Buist[‡]

*Institute of Public Health, University of Cambridge, UK
[†]Department of Global Health, Rollins School of Public Health,
Emory University, USA.
[‡]Pulmonary and Critical Care Medicine,
Oregon Health & Science University, USA

5.1 Introduction

The cataclysmic eruption of Mount St Helens (MSH) on 18 May 1980, was a watershed in modern volcanology and was the first time that US scientists had to confront the fury of an explosive volcano in the conterminous US. This account focuses on one aspect of this seminal eruption, namely the massive ash fall that accompanied it, which blanketed parts of the states of Washington, Idaho and Montana for 5 days and resulted in over a million people being exposed to unprecedented high levels of particulate air pollution from suspended ash. The event triggered an immediate response by the scientific and public health communities, and the finding of relatively high levels of crystalline silica in the ash refocused attention from simply evaluating the acute effects of the ash

73

on respiratory health to including an evaluation of the potential long-term risk of silicosis and other lung conditions that could arise from chronic exposure to the ash that lay over open areas or could be erupted by further volcanic activity.

Far from being merely of historic interest, the acute and chronic health issues raised by the ash fall episode in this eruption have surfaced repeatedly at major eruptions of explosive volcanoes around the world ever since, but St Helens remains the only large eruption to have been extensively studied by virtue of its occurrence in the US and with such a large population affected. Hence, it represents a seminal event for global disaster preparedness and response in volcanic areas. A review of the evidence it provides is therefore timely, especially in view of the remarkable advances in our understanding of the effects of ambient and occupational exposures to particulate air pollution on human health over the last two decades, including the rationale of the new metrics (PM 2.5 and PM 10) used for setting air quality standards on particulate matter.

Another watershed eruption was at the Eyjafjallajökull volcano in Iceland in 2010, when an ash plume was continuously emitted for about 6 weeks during which it blew directly towards Europe, halting air traffic for 6 days and raising concerns in European governments about the acute health effects in citizens of the particulate air pollution reaching their countries (Langmann, 2017). In the UK, in particular, the wider implications of larger open vent eruptions occurring in Iceland in the future raised the level of awareness about the possible impacts of volcanic activity on the health of the population and its economy.[a] For the first time, Icelandic volcanic eruptions became a legitimate interest of European public health officials and respiratory physicians and, in addition, they joined the other major natural and man-made risks on the UK National Risk Register.

---

[a] "Economic Effects of the Mount Saint Helens Eruptions", USITC Publication 1096 September 1980: a Report to the Committee on Ways and Means of the U.S. House of Representatives on Investigation No. 332-110 Under Section 332 of the Tariff Act of 1930. Available at: https://volcanoes.usgs.gov/vsc/file_mngr/file-78/pub1096.pdf (accessed 11 December 2016).

5.2 Background to the Eruption

Located in the Pacific North–West in the State of Washington, Mount St Helens (MSH) is the most active volcano in the Cascade Range, which comprises a string of volcanoes that stretches from northern California to British Columbia, Canada, and where a major eruption happens on average once every 100 years. Surrounded by forest in an area traditionally providing employment in logging as well as recreation for hikers and campers, the nearest large population centres are the city of Portland, Oregon and Vancouver, Washington, 30–40 km towards the southwest.

By a fortunate coincidence, the US Geological Survey (USGS) had undertaken a study of the volcano's eruptive hazards, including past heavy ash falls, based on reconstructing its past eruptions from field evidence. They published their findings in an official report (the Blue Book) in 1978 (Crandell and Mullineaux, 1978), and this was to prove invaluable to the emergency planners. The authors, Crandell and Mullineaux, had even included a prescient statement to the effect that the next eruption could occur in a few years and before the end of the century. With the alarm that followed the new signs of activity, this text was immediately passed around all the key state and federal agencies involved. The US Forest Service took on the responsibility for emergency planning as the volcano was inside the Gifford Pinchot Forest National Park; other agencies that developed contingency plans included the Washington State Department of Emergency Services, the Federal Aviation Administration and the Washington National Guard (Waitt, 2014).

When MSH began its threatening seismic activity on 20 March, 1980, it had been in repose for 130 years. Anxious seismologists and volcanologists in the USGS, who were responsible for monitoring for tell-tale signs of a developing eruption soon understood they were facing a daunting prospect. Not only had the USGS very limited experience of dealing with crises at active volcanoes that erupted explosively, volcanoes like MSH were also very unpredictable as far as forecasting the timing, type and size of its eruptive behaviour was concerned, even with the latest scientific monitoring equipment at their disposal. This had one classic outcome often encountered in volcanic crises: the public and the media were constantly pressing for clear assurances and informed decisions on the crisis as it continued to grow over a two-month period of mounting volcanic unrest. Meanwhile, the authorities and scientists were battling at the edge of scientific knowledge to come up with evidence-based answers. Inevitably, the emergency planning undertaken by the authorities, with technical guidance from the USGS, was limited in some important respects, including preparing communities for the extent of the ash fall described in the Blue Book. The fact that the volcano was very much an unknown quantity greatly added to the danger close to the volcano and contributed to the uncertainty in the population around the whole region over what could unfold (Waitt, 2014).

5.3 The Eruption on 18 May 1980

As the crisis escalated, the seismic activity was building when a small eruption of steam and ash happened on 27 March, with a small crater appearing at the ice-covered summit. This event shifted the emergency response to another level and for the first time access to areas around the volcano began to be restricted. Small eruptions continued with ash being blown downwind, but the portents got worse when the seismicity included harmonic tremor, a sign of magma on the move, from 3 April onwards. A further dramatic escalation was on 30 April, with the appearance of a bulge on the north flank which grew outwards by a metre a day and fueled fears of the flank collapsing in a massive avalanche. As the scientists and the authorities were following these developments with increasing apprehension, further restrictions on access were being introduced against substantial local opposition. On the fateful day of 18 May, a group of cabin owners was preparing to enter the hazard zone to check on their properties in the mistaken belief that there was a lull in the volcanic activity when the massive eruption began at 08.32 without any warning that could be detected by USGS volcanologists despite the use of the most up-to-date monitoring systems (Waitt, 2014; Sarna-Wojcicki *et al.*, 1981).

There were five major aspects of the eruption on 18 May: the flank avalanche; a lateral blast with a pyroclastic surge overriding this, akin to a massive directed explosion (killing 57 people); destructive mudflows and floods along river channels; and an extensive ash plume that dumped ash across central and eastern Washington State, northern Oregon State and beyond as illustrated in Figs. 5.1–5.4 (Sarna-Wojcicki *et al.*, 1981). The extent of the hazard area around the volcano had been based on the Blue Book and had actually been drawn too small as planners had not allowed for a lateral blast and surge. But, unaccountably to us today, the population had not been involved in any planning at all for a prospective ash fall over extensive areas east of the volcano, and, to add to their woes, no warning system had been put in place. Within minutes of the lateral blast, the top of the volcano was blown away and a powerful vertical eruption column was thrust upwards for 20 km

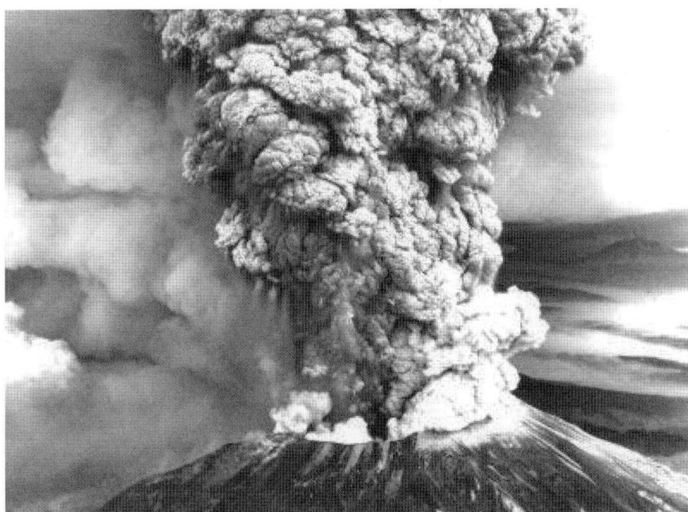

Fig. 5.1.   Plinian column from 18 May 1980, eruption of MSH. Aerial view from southwest. Mount Adams is in the background (right).

Source: https://volcanoes.usgs.gov/volcanic_ash/mount_st_helens_health.html (accessed 18 September 2016).

Fig. 5.2.   The slopes of Smith Creek valley, east of MSH, show trees blown down by the 18 May 1980, lateral blast in the Weyerhauser Corporation logging area. Two USGS scientists can be seen in the lower right, giving the scale.*

Source: https://volcanoes.usgs.gov/volcanic_ash/mount_st_helens_agriculture.html (accessed 18 September 2016).

Fig. 5.3.   Left: Ash plume from felled tree in MSH logging area; Right: mowing a lawn in Portland, June 1980.

Sources: Buist *et al.* (1986). The development of a multidisciplinary plan for evaluation of the long-term health effects of the MSH eruptions, *American Journal of Public Health* **76**(Supplement), 39–44.

https://volcanoes.usgs.gov/volcanic_ash/mount_st_helens_agriculture.html   (accessed 18 September 2016).

Fig. 5.4.   Pre-eruption and post-eruption views of MSH.

Source: The 1980 eruptions of MSH, Washington, available at http://pubs.usgs.gov/pp/1250/report.pdf (accessed 18 September 2016).

above the new, vast crater. This vertical eruption continued for the next 6 h, forming a vast plume that drifted eastwards. The first indication that this was happening for most people was when they spotted a huge and ominous, black ash cloud approaching overhead (Saarinen and Sell, 1985). Figures 5.2 and 5.3 indicate the impacts of the tree blow-down and ash fall exposures on the MSH logging operations of the Weyerhauser Corporation that resumed in the weeks after the eruption.

5.3.1 *The ash fall and its immediate impacts*

On 18 May, helicopters of the Washington Army National Guard were temporarily stationed near Yakima about 90 miles east of MSH. No timely alerts had been issued by the Washington Department of Emergency Services (DES) and thus most officials, as well as residents who were destined to be under the plume, did not know that the eruption was underway. At about 09:45, the ash started falling in Yakima and the pilots scrambled to get airborne, with 19 helicopters getting away in time. Later that day, they took part in the search and rescue operations around the volcano (Saarinen and Sell, 1985).

Given the lack of preparedness, even the knowledge that MSH had erupted would not have perturbed the people living in Eastern Washington, as they had no previous experience of volcanic activity (there had been none in the Cascade Range since Lassen Peak's mild eruption in 1914, which had been long forgotten) and they thought that by living as far as 90 or 300 miles away removed them from any of the eruptive threats from the volcano. The event is, therefore, illustrative of communities totally unprepared for the volcanic air pollution and its immediate effects on all aspects of life over the hours and days that followed. Not even the communities that were first impacted passed on alerts to those further down the plume's path, but then they may not have known about the extent of the plume moving overhead and were in any case preoccupied by their own problems and concerns that the ash was causing. Even so, this lack of awareness is inconceivable today with all the connectivity of social media. No warnings, with advice on what to do, were sent out by the DES to any of the communities subjected to ash fallout (Saarinen and Sell, 1985).

The course of the plume was east–northeast and for the first 13 miles moved at 100–150 miles an hour, but over the next 600 miles, its average speed was nearer 60 miles per hour (Saarinen and Sell, 1985). The plume was tracked by satellite, radar and ground observations. The ash fell mainly on eastern Washington (see Fig. 5.5), northern Idaho and western Montana. Ritzville, about 100 miles northeast of Yakima, saw the ash cloud coming around noon and, as at Yakima, visibility fell to "zero" in the fallout. The same thing happened at Moses Lake and Ellensburg. The bright, sunny day was transformed into darkness as the black cloud of ash enveloped the land on an almost biblical scale. It was the same experience at Spokane and in Missoula, Idaho, where the edge of the cloud arrived some 6–9 h after the eruption had begun (Saarinen and Sell, 1985). In the blackout, some people in Yakima who had been out on the town the night before (Saturday) were reported to us to have woken up, looked outside, and believing it was still nighttime, went back to bed.

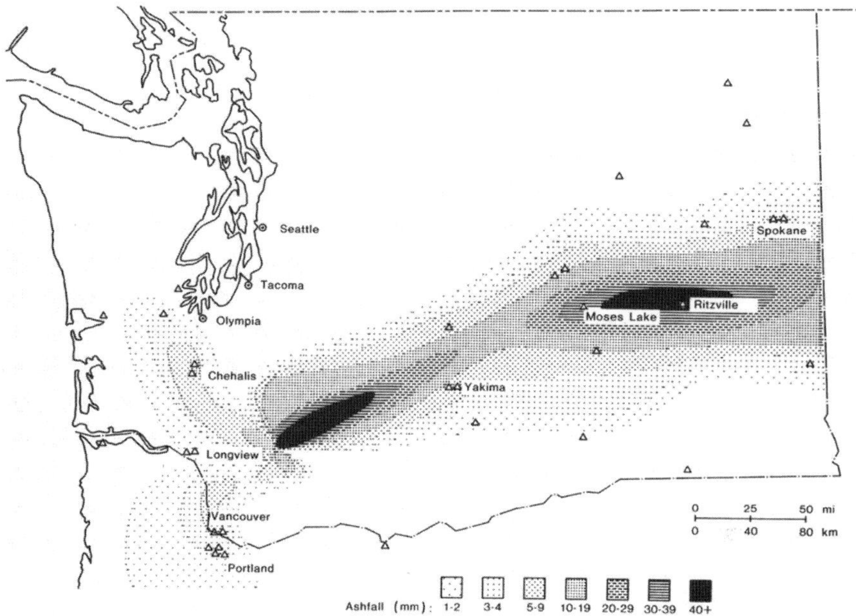

Fig. 5.5.   Ash fall distribution after the three major ash eruptions of MSH and locations of Washington and Oregon Hospitals under CDC surveillance (USGS).

5.3.2 *Impacts of the ash fall*

What happened in the ash fall areas thereafter was outside anyone's previous experience, including most volcanologists. The ash deposit along its central axis was at least 1 cm in depth and became 4–5 cm in two main areas — close to the volcano and, more critically for human exposure, around Moses Lake and Ritzville (Sarna-Wojcicki *et al.*, 1981; Baxter *et al.*, 1981, 1983). The ash that fell on Yakima had a visibly larger coarse component than that which blanketed Moses Lake and Ritzville, which was very fine and easily became airborne again if disturbed. It was this property of the ash that led to the continuation of the crisis for the next 5 days and the fact that the fine ash only settled very slowly.

A state of emergency was declared along with the closure of schools, businesses and government offices in affected areas of the states of Washington and Oregon where people were told to stay in their homes (Saarinen and Sell, 1985). On 21 May 1980, the President declared the State of Washington and part of Idaho major disaster areas following the catastrophic eruption of MSH and the U.S. Federal Emergency Management Agency (FEMA) initiated efforts to coordinate the provision of more than $900 million in financial and technical assistance from 12 federal agencies to the states, counties, and populations in communities in affected areas.[b]

The main task for the population was the removal of the ash from rooftops, residences and the main thoroughfares, but this was initially a slow and massive undertaking, especially dealing with the dry, fine ash when there was a lack of specialised equipment and manpower. They used wetting methods, drawing on limited municipal water supplies, to suppress the ash or shovel it up, and wore improvised masks from wet or dry cloths and surgical masks, as well as industrial respirators (Bernstein *et al.*, 1986). There were road accidents and some falls from roofs as people tried to clear the ash off their houses. Another unexpected problem was moist ash short circuiting electrical equipment. The ash was so abundant that sewage and waste water pipes became clogged and sewage treatment plants were put out of action in Yakima, Moses Lake and Spokane (Saarinen and Sell, 1985; Baxter *et al.*, 1983).

The initial communication failures had one immediate consequence in Eastern Washington which should have been avoidable. The confusion resulted in delays in stopping travel into the areas before the plume arrived. As a result, it was estimated that up to 10,000 people, mainly road users, became stranded when the ash brought all normal activities to a halt, and many motorists had to be rescued from their vehicles (Saarinen and Sell, 1985). The initial pitch-black darkness prevented almost all road, air and railroad movements over the next three days while conditions only very

---

[b]See the message from President Jimmy Carter to Governor Dixy Lee Ray, available at: https://www.csub.edu/~rdaniels/Mt.%20St.%20Helens%20Disaster%20 Declaration,%201980.pdf (accessed 28 March 2017).

slowly improved. Wind and outdoor activities, particularly driving, would resuspend any ash that had settled on the ground, and ash would clog the air filters of the engines of emergency vehicles. The stranded travellers found refuge in Red Cross emergency shelters, or checked into motels, or stayed with friends and relatives. When conditions started to improve by 22 May, a few trains and interstate buses began to move, and motorists who had been stranded could be moved out from the remotest places. The Interstate I-90 that passed through the heart of the fallout area, however, was not opened until one week after the eruption. Normal living was not to resume for 10 days (Saarinen and Sell, 1985).

It was not until 26 May that the conditions transformed after an unseasonal, steady rainfall that allowed life to return to a daily routine (Baxter *et al.*, 1983; Bernstein *et al.*, 1986). The fine ash was dramatically cleared from the air by the rain and consolidated into firm deposits that could be more easily removed from the ground. Had rainfall occurred earlier after the eruption, this whole protracted episode would have ended much earlier, and experience at volcanoes since has repeatedly shown this crucial role of rain in clearing the air of ash. Yet, the ash on the ground was destined to remain in most open areas, road verges and fields, as these were not possible to clear and wind could drive clouds of fine ash into the air whenever the deposits dried out. In towns, thin layers of ash remaining on the surfaces of roads led to the resuspension of ash as traffic passed, making substantial contributions to the ambient air levels of fine ash during the daytime in densely populated areas.

The mantle of ash that had fallen over areas of Washington and three other states (Oregon, Idaho and Montana) was all the more disruptive because of the huge anxiety and uncertainty, it provoked over the possible health effects. In the absence of any previous publications or detailed studies on the health effects of volcanic eruptions, these fears could not be easily allayed by health officials without further investigation. A major effort began within days of the eruption by the US Centers for Disease Control (CDC) in close collaboration with the USGS and other federal agencies to elucidate the hazards and the risks to human respiratory health of the ash (Baxter *et al.*, 1981, 1983).

5.4 Investigating the Health Hazards of Volcanic Ash

5.4.1 *What is volcanic ash?*

For health scientists, including clinicians, the volcanic ash from MSH or, for that matter, from any other volcano, was an unknown quantity. On the day after the eruption, the CDC was contacted by the Idaho State Health Department after people seeking reassurance over possible health effects of the ash besieged hospitals and health centres in the northern part of the State covered by a layer of ash. The Washington State Department of Social and Health Services requested CDC assistance and a CDC team arrived on 21 May to give scientific support. Meanwhile, communities along the line of the ash fall were contending with the shock of the choking ash in the air and the problems it was causing, not only to human health, but every aspect of their daily activities which had become severely curtailed. The following summary on volcanic ash in general and the respirable fraction of MSH ash, in particular, was not available to the population at the time of the 18 May eruption, and only slowly emerged from the investigations conducted by CDC and the Division of Respiratory Disease Studies of CDC's National Institute for Occupational Safety and Health (NIOSH). The following summary is presented here to assist the reader in understanding the nature of the air pollution encountered during and after the eruptions of MSH.

5.4.2 *What is volcanic ash? Composition and morphology of the ash*

Volcanic ash is a specific type of natural mineral particle that is encountered in many different forms. There is no such thing as generic volcanic ash, and the potential health effects will depend on the exposure context as well as the very variable factors responsible for its physical and chemical make-up, as conferred on it by the volcano and its mode of eruption. Hence, the mineral composition of ash reflects the type of the magma, or molten rock, from which it derives, which in turn determines the type of explosive volcano and its style of eruption. Thus, for the geologist, the mineralogy of volcanic ash and rock is a clue to the pre-eruption physical and chemical conditions of the magma in its ascent to the surface. Geologists define ash as the erupted material less than 2 mm in diameter, which for health professionals denotes little about the respiratory health hazard a particular type of ash may present. Thus, the immediate health concerns of the people living in the ash fall areas could not be assuaged without much more information on the ash particles themselves, including the amount of fine, inhalable material that could be suspended in the ambient air and whether the ash could contain toxic minerals, in particular crystalline silica and asbestiform particles. Without any official prompting, laboratories all over the region began to undertake their own studies on ash samples that they collected indiscriminately from nearby ash, with all kinds of different results disseminated in the media to add to the general confusion.

As a default position in the initial knowledge vacuum, many public health officials and clinical doctors tried to reassure the population that the ash posed little risk, while respiratory physicians and others were generally cautious and sought more reliable information from health laboratories with experience in undertaking mineralogical analyses. Another issue that presented, and indeed has since arisen at every ash eruption around the world, was whether the ash could contaminate drinking water and vegetation with dangerously high levels of toxic elements that might add to the potential health risks for both humans and livestock.

Volcanic ash is comprised of silicon as silicate in the form of glass in the presence of an assemblage of certain other minerals. At MSH, the glass was dacitic in composition, reflecting the silicon dioxide concentration in the magma. The mineral assemblage is variable depending upon the eruption conditions, but comprises about two dozen minerals. The most abundant of these are silicates such as feldspars, feldspathoids and ferromagnesian minerals. The fallout may also contain, as it did at MSH, pumice and a varying percentage of crystals, along with pulverised material abraded from the walls of the volcanic conduit or the vent. The morphology of the finer particles in different *magmatic* eruptions can be superficially similar, and larger particles may be rounded by the eruption process, but the shape becomes more variable as the size of the particles increases. Ash from MSH also contained numerous dacitic glass shards, most visible in the sub-10 micron fraction. If the magma comes into contact with water, either from groundwater in the volcano or, for example, by an eruption driving through a lake, the explosiveness is increased and the morphology of the ash will reflect this and be characteristic of *phreato-magmatic* eruptions (Newhall and Fruchter, 1986).

The textures of the surface of the particles can also vary widely between eruptions. Some surfaces are smooth, others are pitted, stepped, or otherwise rough. Rough surfaces retain more moisture and insoluble salts and weather faster than do smooth surfaces. The main volatiles released from the magma on its ascent inside the volcano are the volcanic gases, and these and other chemicals can readily adhere to the particle surfaces. Even before the MSH eruption, many years of experience from eruptions in Iceland had already alerted the scientific community to the possibility of fluoride on the volcanic ash, which may present a lethal hazard to livestock grazing in ash-covered fields (Newhall and Fruchter, 1986). The presence of significant amounts of toxic elements on the surface of the ash grains was, however, rapidly excluded on laboratory testing.

The volcanic gases, which included sulphur dioxide, a common air pollutant in fossil fuel combustion, vented harmlessly to the upper atmosphere during the main eruption, and thereafter were only emitted in small quantities and played no significant part in air pollution in communities.

This brief description should be enough to show that volcanic ash cannot be simply regarded as similar to ordinary rock that has been pulverised into fine fragments, or any other more familiar forms of natural mineral particles. In addition, the surfaces on freshly erupted ash grains will not have had time to be exposed to weathering processes, but instead will have been involved in chemical reactions at different stages of magma ascent and eruption, and also while the ash is being transported in the atmosphere as a plume of ash and gases. It used to be thought that most of this chemical "soup" on the ash surface was water soluble and adsorbed on the grain surfaces, but recent studies point to important chemical reactions with the surfaces as well (Olsen and Fruchter, 1986).

These are some reasons why freshly erupted ash might well have very different acute effects on the respiratory system compared with air pollution episodes from other types of natural mineral particles, such as desert dusts or loess, or compared to weathered ash that was suspended by winds and human activity in the weeks and months after the ash fall. Also, we should be cautious about extrapolating the health consequences of ash from one volcano to another, or even between different eruptions at the same volcano.

5.4.3 *Particle size of the ash*

At MSH, the investigation by the CDC/NIOSH and reliable laboratories in the region found that 90% of the particles by count were in the PM 10 fraction — less than 10 μm in aerodynamic diameter — or 16–18% by weight, indicating that most ash was readily inhalable into the upper airways (Baxter *et al.*, 1981; Olsen and Fruchter, 1986). More pertinent was the proportion of ash in the respirable range (PM < 4 μm) and so able to reach the alveoli. Analyses of the bulk ash samples (collected from horizontal surfaces) were remarkably consistent in finding 1–2 wt% in the 3.5 μm fraction across the ash fall zone (Olsen and Fruchter, 1986). Although, it is typical that the mean diameter of grains decreases with distance from a vent, the percentage of the finer ash (<10 μm) did not necessarily increase downwind. One explanation may be that the finer particles that initially are aggregated become less aggregated downwind (Olsen and Fruchter, 1986).

5.4.4 *Crystalline silica content*

The crystalline silica content of the ash was eventually confirmed by CDC/NIOSH to be around 3–8 wt% in the form of cristobalite and quartz (Baxter *et al.*, 1981). This level of the toxic mineral was unlikely to have any bearing on the health effects of the acute episode that we are focusing on this chapter, but was significant enough to warrant further evaluation of the long-term risks of silicosis in workers and the general public from chronic exposure to ash arising from the widespread deposits that could persist in the general environment for months, and even possibly for years.

Concerns of the Weyerhauser Corporation and their loggers about the potential for respiratory disease led to a CDC/NIOSH-funded 4-year study (Expert Panel on Air Quality Standards, 2001) of 712 loggers exposed over an extended period to varying levels of volcanic ash from the 1980 eruptions of MSH. Baseline testing as begun in June 1980, one month after the major eruption, and follow-up testing continued on an annual basis through 1984. Analysis of lung function data showed that a significant, exposure-related decline in forced expiratory volume in one second ("FEV1", a measure of obstruction of the airways), occurred during the 1st year after the eruption. But, the decline was short-lived, and by 1984 the FEV1 differences between high-, medium-, and low-exposure groups of loggers were no longer significant. Self-reported symptoms of cough, phlegm, and wheeze showed a similar pattern. No ash-related changes were seen in chest roentgenograms taken in 1980 and 1984. The findings are consistent with the hypothesis that the inhaled ash caused acute mucus hypersecretion and/or airway inflammation that reversed when the exposure levels decreased. The ash levels to which the loggers were exposed were low compared with permissible occupational levels for nuisance dusts, but generally higher than the total suspended particulate (TSP) levels permissible in ambient air. Loggers' breathing zone exposures to the respirable crystalline silica component of MSH volcanic ash were controlled by the use of NIOSH-recommended masks, and fell below hazardous levels over time.

5.5 Air Monitoring

Around the time of the MSH eruption, the main air pollutants of concern to public health in advanced economy countries were those arising from the combustion of fossil fuels. The principal agents that were regarded as contributing to the adverse respiratory health effects of air pollution were particulate matter and sulphur dioxide. The measurement of particulate matter in the ambient air was undertaken by different techniques quantifying different properties of the airborne particles and hence a number of different methods were available for expressing the concentration of airborne particulate matter. Unfortunately, for the purposes of this paper, the conversion from one metric to another is rarely possible, as the relationships are site and time specific (Buist *et al.*, 1986a). In the UK, until the early 1990s, almost all measurements of particulate matter in the UK atmosphere were derived using the black smoke method. The air sampler that was adopted for black smoke is known to sample only particles smaller than about 4.5 μm and will include almost all particles formed in combustion and exclude coarser particulate matter (Buist *et al.*, 1986a) like volcanic ash. The main alternative to black smoke methods are the gravimetric methods that depend on weighing the mass of particles collected by drawing a known volume of air through the filter paper. For many years, this was carried out in North America using the high-volume sampler method, and with a specific sampler configuration, the measurement is known as TSP matter. In essence, TSP is theoretically measuring all the particles capable of being suspended in the air and this means it has a cut off of a hundred microns. Unfortunately, the inlet characteristics of the TSP sampler had poor efficiency for collecting particles below 10 μm diameter (Expert Panel on Air Quality Standards, 2001).

The poor characteristics of the TSP sampler led the US Environmental Protection Agency (EPA) to promote the development of size selective samplers, introducing the PM 10 sampler as the standard in 1987—after the MSH eruption. In 1994, the International Standards Organisation (ISO) defined three criteria for biologically relevant size selective aerosol sampling, primarily for studies of workers exposed to hazardous dust in industrial situations: inhalable dust is the mass fraction of total airborne particles that is inhaled through the nose and mouth; the thoracic fraction is the mass fraction of aerosol which penetrates beyond the larynx, and the respirable fraction is the mass of inhaled particles penetrating to the non-ciliated, smallest airways of the lung. PM 10 closely approximates to the ISO thoracic sampling criterion while the respiratory criterion is met by a PM 4 sampler (PM 3.5 under an older US standard for respirable dust sampling) (Expert Panel on Air Quality Standards, 2001).

5.5.1 *Air monitoring in the ash fall zones*

The EPA had already established a network of TSP monitors before the MSH eruption which included a monitor in Yakima (on top of the County Hall building in the centre of Yakima) and one in Spokane in Eastern Washington State. On 18 May, the ash started to fall at Yakima from 09.45 and continued for 8 h. TSP levels averaged over 24 h were 33.4 mg m^{-3} on that day but could have been higher because the sampler was not designed to measure such high loads (for comparison, the geometric mean TSP level for Yakima in 1979 was 80 μg m^{-3}). These were the most exceptional particle concentrations by any standards.

The next day TSP levels fell to 13.6 mg m^{-3} and then to 5.8 mg m^{-3} on 20 May. Winds increased during the next 2 days, with gusts up to 20 mph and though there was a trace of rain on 21 May TSP levels still rose to 13.2 mg m^{-3} on 21 May and to 28.4 mg m^{-3} on 22 May, falling again to 9.8 mg m^{-3} on 23 May. The mean wind speed of 15 mph kept TSP levels between 1.7 and is 3.4 mg m^{-3} on 24 and 25 May until rainfall of 0.44 inches in the early morning of 26 May was followed by the TSP to falling dramatically to 248 μg m^{-3}. During the next 3 weeks, TSP levels ranged from 55 to 2504 μg m^{-3}, except on 28 and 29 May when wind speeds of 13–21 mph drove TSP levels to 334 and 689 μg m^{-3}, respectively.

On 25 May and 13 June, ash fell on the opposite side of the volcano in two small eruptions, the first northwesterly towards Chehalis,

Centralia and Longview, and the next southwesterly towards Vancouver, Washington and Portland, Oregon. Interestingly, very high peak levels of TSP comparable to those seen in the east on 18 May were measured by EPA monitors, but over less than 24 h, in spite of only thin deposits (mostly <5 mm) of ash on the ground and without major disruption to daily living. After the 25 May eruption only the area around Centralia and Chehalis received as much as 8–10 mm of ash.

In June, CDC/NIOSH undertook a survey of exposure to total and "respirable" airborne ash in five communities in Washington and Idaho that had been under the ash falls, east and west of St Helens — recalling that "respirable" in the terminology in their paper is today called PM 10 (Merchant *et al.*, 1982). The investigators undertook personal sampling, i.e. using samplers worn by individuals recruited for the survey. The results showed that elevated exposures to PM 10 continued to occur in outdoor workers in a range of occupations. Exposure comparisons between five communities in ash fall areas showed they all continued to have higher overall levels of PM 10 than would be expected in unexposed communities (Merchant *et al.*, 1982).

Of special note in this survey was that area sampling done inside homes, schools, commercial establishments and automobiles showed PM 10 concentrations to be an order of magnitude lower than those measured in individuals outside, and were close to normal levels, which supported the official advice that people should stay inside their homes during periods of high ash resuspension, especially those suffering from pre-existing respiratory disease.

5.6 Health Consequences

The following is a short summary of the hospital surveillance network that was established by CDC to monitor the health of people living in the areas of ash fall (Baxter *et al.*, 1981). Comparisons of attendances at emergency departments (ED) for the same period in the preceding year or in preceding weeks were made and obvious increases in the numbers of patients that closely coincided with ash fall events were evident. In contrast, there were no apparent increases in numbers of deaths in the population of over a million people at risk, but time series analyses, a mainstay of present-day air pollution epidemiology, was not then in the epidemiologist toolkit.

Within days of the 18 May eruption, the surveillance network included 18 main hospitals in the ash fall areas of central and eastern Washington. Nine hospitals in the affected area of Western Washington were added after 25 May eruption. In the week following the 18 May eruption, increases in ED visits and hospital admissions for respiratory complaints (mainly asthma and bronchitis diagnoses) occurred in those areas of central Washington that had received the heaviest ash fall (Moses Lake, Othello and Ritzville, 32–70 mm depth) and these increases persisted for 2 weeks. Similar increases were also seen in Yakima (>8 mm depth) and Spokane (3–4 mm depth), which were larger cities where traffic readily resuspended the ash. However, little if any change was seen in the small communities of Ellensburg (10 mm depth), Soap Lake (10 mm depth) and Pullman (5 mm depth). In the worst affected areas of Moses Lake and Ritzville, increases in the number of hospital visits for non-respiratory illnesses coincided with physicians closing their offices due to severe travel difficulties (Baxter *et al.*, 1981).

After the 25 May eruption, only the area in and around Centralia and Chehalis received as much as 8–10 mm ash. An increase in ED visits but not admissions for respiratory complaints was seen in the cities in the following week. Slight increases in the number of ED visits only were also apparent in Longview (1–2 mm depth) and Aberdeen (0.3–0.5 mm depth) (Baxter *et al.*, 1981).

After the 12th June eruption when about 3–4 mm of ash fell on Portland, six key hospitals in Northwest Oregon were added to the surveillance network. Transient increases in ED visits for mainly asthma and wheezing were seen (Baxter *et al.*, 1981).

A follow-up study by CDC of the attendees at the EDs suggested that the majority of patients suffered from pre-existing asthma or bronchitis which had been exacerbated by exposure to the ash, or were concerned about possible health effects (Baxter *et al.*, 1983).

A group of 97 patients living in Yakima County and Ellensburg with chronic respiratory disease (COPD, asthma or emphysema) were identified through the Washington Lung Association and interviewed. One third of these participants reported that between 18 and 25 May they had suffered exacerbations of their condition starting on 18 May. Four of those interviewed had visited a hospital ED during the week for their complaints and one of them had been admitted to hospital. Over half of the respondents said their condition had not improved by the time of the interviews (in August and September), when ash was still visible on the ground in many places.

5.7 Discussion

Explosive eruptions with widespread heavy ash falls will occur rarely in the populated active volcanic regions around the world, on average, in total, several times a year, their plumes often affecting neighbouring populated areas, as at MSH. Until rainfall clears the air and consolidates all the ash into deposits on the ground, severe air pollution episodes may typically persist for hours or even days. More often than not, just like at MSH, local communities are not prepared for these events and have to take improvised measures to protect themselves from the choking quantities of the fine airborne ash particles ash which penetrate everywhere, including inside houses unless they are well sealed against infiltration to the outside air. The MSH episode of 18 May–26 June was exceptional because of the size of the erupted plume, and with the ash falling into an agricultural area where there was infrequent rainfall and farmers relied on artificial irrigation of their crops.

The extreme pollution conditions caused by MSH over 10 days recalled the notorious London Smog which brought central London to a virtual standstill in a pollution episode on 5–9 December 1952 (Anderson, 1999). In this landmark disaster, windless conditions and unusually low temperatures trapped domestic and industrial smoke emissions in a temperature inversion over the city (Brimblecombe, 2017). Particulate (smoke) levels reached a maximum concentration of 4,460 μg m^{-3} at County Hall and an average over the four days of about 1,600 μg m^{-3} (about five times normal level); for sulphur dioxide a maximum of 1,344 ppb at County Hall and the maximum values at different sites ranging from three to five times normal. During this episode, it was estimated there were 4,000 excess deaths and it was obvious that there was a sharp increase in people falling seriously ill and a steep rise (160%) in hospital admissions for respiratory diseases. Although there was a steep rise in bronchitis diagnoses, there was surprisingly little change for asthma. The Black Smoke metric is the measure of the respiratory fraction of particles (see above), mostly those arising from combustion sources, and may have been in even higher concentrations as the filters became saturated: the quoted figure was actually an indirect estimate from the sulphur dioxide peak (Anderson, 1999).

Our initial concerns at MSH were over the possibility of an acute increase in mortality due to the raised particle concentrations from the ash fall, in a similar way to events in the London Smog. This view was prompted by the blackouts in both incidents, even before our investigations had begun, but our fears were allayed by the hospital surveillance network which began receiving feedback from the hospitals in the region within the first days after the eruption. For reasons mentioned above, we cannot make direct comparisons between the TSP measured in the eruption with black smoke measured in London in 1952. But we do know that the PM 10 fraction constituted about 10 wt% of the total in the bulk ash samples of the fallout from across the region and 90% of the ash particles by count, with 2–3 wt% under 3.5 μg diameter and corresponding to the respirable range (Newhall and Fruchter, 1986; Olsen and Fruchter, 1986). Even if we supposed that the respirable fractions were in a similar range or magnitude, other major differences in the smoke in London formed by coal combustion compared to volcanic particles were likely to account for the differences in the observed health impacts, even though the causal links between the severe health impacts and the individual pollutants in the London Smog remain unexplained to this day. Air quality standards devised after this event, which became a milestone in the campaign against smoke air pollution, were based on both the particulate and sulphur dioxide concentrations in the ambient air and their health effects, as investigated in studies conducted in the following years (Anderson, 1999; Ware *et al.*, 1981).

The 1952 London Smog raised a controversial issue of the definition of an episode, particularly for epidemiological studies, namely when it could be said to have begun and finished (Merchant *et al.*, 1982). In volcanic eruptions, the date of eruption would be the date of onset, but with ash persisting in the environment for months or even years, the date of offset is not clear. In terms of severe episodes, like the one described in this chapter, one endpoint could be when living conditions returned to normal, which here was not until 28 May. By comparison, in the eruption of Cordon Calle in Patagonia, Chile, in 2011, the main ash fall was in Argentina and towns in the semi-arid region along the Linea Sur did not return to normal for 3 months (Wilson *et al.*, 2012). Many volcanic regions do not have a wide coverage of air quality monitors, but the return to guideline ambient air concentrations for PM 2.5 or PM 10 would be another timing of offset.

By around 1990, it became apparent that the decline in smoke emissions in advanced economy nations was being replaced by air pollutants from the petrol combustion engine and vehicle exhausts. Evaluation of the acute health impacts of traffic-related sources of particles has only been made possible by the adoption of time series methods in numerous epidemiological studies that have been undertaken in many countries over recent decades. These have over time focused on PM 2.5 as the main metric of concern, though the PM 2.5–10 fraction, the coarse range, is also linked to adverse respiratory health consequences, ensuring that the PM 10 metric used by some investigators at MSH is relevant in making comparisons with today's studies (and it incorporates PM 2.5 as well).

The latest estimates of the global burden of disease due to air pollution have been based on PM 2.5 (Burnett *et al.*, 2014), now regarded as the most ubiquitous environmental pollutant harmful to humans when all its sources are considered and combined in these estimates. PM 2.5 has measurable acute and chronic impacts on morbidity and mortality, but the causal factors in PM 2.5 operating in the short and long term are unlikely to be the same (Burnett *et al.*, 2014; Lippman, 2014). Because of the short duration of exposure after most ash falls, we are only concerned with the acute effects. The only common particulate pollutant comparable to volcanic ash is that derived from crustal-derived material, like desert dust, yet volcanic ash is not merely pulverised rock — as previously explained — so this comparison is not very valid either. The feature they do have in common is that they are not formed as products of combustion, as debate continues on whether the toxicity of PM 2.5 is associated with other primary or secondary products of combustion, or other particles generated by traffic.

Unfortunately, with the exception of the prospective study of the Weyerhauser loggers, there have been no time series epidemiological studies performed after any volcanic eruption to date, so we have had to rely in the past on issuing reassurance that ash does not seem to have the same serious consequences as traffic-produced PM 2.5, based mainly on our experience of this high exposure episode at MSH and a few limited investigations at other major eruptions since. This sanguine, or mainly reassuring, view held sway right up to the eruption of *Eyjafjallajökull* volcano in Iceland in 2010, as mentioned in the Section 5.1, when European countries braced themselves for fine volcanic ash particles being blown towards them.

However, there is little doubt since then that future health risk assessments for acute episodes of pollution by volcanic ash can no longer ignore the new role of PM 2.5, even if these, at least by analogy, draw upon the growing number of studies on the adverse health effects of PM 2.5 from other natural sources, such as desert dust (Alessandrini *et al.*, 2013; Mallone *et al.*, 2011; Staffogia *et al.*, 2016). It is too early to say, in our current state of knowledge, how applicable, or useful, such risk assessments would be to the populations at risk of exposure to volcanic ash.

References

Alessandrini, E. R., Staffogia, M., Faustini, A., Gobbi, G. P. and Forastiere, F. (2013). Saharan dust and the association between particulate matter and daily hospitalisations in Rome, Italy, *Occupational Environmental Medicine*, **70**, 432–439.

Anderson, H. R. (1999). Health effects of air pollution episodes, in Holgate, S. T., Samet, J. M., Koren, H. S., Maynard, R. L. (eds.), *Air Pollution and Health* (Academic Press, London), pp. 461–481.

Baxter, P. J., Ing, R., Falk, H., French, J., Stein, G. F., Bernstein, R. S., Merchant, J. and Allard, J. (1981). Mount St Helens Eruptions, May 18 to June 12, 1980: An overview of the health impact, *JAMA* **246**, 2585–2589.

Baxter, P. J., Ing, R., Falk, H. and Plikaytis, B. (1983). Mount St Helens eruptions: The acute respiratory effects of volcanic ash in a North American community, *Archives of Environmental Health*, **38**, 138–143.

Bernstein, R. S., Baxter, P. J., Falk, H., Ing, R., Foster, L. and Frost, F. (1986). Immediate public health concerns and actions in volcanic eruptions: Lessons from the Mount St Helens eruptions, May 18 – October 18, 1980, in Buist, A. S. and Bernstein, R. S. (eds.) Health Effects of Volcanoes: An Approach to Evaluating the Health Effects of an Environmental Hazard. *American Journal of Public Health*, **76** (suppl.), 25–37.

Brimblecombe, P. (ed.) (2017). London 1952: An enduring legacy, *Air Pollution Episodes* (World Scientific, Singapore).

Burnett, R. T. *et al.* (2014). An integrated risk function for estimating the global burden of disease attributable to ambient fine particulate matter exposure. *Environmental Health Perspectives*, **122**, 397–403.

Buist, A. S., Vollmer, W. M., Johnson, L. R., Bernstein, R. S. and McCamant, L. E. (1986a). A four-year prospective study of the respiratory effects of volcanic ash from Mt. St. Helens, *American Review Respiratory*, **133** (4), 526–534.

Crandell, D. R. and Mullineaux, D. R. (1978). Potential hazards from future eruptions of Mount St Helens Volcano. Geological Survey Bulletin 1383-C. U.S. Government Printing Office, Washington D.C.

Expert Panel on Air Quality Standards (2001). *Airborne Particles. Department of the Environment, Transport and the Regions (DEFRA)* (Stationary Office, London).

Langmann, B. (2017). North atlantic volcanic ash 2010: Contemporary social vulnerability to a natural event, in Brimblecombe, P. (ed.), *Air Pollution Episodes* (World Scientific, Singapore).

Lippmann M. (2014). Toxicological and epidemiological studies of cardiovascular effects of ambient air fine particulate matter (PM2.5) and its chemical components: coherence and public health implications. *Crit Rev Toxicol*, **44**, 299–347.

Mallone, S., Staffogia, M., Faustini, A., Gobbi, G. P., Marconi A. and Forastiere, F. (2011). Saharan dust and associations between particulate matter and daily mortality in Rome, Italy, *Environmental Health Perspectives*, **119**, 1409–1414.

Merchant, J., Baxter, P., Bernstein, R., McCawley, M., Falk, H., Stein, G., Ing, R. and Attfield, M. (1982). Health implications of the Mount St Helens eruption: Epidemiological considerations, *Annals of Occupational Hygiene*, **26**, 911–919.

Newhall, C. G. and Fruchter, J. S. (1986). Volcanic activity: A review for health professionals, in Buist A. S. and Bernstein R. S. (eds.), Health Effects of Volcanoes: An Approach to Evaluating the Health Effects of an Environmental Hazard. *American Journal of Public Health*, **76** (suppl.), 10–24.

Olsen, K .B. and Fruchter, J. S. (1986). Identification of the physical and chemical characteristics of volcanic hazards, in Buist A. S. and Bernstein R. S. (eds.), Health Effects of Volcanoes: An Approach to Evaluating the Health Effects of an Environmental Hazard. *American Journal of Public Health*, **76** (suppl.), 45–52.

Saarinen, T. F. and Sell, J. L. (1985). *Warning and response to the Mount St. Helens Eruption* (State University of New York, Albany).

Sarna-Wojcicki, A. M., Shipley, S., Waitt, R. B., Dzurisin, D. and Woods, S. H. (1981). Areal Distribution, thickness, mass, volume and grain size of air–fall ash from the six major eruptions of 1980. The 1980 Eruptions of Mount St Helens, Washington. U.S Geological Survey Professional Paper 1250. Washington D.C.: U.S. Government Printing Office, 577–600. Available at: http://pubs.usgs.gov/pp/1250/report.pdf (accessed 18 September 2016).

Staffogia, M., Zauli-Sajani, S, Pey, J., *et al.* and the MED-PARTICLES Study Group (2016). Desert dust out breaks in Southern Europe: Contribution to daily PM 10 Concentration and short term associations with mortality and hospital admissions, *Environmental Health Perspectives*, **124**, 413–419.

Ware, J. H., Thibodeau, L. A., Speizer, F. E., Colome, S. and Ferris, Jr., B. G. (1981). Assessment of the health effects of atmospheric sulfur oxides and particulate matter: Evidence from observational studies, *Environmental Health Perspectives*, **41**, 255–276.

Waitt, R. B. (2014). *In the Path of Destruction: Eyewitness Chronicles of Mount St Helens* (Washington State University, Pullman).

Wilson, T., Stewart, C., Bickerton, H., Baxter, P., Outes, V., Villarosa, G. and Rovere, E. (2012). Impacts of the June 2011 Puyehue-Cordon Calle volcanic complex eruption on urban infrastructure, agriculture and human health. *GNS Science Report* 2012/20. Lower Hutt: GNS Science.

Further Reading

Bernstein, R. S., McCawley, M. A., Attfield, M. D., Green, F. H. Y., Olenchock, S. A. and Dollberg, D. D. (1981). Evaluation of the potential pulmonary hazards of breathing zone exposures to Mount St. Helens volcanic ash in loggers: The results of short-term toxicological, environmental, and epidemiological studies. DHHS (NIOSH) Interim Report No. GHE-80-112. (National Institute for Occupational Safety and Health Morgantown, WV).

Buist, A. S. and Bernstein, R. S. (eds.) (1986). Health effects of volcanoes: An approach to evaluating the health effects of an environmental hazard. *American Journal of Public Health*, **76**(suppl.), 1–90. Available at: http://ajph.aphapublications.org/toc/ajph/76/Suppl. (accessed 20 September 2016).

Buist, A. S. and Bernstein, R. S. (eds.) (1986). Health effects of volcanoes: an approach to evaluating the health effects of an environmental hazard. *American Journal of Public Health*, **76**(suppl.), 10–24.

Buist, A. S. and Bernstein, R. S. (eds.) (1986). Health effects of volcanoes: an approach to evaluating the health effects of an environmental hazard. *American Journal of Public Health*, **76**(suppl.), 45–52.

Buist, A. S., Bernstein, R. S., Johnson, L. R. and Vollmer W. M. (1986b). Evaluation of physical health effects due to volcanic hazards: Human studies, in Buist, A. S. and Bernstein, R. S. (eds.) Health Effects of Volcanoes: An Approach to Evaluating the Health Effects of an Environmental Hazard. *American Journal of Public Health*, **76**(suppl.), 66–75.

USGS (1981). The 1980 Eruptions of Mount St Helens, Washington. U.S. Geological Survey Professional Paper 1250. (U.S. Government Printing Office, Washington D.C).

Chapter 6

Bhopal (1984): Cancer Risk Among Survivors and Opportunities for Translational Environmental Health Research

Pradyumna Kumar Mishra*, Neha Bunkar[†],
Arpit Bhargava*, Subodh Kumar Jain[†],
Naveen Kumar Khare[‡] and Neelam Pathak[‡]

*National Institute for Research in Environmental
Health (ICMR), Bhopal, India
[†]School of Biological Sciences, Dr. H. S. Gour Central University, Sagar, India
[‡]Division of Translational Research, ACTREC, Tata Memorial Centre,
Navi Mumbai, India

6.1 Bhopal Gas Tragedy

On the intervening night of 2–3 December 1984, at around 11:00 pm in Bhopal, an initial leakage of deadly methyl isocyanate (MIC) gas in Union Carbide's pesticide plant and increasing pressure inside a storage tank was noticed. Unfortunately, on the day of tragedy the three safety devices: (i) vent-gas scrubber; (ii) refrigeration unit; and (iii) gas flare safety system, primarily responsible for neutralising the toxic MIC discharge were non-functional. The refrigeration system to keep MIC at $0°C$ was turned off to save costs.

The flare had been out of order for couple of months to replace a corroded pipe and the gas scrubber was not working. A malfunctioned valve along with a series of errors allowed high volume of water to contaminate with 40 tonnes of MIC, resulting in forceful exothermic reactions. Moreover, the operators of the plant failed to shift MIC from the problem tank to a storage tank, as the latter was not empty. Due to high vapour pressure and volatile nature of MIC, it quickly reacted with air and water, and the reaction went out of control, as a result, the toxic gas and its breakdown products spilled across 75–100 km^2. Acute exposure to this toxic gas caused severe breath discomfort, nausea, eye and skin irritation. A large number of inhabitants in the township of Bhopal were exposed to different degrees, depending on their proximity to the plant and atmospheric factors. Approximately 3,800 immediate deaths were recorded and 7,000 persons died within a couple of days of the accident. Although number of deaths declined rapidly, the extent of the impact on the survivors' health was compounded by the fact that nothing was known about the toxic effects of MIC, hence there was no clue about a possible antidote to minimise the impact. In fact, there are more than 570,000 registered victims that survived this tragedy who continue to suffer from various chronic ailments not limited to bronchial asthma, pulmonary fibrosis, chronic obstructive pulmonary disease, emphysema, keratopathy, recurrent chest infections, corneal opacities, impaired immune competence, gastro-intestinal ailments, neurologic and psychiatric problems, and reproductive health problems (Sriramachari, 2004; ICMR, 2010a; Mishra *et al.*, 2009a; Mishra, 2012).

The exposure levels such as "acute & severe", "sub-acute & moderate" and "chronic & mild" led to the development of diverse pathological conditions in the survivors that correspond to the distinct patterns of the immediate, early and late mortality and morbidity (Table 6.1). Realising the far-reaching impact of this tragedy, Indian Council of Medical Research (ICMR), an apex national body in biomedical research in India, initiated several clinical research programmes aimed to delineate the long-standing effect of toxic gas on human health including genetic risk. The main reason for death

Table 6.1.   Summary of health effects observed in MIC exposed population.

Systems affected	Clinical signs and symptoms		References
	Acute and sub-acute phase (1–6 months)	Chronic phase (6 months onwards)	
Ocular system	Extreme irritation, burning, photophobia, blurred vision, corneal ulcer, conjunctiva and circum-corneal congestion	Injury in posterior ocular chamber, corneal opacity, conjunctivitis, chronic lesions, and loss of tear secretion	Andersson *et al.*, (1988); ICMR (2010a)
Respiratory system	Pulmonary edema, breathlessness, chest pain, cough, pneumonitis	Damaged lung functioning, chest pain, cough, dyspnea, wheezing, obstructive airway disease, acute allergic bronchio-alveolitis	Irani and Mahashur (1986); Kamat *et al.*, (1992); Vijayan and Sankaran (1996); Dhara *et al.*, (2002); ICMR (2010a)
Neurological system	Depression, anxiety, impaired auditory and visual memory	Growth retardation, flawed standard progressive matrices, motor speed precision test	Irani and Mahashur (1986); Misra and Kalita (1997); Sharma (2005); ICMR (2010a)
Gastro-intestinal system	Diarrhea, anorexia, abdominal pain	Disorders in upper and lower gastrointestinal tract	Shilotri *et al.*, (1986); Bhandari *et al.*, (1990); ICMR (2004)

(Continued)

Table 6.1.   (*Continued*)

Systems affected	Clinical signs and symptoms		References
	Acute and sub-acute phase (1–6 months)	Chronic phase (6 months onwards)	
Immune system	Down-regulated immune response, decreased T cell count and phagocytic response	Hyper-responsiveness among *in-utero* exposed individuals	Deo *et al.*, (1987); Karol and Kamat (1987); Saxena *et al.*, (1988); Mishra *et al.*, (2007, 2008, 2009b); Bhargava *et al.*, (2010)
Reproductive system	Miscarriages, irregular menstrual cycle, perinatal and neonatal mortalities	Increased abortion rate, pregnancy loss, infant mortality, reduced fetal weight	Bhandari *et al.*, (1990); Ranjan *et al.*, (2003); Sharma (2005)
Others	Muscle weakness, sleepiness, loss of appetite, nausea, vomiting, fever	Increased chromosomal abnormalities	Goswami (1986); Acquilla *et al.*, (1996)
Cancer	—	Increase in cancer patterns	Dikshit and Kanhere (1999); ICMR (2010b)

during initial years (between 1986 and 2010) was reported to be the disorders of respiratory tract followed by digestive and cerebrovascular abnormalities. The primary morbidity (96–99%) during the acute phase in severely affected area was eyes and lungs symptoms, followed by gastrointestinal symptoms (74%). While in moderate and mild affected areas, the morbidity for eyes-lungs, gastrointestinal tract and skin were 48%, 14% and 14%, respectively. Initial acute phase observation also showed gastrointestinal morbidities among 73.53%, 26.36% and 15% in severely affected, moderate and mild areas which decreased to 7.99%, 6.52% and 5.88% within 5 years (ICMR, 2013). Interestingly, a 19-year-long study *Cancer in Bhopal: Comparison of Cancer Patterns in MIC Affected and Non-affected Areas* (1988–2007), conducted by ICMR reported that overall cancer cases in Bhopal have tripled among men and more than doubled among women between 1988 and 2007 (ICMR, 2010b).

6.2 Clinical Reports

The initial treatment after MIC exposure was limited to symptom management, as it was still uncertain whether the observed consequences were due to the toxicity of MIC *per se* or other reaction products. In addition, a lack of the data about the parent compound toxicity was a major obstacle in therapeutic intervention and management of gas victims. The exposed individuals were reported to suffer from different chronic health effects such as febrile illnesses, pulmonary fibrosis, bronchial asthma, chronic obstructive pulmonary disease, emphysema, recurrent chest infections, gastro-intestinal ailments, keratopathy and corneal opacities (Naik *et al.*, 1986a, 1986b; Mehta *et al.*, 1990; Dhara *et al.*, 2002). The survivors also experienced higher incidence neurological and psychiatric problems along with an impaired immune competence (Dhara and Dhara, 2002; Mishra *et al.*, 2007, 2009b). An increased rate of spontaneous abortion also reported in a pregnancy outcome study among women from severely affected areas (Cullinan *et al.*, 1996). Initial autopsy studies (first 4 weeks) showed a futuristic discolouration of the lungs. During acute phase i.e. approximately 0–1 month of the toxic gas exposure, a study observed respiratory symptoms in 79%, while ophthalmic symptoms were reported in 74% of the total studied population. Similarly, it was also observed that breathlessness (99%), cough (95%), choking and irritation (46%), chest pain (25%), nausea and vomiting (52%), tachycardia (52%) and coma (7%) were among the common symptoms reported in studied population during acute phase. Significant lung abnormalities (interstitial and alveolar lesions) were observed among 98% of the observed chest radiographs within 72 h of the tragedy. Moreover, hematological profiling of exposed individuals after 2 weeks tragedy showed increased polymorphonuclear cells (35%), lymphocytosis (52%) and eosinophilia (19%) (ICMR, 2013).

6.3 Respiratory Health Effects

Lungs comprise one the first organs of contact for most environmental acute toxic exposures. Inflammatory stimulation in lungs accounts for exacerbations that significantly affect cellular respiration and lead to several chronic respiratory diseases including chronic obstructive lung disease. MIC-induced chronic inflammation is attributed to increased prevalence of respiratory morbidity among the surviving population of Bhopal gas tragedy (ICMR, 2004; Dhara *et al.*, 2002; Kamat *et al.*, 1992). Autopsy studies in MIC-exposed persons revealed severe necrotising lesions in upper respiratory tract, which suggested that the respiratory system was most severely affected. Documented reports have revealed that asphyxia arising from acute lung injury or acute respiratory distress syndrome as major reasons for the observed mortality. Abnormalities in lung function test and lower respiratory tract (alveolitis) was also shown in exposed individuals (Vijayan and Sankaran, 1996).

6.4 Ocular Health Effects

The acute ocular lesion caused by MIC was comparable to exposure caused by weak acid. The immediate symptoms comprised, of severe burning, foreign body sensation, lacrimation, inability to open eyes and blurred vision. The clinical manifestations appeared to be proportional to the duration of exposure and concentration of MIC gas. Severe ocular burning, watering, pain and photophobia were reported in the individuals after initial 2 months of exposure (Andersson *et al.*, 1988). However in some cases the damage was to the posterior ocular chamber and required surgical replacement of corneal tissue. Upon comparative frequency analysis of ophthalmic symptoms in people living at different distances from the factory, the reported ocular problems varied from 80% in people residing within 0.5 km area to 40% in people residing in 8 km area from the factory. The differences in the vision of populations living in moderately affected was between 6/12 and 6/60 while those living in more severely affected are the vision was lower than 6/60 (Maskari, 1986). Some patients continue to complain of irritation in eyes and mild degree in reduction of vision which may be attributed to persistent corneal edema and deficiency in tear secretion (ICMR, 2013).

6.5 Psychological and Neurological Health Effects

Neurons have a high metabolic rate, therefore, are at greatest risk for acute toxic exposure, followed by oligodendrocytes, astrocytes, microglia and cells of the capillary endothelium. Acute exposure to large amount of toxins can alter protein form, fluid content, ionic exchange capability of membranes, leading to swelling of neurons, astrocytes and damage to the delicate cells lining blood capillaries. Moreover, disruption of neurotransmitter mechanisms block access to post-synaptic receptors, produce false neurotransmitter effects, and alter the synthesis, storage, release, reuptake or enzymatic inactivation of natural neurotransmitters. Changes in cellular membrane structure impair excitability and impede impulse transmission. Significant neurological, neurobehavioural and psychological effects were also observed among the survivors. In a randomised study, 3–5 months after the disaster, 22.6% were found to be suffering from psychological disorders, neurotic depression, anxiety and social adjustment problems. Neuromuscular problems, such as tingling numbness, sensation of pins, and muscle aches, were commonly observed among the victims (ICMR, 2010a, 2013). In a study conducted on 208 exposed persons suffering from psychological problems, 45% suffered from neuroses, 35% from anxiety states and 9% from exacerbation of pre-existing adjustment reactions (Sethi *et al.*, 1987). Impairment of auditory and visual memory, attention response speed, and vigilance was also reported among exposed individuals (Gupta *et al.*, 1988). In follow-up study, conducted after one year, associate learning and precision were observed to be significantly disturbed in affected victims (Misra and Kalita, 1997). In addition, definite psychological problems such as apprehensiveness, jitteriness, depression and verbosity were also observed in children (Irani and Mahashur, 1986).

6.6 Reproductive Health Effects

The biological effect of acute toxic exposure depends primarily on the dose-time relationship. The ability to inflict systemic damage to reproductive organs and tissues as a result of one-time toxic exposure can range from mild and reversible to serious and irreversible. Menstrual abnormalities, vaginal discharge and premature menopause were the common reproductive problems observed among MIC exposed women and their children (Sharma, 2005). A significantly higher incidence of spontaneous miscarriages (24.2%) was observed in the women exposed to the toxic gas during pregnancy. Studies have also shown higher perinatal and neonatal mortalities in the MIC-affected area (6.9% and 6.1%, respectively), in comparison to the control area (5.0% and 4.5%, respectively) (Bhandari *et al.*, 1990). Higher incidences of abnormal uterine bleeding and abnormal pap smears amongst exposed women were also reported. An anthropometric study, carried out almost 16 years after the disaster, showed selective retardation in boys, who had been exposed to MIC during their toddler age or pregnancy (Ranjan *et al.*, 2003).

6.7 Immunotoxic Consequences

Immunotoxic assessments has been of lesser concern as a standalone health effect but more as a mechanism contributing to many, if not all, manifestations of toxicity of chemicals. The repercussion is even more difficult to measure as it is not a functional evaluation of a single organ, but bone marrow, thymus, spleen, about 1,000 lymph nodes and a similar number of Peyers' patches in the gut, the lymphoid tissue associated with skin, mucosa, bronchi, gut and the genitourinary tract as well as the peripheral leukocytes. Several studies on Bhopal disaster survivors suggest long-term immunological effects, including the potential of MIC to produce hypersensitivity reactions. A significant delay of the cell cycle and decreased response to mitogen-activated stimulation of proliferative lymphocytes was reported among cells of exposed individuals (Deo *et al.*, 1987). Studies have also suggested that the single large exposure to MIC elicited a definite immunologic response and was concomitant with chronic respiratory effects of MIC exposure (Karol and Kamat, 1987). Upon immune function study after 2.5 months of exposure the T-cell population (28%) among exposed individuals was observed to be less than half that normally found in the Indian population (65%). In addition, significantly decreased phagocytic activity of lymphocytes among exposed persons indicated a potentially compromised cell-mediated immune functioning in these individuals (Saxena *et al.*, 1988). In a significant study conducted by our group, we reported that *in-utero* MIC exposure has caused a persistently hyper-responsive cellular and humoral immune state in affected individuals (Mishra *et al.*, 2007, 2009b). In a separate study, we aimed to decipher any persistent and subtle immunotoxic effects of MIC in the survivors of Bhopal gas tragedy. The study was divided into three groups i.e. group I ($n = 40$); age and gender matched non-exposed healthy controls recruited from places within the geographical region of Bhopal but from unaffected zones, group II ($n = 40$); age and gender matched nonexposed healthy controls recruited from places well outside geographical region of Bhopal and group III ($n = 40$); age and gender matched MIC exposed subjects from affected zones inside geographical region of Bhopal and the status of inflammatory biomarkers (IL-8, IL-1β, IL-6, TNF, IL-10, IL-12p70 cytokines and C-reactive protein) were analysed. Our results displayed a significant increase in the levels of all circulating inflammatory biomarkers in the MIC exposed group in comparison to nonexposed cohorts (Bhargava *et al.*, 2010).

6.8 Toxico-genomic Outcomes

In an initial study, increased frequency of sister-chromatid exchanges frequency in lymphocytes of MIC-exposed persons were observed (Goswami, 1986). Chromosomal breaks were also observed in 71.4% of the studied population in comparison to 21.4% observed among controls. Surprisingly, some of the survivors also reported even chromatin bodies in addition to the normal 46 chromosomes. Moreover, the repeatedly developed chromosomal aberration among exposed subjects suggested of potential MIC induced DNA damage in these individuals (Goswami *et al.*, 1990). Further, these abnormalities were observed to be linked with different pathological conditions such as tumours, recurrent miscarriage or transmission of defects to their offspring in among 50% of the exposed subjects. A unique study conducted in 1990 clearly establishes genetic link of cancer patterns among gas victims of the tragedy with MIC exposure (Dikshit and Kanhere, 1999). Such studies were not conducted during the late recovery phase that would have helped identify people with chromosomal aberrations and at high risk of developing cancer.

6.9 Carcinogenic Risk Assessment of Isocyanates

Environmental toxic exposures are now known to be significantly associated with increased cancer associated risks. However, there remain many missing links which have not been adequately addressed. Moreover, identification of biomarkers, toxicogenomic effects and gene susceptibilities in the post-genomics era offer immense opportunities to design pragmatic risk assessment strategies which may be of great clinical importance. Cancer risk assessment among the 570,000 MIC exposed individuals of Bhopal gas tragedy provides an unparallel opportunity to recognise the long-term consequences of such massive toxic exposures (Mishra, 2012;

Mishra *et al.*, 2015a). More so, for a developing economy like India which is over-loaded with environmental and economical threats, examining such a carcinogenic risk-assessment model will be vital to formulate a suitable regulatory measure for effective management of occupational and environmental carcinogenesis.

In past 10 years, work from our laboratory has provided mechanistic insights in understanding the intricate molecular mechanisms of toxico-genomic and epigenomic implications invoked upon exposure to environmental stress signals and bio-transforming agents including isocyanates. Using germ-line stem cells and cultures of epithelial and fibroblast cells representing vital organs [human lymphocytes and neutrophils; IMR-90 (human lung fibroblasts), FHC (human colon epithelial), HEK-293 (human embryonic kidney epithelial), HPAE-26 (human pulmonary arterial endothelial), GC-1 spg (mouse spermatogonial), B/CMBA.Ov (mouse ovarian epithelial), MM55.K (mouse kidney epithelial), and NCTC-1469 (mouse liver epithelial) cells], we have delineated the sequence of events that lead to DNA damage repair, apoptosis, oxidative stress, inflammation, genomic instability and oncogenic transformation following induction of environmental stress stimuli. Our findings strengthened the current understanding of consequences of environmental stress signals at genome, proteome and cytome level (Mishra *et al.*, 2007, 2008, 2009a, 2009b, 2009c, 2009d, 2009e, 2009f, 2010, 2014, 2015b; Bhargava *et al.*, 2010; Raghuram *et al.*, 2010a, 2010b; Raghuram and Mishra, 2014; Khan *et al.*, 2011, 2014; Panwar *et al.*, 2014). These systematic laboratory investigations coupled with molecular surveillance studies have comprehensively demonstrated that the risk of developing an environmental associated aberrant phenotype involves a complex interplay of genomic and epigenetic reprogramming (refer to Fig. 6.1).

6.9.1 *Chronic inflammation*

Aberrant inflammatory response plays a crucial role in the induction of cellular proliferation, genetic instability and an increased

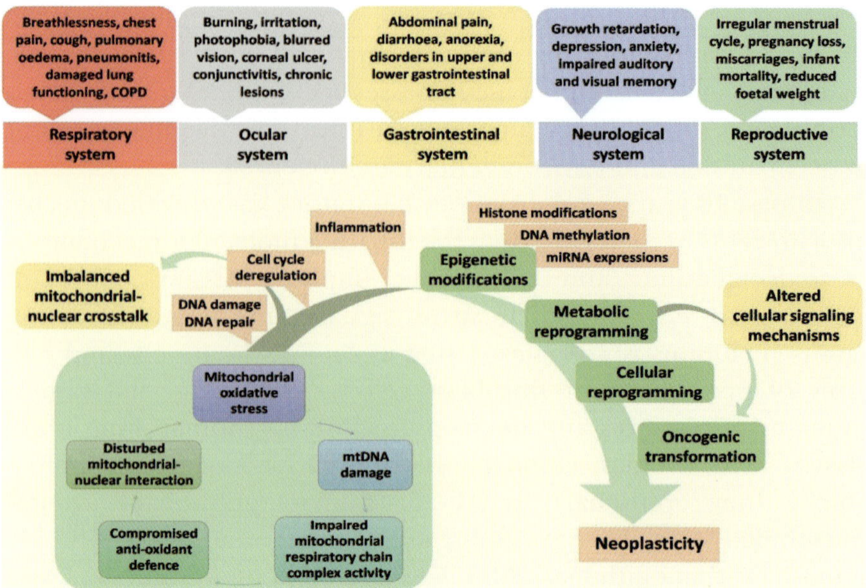

Fig. 6.1.   A molecular framework showing the exposure-response relationship of isocyanate-induced cellular perturbations possibly responsible for the long-term health effects observed in the population that survived the Bhopal gas tragedy.

risk of carcinogenesis (Khan *et al.*, 2011). In our study, we observed a time- and dose-dependent escalation in the levels of the inflammatory cytokines (IL-6, TNF-α, IFN-γ, IL-8, IL-1β, NF-$\kappa\beta$ and IL-12p70) in culture supernatants of isocyanate-treated cells (Mishra *et al.*, 2008, 2009b, 2009c, 2010). A parallel study performed in a strictly selected surviving cohort of the Bhopal gas tragedy not only manifested significantly enhanced levels of circulating inflammatory mediators, but presented a persistent and subtle toxic effect that well corroborated with the results of *in vitro* experiments (Mishra *et al.*, 2009d; Bhargava *et al.*, 2010). Therefore, well-designed genome-wide association studies in a restricted cohort might help to understand the precise role of inflammatory mediated cancer risk and also identification of a suitable biomarker for wide spectrum population screening.

6.9.2 *Mitochondrial retrograde signaling*

One of the putative mechanisms of a pro-inflammatory cytokine response is mitochondrial oxidative stress (Bhargava *et al.*, 2011; Khan *et al.*, 2011, 2014). In our study, we observed an early and continual increase in the generation of reactive oxygen species in the treated cells. In addition, up-regulation of pro-apoptotic bax gene along with increased activation of annexin V and caspase-3 proteins, further depicted considerable amount of apoptosis due to isocyanate exposure (Mishra *et al.*, 2009e; Raghuram *et al.*, 2010a; Panwar *et al.*, 2013). These observations imply that excessive accumulation of H_2O_2 in isocyanate exposed cells could lead to pronounced oxidative damage and instigate a variety of cellular responses.

6.9.3 *Oxidative DNA damage*

With over a hundred oxidative DNA adducts, oxidative DNA damage is also major source for the mutation load in living organisms. Our investigations in isocyanate treated cells revealed significant induction of oxidative stress through accumulation of 8-oxo-dG bases and ARP (aldehyde reactive probe) sites in the DNA and inhibition of (SOD) super oxide dismutase and (GR) glutathione reductase activities (Mishra *et al.*, 2009c; Jain *et al.*, 2011; Pathak *et al.*, 2014; Bunkar *et al.*, 2016). These results suggested that isocyanate confer cells to be more vulnerable to free radicals induced oxidative damage, increased protein thiolation or oxidation of thiol groups and indirectly contribute to the ineffective repair of DNA damage and acquisition of damaged genotype.

6.9.4 *DNA double strand breaks*

Toxic oxygen metabolites such as the hydroxyl radical can covalently attack nucleophilic sites and may cause DNA strand breakage which could lead to mutagenic effects (Bhargava *et al.*, 2010). Qualitative immunofluorescence microscopy in cells after treatment with MIC showed elevated expression of pATM, γ-H2AX and p-p53 proteins with interspersed and punctuated nuclear pattern. While, quantitative analysis through flow cytometry displayed an analogous upshot in phosphorylation time kinetics of p-ATM/γ-H2AX and p-p53 proteins expression (Mishra *et al.*, 2009e; Raghuram *et al.*, 2010b; Panwar *et al.*, 2014). Our results imply that isocyanate treatment cause the formation of carcinogenic-DNA adducts and induce alterations to DNA ultra-structure.

6.9.5 *DNA repair response*

Double stranded breaks are potentially lethal, and cells have evolved a mechanism leading to recruitment of repair proteins. Analysis of damage repair and arrest factors Mre11, Rad50, Nbs1 and GADD45α in our studies with isocyanate treated cells revealed abrupt augmentation in their expression pattern suggesting the active recruitment of these proteins in response to severity of stress-induced DNA damage and cycle arrest due to active accumulation and trans-activation of check-point genes suggesting a compromised state of genome and increased susceptibility to initiation of carcinogenesis (Mishra *et al.*, 2009e; Panwar *et al.*, 2014).

6.9.6 *Cell cycle deregulation*

Moreover, excess accumulation of DNA damage responsive factors can cause early cell cycle arrest (Mishra *et al.*, 2009f, 2011a). In line with this, our studies in cells treated with isocyanate showed a significant early halt in G1 phase of their growth cycle than expected and sustained the perturbed trend thereafter with significant G2/M phase arrest eventually showing inclination towards aneuploidy (Mishra *et al.*, 2009b, 2009c; Raghuram *et al.*, 2010b). These results indicated imbalance in ploidy levels and are probably the response of check-point proteins and interruption of cell cycle progression, due to toxic isocyanate treatment.

6.9.7 *Cell cycle checkpoints*

Upon examining the role of cell cycle regulatory proteins after isocyanate treatment in cells, we noted an abrupt up-regulation in the levels of both tumour suppressor p53 and cyclin-dependent kinase inhibitor p21WAF/Cip1 proteins. In addition, a temporal upshot of p16INK4A and sequential up-regulation of pRb proteins along with down regulated expression of cyclin D1 in isocyanate exposed cells alluded to the deregulated cell cycle with significant impediment in G1/S transition phase caused by isocyanate exposure. Furthermore, the transitions between late G1, and G1-S and G2/M cell cycle phases are regulated by cyclins A/E in conjunction with their catalytic partners CDK2 through their nuclear translocation to control cell cycle progression, DNA synthesis and mitosis. Interestingly, we also observed an uncontrolled activation/over-expression of cyclin A, cyclin E and CDK2 proteins at different passages 1–5 post isocyanate treatment suggestive of deregulated G1/S transition machinery with sub-optimal S phase allowing mitotic slippage and initiation of genomic instability (Mishra *et al.*, 2009a, 2010; Panwar *et al.*, 2014).

6.9.8 *Genomic instability*

Defective intra-cellular pathways of cell cycle and differentiation and / or signaling molecules might underlie the undeterred growth and genetically unstable phenotype (Mishra *et al.*, 2009e). Our experiments with isocyanate exposed cells showed an aberrant amplification of aurora kinase1 gene and over expression of aurora kinase B protein indicating mitotic slippage by override of the checkpoint arrest through rapid exit from mitosis and defective spindle assembly checkpoint and aneuploidy resulting in genomic instability. Further investigations in isocyanate treated cells showed the over-expression of centrosomal protein pericentrin levels suggestive of an abnormal centrosome functioning thereby generating supernumerary centrioles and improper segregation of chromosomes during cell division. Moreover, metaphase spreads obtained from treated cells showed an array of structural and numerical chromosomal aberrations indicating the occurrence of chromo-somal instability due to isocyanate exposure. Upon centromeric fluorescence *in situ* hybridisation analysis, we observed centromeric amplification of α-satellite repeats in isocyanate treated cells suggesting that this portion of the genome may be destabilised and might lead to "neo-centromerisation." While, telomeric and immune FISH analysis in isocyanate exposed cells further confirmed the loss of telomeric integrity (absence of signals at telomeric sites) of chromo-somes. Concurrently, our spectral karyotyping studies in isocyanate treated cells, respectively, showed incidence of cryptic and multiple translocations implying the greater susceptibility of telomeres to isocyanate exposure, which, consecutively, might play a part in transfer of instability from one chromosome to another through abnormal telomeres and generate many types of rearrangements. Collectively, our investigations confirmed that there is a synergistic inception of chromosomal instability in the genome as indicated by various unstable-type chromosome aberrations. Moreover, ISSR PCR analysis in isocyanate treated cells, presented the extension of genetic instability to microsatellites (Mishra *et al.*, 2008, 2009b, 2009c, 2009d, 2009e; Raghuram *et al.*, 2010b; Panwar *et al.*, 2014).

6.9.9 *Oncogenic transformation*

Genomic instability plays a major part in the cellular drift towards phenotypic transformation. Experiments performed in cells exposed to isocyanate further substantiates these observations by means of the morphological variations and increased uptake of β-gal stain by these, probably transformed, senescent cells manifested at higher time course and passage levels (Mishra *et al.*, 2009c, 2015b; Raghuram and Mishra, 2014).

6.9.10 *Stress-induced premature senescence*

Induction of a stress induced senescence-like phenotype occurs after environmental and genotoxic insults can stimulate a senescence-like state prematurely in young cells. This forces cells to acquire multiple phenotypic changes and thereby compromising the tissue structure and function (Raghuram and Mishra, 2014). Consistent with this, the present investigation showed an increase in the cell size, a characteristic feature of senescence with early induction of positive staining for β-galactosidase in isocyanate exposed cells in-advance with large granular nucleus and focal enrichment of lysosome-related β-galactosidase activity at vacuoles which sustained progressively along the incremental time course. These outcomes further support the notion that an increase in lysosomal mass is responsible for the increase in β-galactosidase activity observed in senescent cells (Raghuram *et al.*, 2010b; Raghuram and Mishra, 2014; Mishra *et al.*, 2015b).

6.9.11 *Neoplasticity*

In some cells, senescence associated DNA damages due to mutagenic exposure could affect a cocktail of oncogenes and tumour-suppressor genes, favouring an evolution toward transformation or immortalisation. Consistent with this in our study soft-agar culture of isocyanate-treated cells exhibited morphological variations with emergence of cells overcoming senescence. *In vitro* cellular transformation assay after 8 days of culture presented the striking appearance of anchorage-independent growth of treated cells in clusters having retractile property along with a significant loss of β-gal activity suggesting that cells, already partially transformed, are highly prone to further neoplastic development upon exposure to isocyanates (Raghuram *et al.*, 2010b; Raghuram and Mishra, 2014; Mishra *et al.*, 2015b).

6.9.12 *Epigenetic signatures*

Chromatin changes are intimately linked with DNA damage response. Over-accumulation of defectively repaired adducts, due to impaired mitochondrial-nuclear cross talk, alters epigenetic programming through introduction of a negative charge on the chromatin structure. We observed significant post-translational histone modifications including hypo-acetylation of AcH3 and AcH4, hyper-methylation of H3K9me1 and H4K20me3, ubiquitinated uH2A/uH2B and increased phosphorylation of H2AX and H3 in the isocyanate exposed cells. In addition, the altered expression of miRNAs denoted possible location of corresponding genes at oxidative-damaged fragile sites. Altered DNA methylation profile (p. 16) reported in cells after MIC exposure suggested dis-regulation of vital cellular mechanisms which may initiate different multifaceted human disorders, including cancer. This is important because changes in DNA methylation patterns can be heritably transferred across upcoming generations and may act as a critical player for cancer risk assessment among MIC exposed individuals (Raghuram *et al.*, 2010b; Raghuram and Mishra, 2014; Mishra *et al.*, 2014, 2015b).

Results of our studies delineated the underlying molecular link between MIC exposure and potential risk of tumour development (see Fig. 6.2). For instance, preliminary evidences emerging from our work on gall bladder carcinoma suggested that Rad-50 and cyclin-E may be one of the potential candidates utilised for risk assessment of gall bladder cancers (Mishra *et al.*, 2009d, 2011b). Moreover, relationship of oxidative stress and cancer initiation may be another strategy which can be exploited to risk assessment exposed populations. Similarly, identifying the status of mitochondrial genome and factors affecting its biogenesis could be of potential significance for cancer risk assessment. Further, investigations specifically focusing on designing risk assessment strategies will certainly prove to be fundamental and will be helpful in assortment of heritable epigenetic traits likely to be transmitted to future generations.

However, these studies need to be critically designed as drafting an appropriate cancer risk assessment model that will require: (i) appropriate biological samples, (ii) selection of study parameters capable to relate exposure and cancer risks and (iii) efficient interpretation of the obtained data. Since a long time has elapsed after the exposure, recruitment of the MIC exposed subjects should be done with immense care (Chiu *et al.*, 2013). Besides recruiting subjects

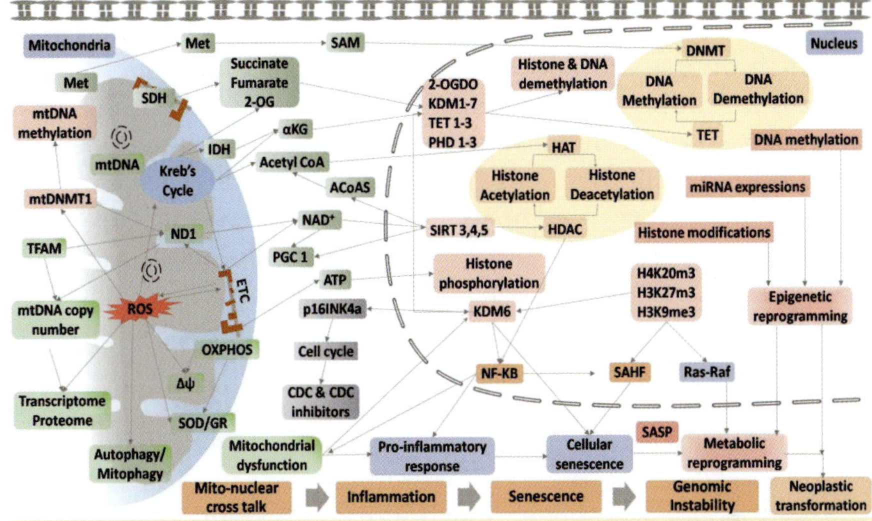

Fig. 6.2. A comprehensive molecular understanding of mitochondrial retrograde signaling induced genomic and epigenomic changes following isocyanate exposure.

from the affected area (within 2.5 km from the plant), non-exposed healthy controls (age and gender matched) from unaffected zones within the city (more than 25 km from the plant) and from outside Bhopal (more than 200 km from the plant) should also be recruited. In addition, different confounding factors such as history and duration of stay, socio-economic conditions, alcohol/tobacco use, and any medical history should be assessed carefully. Moreover, MIC exposure to the subjects should be done through a set of points: (i) physical presence and distance from the plant on the night of the tragedy; (ii) the shielding measures taken during MIC leak, if any; (iii) outdoor or indoor exposure; (iv) activity profile, i.e. running or walking, etc. and (v) post-exposure symptom profile to corroborate exposure (Mishra *et al.*, 2011c).

6.10 Clinical and Translational Perspective

A shift in disease burden from a predominance of infectious diseases to a growing contribution of non-communicable diseases is evident in low- and middle-income countries comprising >80% of the world population. Among the non-communicable diseases, cancer alone caused over eight million deaths worldwide and is the second leading cause of death behind cardiovascular disease in 2013. Of late, there has been substantial progress with regard to prevention and treatment options for certain cancers, but despite this cancer, burden is increasing owing to a growing number of environmental risk factors. In general, for carcinogenic risk assessment for chemicals, a battery of toxicological investigations are conducted based on regulatory test guidelines with a key objective of providing a dose-response assessment that estimates a point of departure lowest-observed adverse-effect level, LOAEL). Subsequently, findings of these studies are extrapolated to provide an exact quantity of substance above which adverse effects can be expected in humans. But the complex nature of the biology of cancer and an advanced understanding of the disease has added several layers of complexities for the risk assessment studies. Moreover, it is difficult to ascertain the carcinogenic outcome on humans of a single acute toxic chemical exposure through conventional risk assessment practices. Interestingly, an increased association between acute toxic exposure and early biomarkers related to carcinogenesis has been reported in several molecular epidemiological studies. Monitoring genomic instability through inter-linked mechanisms such as, DNA damage repair response, mitochondrial function, epigenetic modifications and telomeric integrity are now used for human bio-monitoring of environmental carcinogenesis (Raghuram and Mishra, 2014; Mishra *et al.*, 2014, 2015b).

Organic isocyanates contain one or more isocyanate groups (–NCO) and are widely used in the manufacture of flexible and rigid foams, fibres, paints and varnishes, elastomers and building insulation materials. Due to inadequate evidence of carcinogenicity in humans, the International Agency for Research on Cancer classified isocyanates as "potential human carcinogens", known to cause cancer in animals. A formal risk assessment study to address MIC-induced carcinogenesis in the surviving population of Bhopal gas tragedy is fundamental as it would revalidate the inherent toxicity of the compound and unravel the molecular pathways related with acute exposure. Three types of investigations: (i) exposure-response studies, to confirm the relationship between MIC exposure and carcinogenic effect; (ii) descriptive studies, to compare the incidence of cancer between an exposed population with that in a reference population and (iii) studies of incidence of cancer in registered survivors of the tragedy will be necessary to provide the least confounded factors of an association between MIC exposure and cancer risk. Of course, a central question for risk assessment needs to be derived from understanding the mode of action of MIC at an environmentally-relevant dose. In this regard, descriptions of the molecular mechanisms of the carcinogenic process provided by our laboratory might serve as the basis of biologically based dose-response models. Given the long latency period in onset of most cancer forms and types, a pragmatic carcinogenic-risk assessment study will be a key for the development of effective strategies for the thousands of ailing survivors of the Bhopal gas tragedy.

References

Acquilla, S. D., Cullinan, P. and Dhara, V. R. (1996). Long term morbidity in survivors of the 1984 Bhopal gas leak, *National Medical Journal of India*, **9**, 5–10.

Andersson, N., Kerr, M. M., Mehra, V. and Salmon, A. G. (1988). Exposure and response to methyl isocyanate: Results of a community based survey in Bhopal, *British Journal of Industrial Medicine*, **45**, 469–475.

Bhandari, N. R., Syal, A. K., Kambo, I., Nair, A., Beohar, V., Sexena, N. C., Dabke, A. T., Agarwal, S. S. and Saxena, B. N. (1990). Pregnancy outcome in women exposed to toxic gas at Bhopal, *Indian Journal of Medical Research*, **92**, 28–33.

Bhargava, A., Punde, R. P., Pathak, N., Dabadghao, S., Desikan, P., Jain, A., Maudar, K. K. and Mishra, P. K. (2010). Status of inflammatory biomarkers in the population that survived the Bhopal gas tragedy: A study after two decades, *Industrial Health*, **48**, 204–208.

Bhargava, A., Punde, R., Varshney, S., Pathak, N. and Mishra, P. K. (2011). A novel FRET probe-based approach for identification, quantification, and characterization of occult HCV infections in patients with cryptogenic liver cirrhosis, *Indian Journal of Pathology & Microbiology*, **54**(2), 420–421.

Bunkar, N., Bhargava, A., Khare, N. K. and Mishra, P. K. (2016). Mitochondrial anomalies: Driver to age associated degenerative human ailments, *Frontiers in Bioscience (Landmark Ed.)*, **21**, 769–793.

Chiu, W. A., Euling, S. Y., Scott, C. S. and Subramaniam, R. P. (2013). Approaches to advancing quantitative human health risk assessment of environmental chemicals in the post-genomic era, *Toxicology Applied Pharmacology*, **271**(3), 309–323.

Cullinan, P., Acquilla, S. D. and Dhara, V. R. (1996). Long term morbidity in survivors of the 1984 Bhopal gas leak, *National Medical Journal of India*, **9**, 5–10.

Deo, M. G., Gangal, S., Bhisey, A. N., Somasundaram, R., Balsara, B., Gulwani, B., Darbari, B. S., Bhide, S. and Muru, G. B. (1987). Immunological, mutagenic and genotoxic investigations in gas exposed population of Bhopal, *Indian Journal of Medical Research*, **86**, 63–76.

Dhara, V. R. and Dhara, R. (2002). The Union Carbide disaster in Bhopal: A review of health effects, *Archives of Environmental Health*, **57**, 391–404.

Dhara, V. R., Dhara, R., Acquilla, S. D. and Cullinan, P. (2002). Personal exposure and long-term health effects in survivors of the Union Carbide disaster at Bhopal, *Environmental Health Perspectives*, **110**, 487–500.

Dikshit, R. P. and Kanhere, S. (1999). Cancer patterns of lung, oropharynx and oral cavity cancer in relation to gas exposure at Bhopal, *Cancer Causes and Control*, **10**, 627–636.

Goswami, H. K. (1986). Cytogenetic effects of methyl isocyanate exposure in Bhopal, *Advances in Human Genetics*, **74**, 81–84.

Goswami, H. K., Chandorkar, M., Bhattacharya, K., Vaidyanath, G., Parmar, D., Sengupta, S., Patidar, S. L., Sengupta, L. K., Goswami, R. and Sharma, P. N. (1990). Search for chromosomal variations among gas-exposed persons in Bhopal, *Human Genetics*, **84**, 172–176.

Gupta, B. N., Rastogi, S. K., Chandra, H., Mathur, A. K., Mathur, N., Mahendra, P. N., Pangtey, B. S., Kumar, S., Kumar, P., Seth, R. K., Dwivedi, R. S. and Ray, P. K. (1988). Effect of exposure to toxic gas on the population of Bhopal. Part I. Epidemiological, clinical, radiological and behavioural studies, *Indian Journal Experimental Biology*, **26**, 149–160.

Indian Council of Medical Research (ICMR). (2010a). Health effects of the toxic gas leak from union carbide methyl isocyanate plant in Bhopal, Bhopal, India.

Indian Council of Medical Research (ICMR). (2010b). Cancer in Bhopal: Comparison of cancer patterns in MIC affected and unaffected areas (1988–2007), Bhopal, India.

Indian Council of Medical Research (ICMR). (2013). Technical report on population based long term epidemiological studies Part-II. Bhopal, India.

Irani, S. F. and Mahashur, A. A. (1986). A survey of Bhopal children affected by methyl isocyanate gas, *Journal of Postgraduate Medicine*, **32**(4), 195–198.

Jain, D., Pathak, N., Khan, S., Raghuram, G. V., Bhargava, A., Samarth, R. and Mishra, P. K. (2011). Evaluation of cytotoxicity and anti-carcinogenic potential of Mentha leaf extracts, *International Journal of Toxicol*ogy, **30**(2), 225–236.

Kamat, S. R., Patel, M. H., Pradhan, P. V., Taskar, S. P., Vaidya, P. R., Kolhatkar, V. P., Gopalani, J. P., Chandarana, J. P. Dalal, N. and Naik, M. (1992). Sequential respiratory, psychologic, and immunologic studies in relation to methyl isocyanate exposure over two years with model development, *Environmental Health Perspectives*, **97**, 241–253.

Karol, M. H. and Kamat, S. R. (1987). The antibody response to methyl isocyanate: experimental and clinical findings, *Bulletin of European Physiopathology of Respiration*, **23**, 591–597.

Khan, S., Bhargava, A., Pathak. N., Maudar, K. K., Varshney, S. and Mishra, P. K. (2011). Circulating biomarkers and their possible role in pathogenesis of chronic hepatitis B and C viral infections, *Indian Journal of Clinical Biochemistry*, **26**(2), 161–168.

Khan, S., Raghuram, G. V., Pathak, N., Jain, S. K., Chandra, D. H. and Mishra, P. K. (2014). Impairment of mitochondrial-nuclear cross talk in neutrophils of patients with type 2 diabetes mellitus, *Indian Journal of Clinical Biochemistry*, **29**(1), 38–44.

Maskari, Q. B. (1986). Ophthal survey of Bhopal victims 104 days after the tragedy, *Journal of Postgraduate Medicine,* **32**(4), 199–202.

Mehta, P. S., Mehta, A. S., Mehta, S. J. and Makhijani, A. B. (1990). Bhopal tragedy's health effects. A review of methyl isocyanate toxicity, *JAMA,* **264**, 2781–2787.

Mishra, P. K., Dabadghao, S., Modi, G., Desikan, P., Jain, A. and Mittra, I. (2007). Immune status in individuals exposed *in-utero* to methyl isocyanate during the Bhopal gas tragedy, a study after two decades, *Proceeding of 13th International Congress of Immunology,* Medimond Intern. Proc., pp. 341–344.

Mishra, P. K., Panwar, H., Bhargava, A., Gorantla, V. R., Jain, S. K. Banerjee, S. and Maudar, K. K. (2008). Isocyanates induces DNA damage, apoptosis, oxidative stress, and inflammation in cultured human lymphocytes, *Journal of Biochemical and Molecular Toxicology,* **22**, 429–440.

Mishra, P. K., Samartha, R. S., Pathak, N., Jain, S. K., Banerjee, S. and Maudar. K. K. (2009a). Bhopal gas tragedy: Review of clinical and experimental findings after 25 years, *International Journal of Occupational Medicine and Environmental Health,* **22**, 193–202.

Mishra, P. K., Dabadghao, S., Modi, G., Desikan, P., Jain, A., Mittra, I., Gupta, T., Chauhan, C., Jain, S. K. and Maudar, K. K. (2009b). *In-utero* exposure to methyl isocyanate in the Bhopal gas disaster: evidence of persisting hyper-activation of immune system two decades later, *Occupational Environmental Medicine,* **66**, 279.

Mishra, P. K., Raghuram, G. V., Panwar, H., Jain, D., Pandey, H. and Maudar, K. K. (2009c). Mitochondrial oxidative stress elicits chromosomal instability after exposure to isocyanates in human kidney epithelial cells, *Free Radical Research,* **43**, 718–728.

Mishra, P. K., Bhargava, A., Raghuram, G. V., Gupta, S., Tiwari, S., Upadhyaya, R., Jain, S. K. and Maudar, K. K. (2009d). Inflammatory response to isocyanates and onset of genomic instability in cultured human lung fibroblasts, *Genetics and Molecular Research,* **8**, 129–143.

Mishra, P. K., Bhargava, A., Raghuram, G. V., Jatawa, S. K., Akhtar, N., Khan, S., Tiwari, A. and Maudar, K. K. (2009e). Induction of genomic instability in cultured human colon epithelial cells following exposure to isocyanates, *Cell Biology International,* **33**, 675–683.

Mishra, P. K., Gorantla, V. R., Akhtar, N., Tamrakar, P., Jain, S. K. and Maudar, K. K. (2009f). Analysis of cellular response to isocyanate exposure in cultured mammalian cells, *Environmental and Molecular Mutagenesis*, **50**, 328–336.

Mishra, P. K., Khan, S., Bhargava, A., Panwar, H., Banerjee, S., Jain, S. K. and Maudar, K. K. (2010). Regulation of isocyanate-induced apoptosis, oxidative stress, and inflammation in cultured human neutrophils: Isocyanate-induced neutrophils apoptosis, *Cell Biology and Toxicology*, **26**, 279–291.

Mishra, P. K., Raghuram, G. V., Bhargava, A., Ahirwar, A., Samarth, R., Upadhyaya, R., Jain, S. K. and Pathak, N. (2011a). *In vitro* and *in vivo* evaluation of the anticarcinogenic and cancer chemopreventive potential of a flavonoid–rich fraction from a traditional Indian herb *Selaginella bryopteris*, *British Journal of Nutrition*, **106**(8), 1154–1168.

Mishra, P. K., Raghuram, G. V., Jatawa, S. K., Bhargava, A. and Varshney, S. (2011b). Frequency of genetic alterations observed in cell cycle regulatory proteins and microsatellite instability in gallbladder adenocarcinoma: A translational perspective, *Asian Pacific Journal of Cancer Prevention*, **12**(2), 573–574.

Mishra, P. K., Raghuram, G. V., Bhargava, A. and Pathak, N. (2011c). Translation research in molecular disease diagnosis: Bridging gap from laboratory to practice, *Journal of Global Infections Diseases*, **3**(2), 205–206.

Mishra, P. K. (2012). A pragmatic & translational approach of human biomonitoring to methyl isocyanate exposure in Bhopal, *Indian Journal of Medical Research*, **135**, 479–484.

Mishra, P. K., Raghuram, G. V., Jain, D., Jain, S. K., Khare, N. K. and Pathak, N. (2014). Mitochondrial oxidative stress-induced epigenetic modifications in pancreatic epithelial cells, *Indian Journal of Toxicology*, **33**(2), 116–129.

Mishra, P. K., Raghuram, G. V., Bunkar, N., Bhargava, A. and Khare, N. K. (2015a). Molecular bio-dosimetry for carcinogenic risk assessment in survivors of Bhopal gas tragedy, *International Journal of Occupational Medicine and Environmental Health* **28**(6), 921–939.

Mishra, P. K., Bunkar, N., Raghuram, G. V., Khare, N. K., Pathak, N. and Bhargava, A. (2015b). Epigenetic dimension of oxygen radical injury in spermatogonial epithelial cells, *Reproductive Toxicology*, **52**, 40–56.

Misra, U. K. and Kalita, J. (1997). A study of cognitive functions in methylisocyanate victims one year after Bhopal accident, *Neurotoxicology*, **18**, 381–386.

Naik, S. R., Acharya, V. N., Bhalerao, R. A., Kowli, S. S., Nazareth, H. H., Mahashur, A. A., Shah, S., Potnis, A. V. and Mehta, A. C. (1986a). Medical survey of methyl isocyanate gas affected population of Bhopal. Part I. General medical observations 15 weeks following exposure, *Journal of Postgraduate Medicine*, **32**, 175–184.

Naik, S. R., Acharya, V. N., Bhalerao, R. A., Kowli, S. S., Nazareth, H. H., Mahashur, A. A., Shah, S., Potnis, A. V. and Mehta, A. C. (1986b). Medical Survey of methyl isocyanate gas affected population of Bhopal. Part II. Pulmonary effects in Bhopal victims as seen 15 weeks after MIC exposure, *Journal of Postgraduate Medicine*, **32**, 185–191.

Panwar, H., Jain, D,, Khan, S., Pathak, N., Raghuram, G. V., Bhargava, A., Banerjee, S., Mishra, P. K. (2013). Imbalance of mitochondrial-nuclear cross talk in isocyanate mediated pulmonary endothelial cell dysfunction, *Redox Biology*, **1**, 163–171.

Panwar, H., Raghuram, G. V., Jain, D., Ahirwar, A. K., Khan, S., Jain, S. K., Pathak, N., Banerjee, S. and Maudar, K. K. (2014). Cell cycle deregulation by methyl isocyanate: Implications in liver carcinogenesis, *Environmental Toxicology*, **29**(3), 284–297.

Pathak, N., Khan, S., Bhargava, A., Raghuram, G. V., Jain, D., Panwar, H., Samarth, R. M., Jain S. K., Maudar K. K., Mishra, D. K. and Mishra P. K. (2014). Cancer chemopreventive effects of the flavonoid-rich fraction isolated from papaya seeds, *Nutrition Cancer*, **66**(5), 857–871.

Raghuram, G. V., Pathak, N., Jain, D., Panwar, H., Jain, S. K. and Mishra, P. K. (2010a). Molecular mechanisms of isocyanate induced oncogenic transformation in ovarian epithelial cells, *Reproductive Toxicology*, **30**(3), 377–386.

Raghuram, G. V., Pathak, N., Jain, D., Pandey, H., Panwar, H. and Mishra, P. K. (2010b). Molecular characterization of isocyanate–induced male germ-line genomic instability, *Journal of Environmental Pathology, Toxicology and Oncology*, **29**, 213–234.

Raghuram, G. V. and Mishra, P. K. (2014). Stress induced premature senescence: A new culprit in ovarian tumorigenesis?, *Indian Journal of Medical Research*, **140**, 99–108.

Ranjan, N., Sarangi, S., Padmanabhan, V. T., Holleran, S., Ramakrishnan, R. and Varma, D. R. (2003). Methyl isocyanate exposure and growth patterns of adolescents in Bhopal, *JAMA*, **290**, 1856–1857.

Saxena, A. K., Singh, K. P., Nagle, S. L., Gupta, B. N., Ray, P. K., Srivastav, R. K., Tiwari, S. P. and Singh, R. (1988). Effect of exposure to toxic gas on the population of Bhopal. Part IV. Immunological and chromosomal studies, *Indian Journal of Experimental Biology*, **26**, 173–176.

Sethi, B. B., Sharma, M., Trivedi, J. K. and Singh, H. (1987). Psychiatric morbidity in patients attending clinics in gas affected areas in Bhopal, *Indian Journal of Medical Research*, **86**, 45–50.

Sharma, D. C. (2005). Bhopal: 20 years on, *Lancet*, **365**, 111–112.

Shilotri, N. P., Raval, M. Y. and Hinduja, I. N. (1986). Gynaecological and obstetrical survey of Bhopal women following exposure to methyl isocyanate, *Journal of Postgraduate Medicine*, **32**, 203–205.

Sriramachari, S. (2004). The Bhopal gas tragedy: An environmental disaster, *Current Science*, **86**, 905–920.

Vijayan, V. K. and Sankaran, K. (1996). Relationship between lung inflammation, changes in lung function and severity of exposure in victims of the Bhopal tragedy, *European Respiratory Journal*, **9**, 1977–1982.

Chapter 7

Lake Nyos (1986): The Geology of Killer Lakes

Lakshmi Kantha

Aerospace Engineering Sciences, University of Colorado, USA

7.1 Introduction

The massive outgassing at Lake Nyos, Cameroon on 21 August 1986, involved a colourless, odourless gas that people and animals exhale, and plants use in photosynthesis vital to the very existence of life on Earth. The gas is the life-permitting carbon dioxide (CO_2), a greenhouse gas, whose presence in the atmosphere in minute quantities is responsible for the temperature range that makes life possible on Earth. So carbon dioxide is quite harmless, right? After all we live with it everyday[1]!

In small concentrations, CO_2 is indeed harmless. We routinely breathe in air with CO_2 concentrations of around 400 ppm (parts per million, 0.04 wt%), and breathe out air with higher concentrations of around 4%, the increase in concentration being due to generation of CO_2 by cellular metabolism of glucose for our daily

---

[1]CO_2 is also the gas most frequently talked about in anthropogenic emissions from fossil fuel burning, which are causing global warming that could lead to profound and potentially disruptive climatic changes in the coming decades.

129

energy needs. However, in high concentrations, say above about 8%, it is dangerous. Above 15%, CO_2 is lethal. It asphyxiates. This is what happened on that fateful day. Some trigger released a massive amount of CO_2 dissolved in the deep waters of the lake, which then flowed down the valley like a river asphyxiating and killing all people, livestock, animals and even insects in its path.

7.2 Physical Setting

Lake Nyos, located in Cameroon, Central Africa (6°26′17″N, 10°17′56″E), at an elevation of 1,090 m, is a fresh water crater lake, formed by a volcanic eruption several hundred years ago. Magma from a chamber below came in contact with groundwater-saturated rock near the surface and blasted a crater that then filled with water to form a crater lake 208 m deep. Its uniqueness lies in the fact that it is one of the very few volcanic lakes that contain enormous quantities of dissolved CO_2 in its hypolimnium; the only others in Africa being Lake Monoun, also in Cameroon along the Cameroon Volcano Line, and Lake Kivu on the border between the Republic of Congo and Rwanda. A 45-m wide, 40 m high natural dam formed from volcanic rock impounds the lake waters. The lake is deep unlike its much smaller companion Monoun, which at 103 m is only half as deep. This depth difference is important as we will see. With a length and width of about 2.0 and 1.2 km, respectively, the lake is relatively small with a surface area of approximately 1.58 km^2. The total volume of water is only 0.18 km^3. Except for near-surface waters, where convective cooling can occur leading to a homogeneous mixed layer as deep as 50 m, the lake waters are stably stratified because of dissolved solids and CO_2. This means, the lake does not overturn periodically and is therefore a meromictic fresh water lake, as are Lake Monoun and Lake Kivu. This allows any dissolved gases introduced at the lakebed to accumulate in the hypolimnium year after year.

Lake Nyos is situated above an extinct volcano, but a pool of magma exists at a depth of 80 km below the lake and releases CO_2 and other gases, which rise to the surface, dissolve in the spring waters surrounding the lake. This CO_2-rich "soda water" with a concentration of about 360 mmol kg^{-1} (15.84 g kg^{-1}) seeps through the lakebed at a rate of about 0.12 gmol a^{-1} (5,280 mt a^{-1} or 2.72 × 10^6 m^3 a^{-1} at STP). The thickness of this CO_2-rich water column above the lakebed increases gradually with time. Because of the pressure exerted by the water column above, the gas stays dissolved. However, if the top of these waters were to reach a depth of 113 m below the surface, the CO_2 saturation level, or the waters were to be raised somehow to that depth by some disturbance such as a landslide or cold rain water sinking to the bottom displacing the CO_2-laden waters upward, then gas bubbles begin to form creating a jet of less buoyant bubbly water that raises to the surface. This process is self-sustaining and the positive feedback could lead to explosive release of a major fraction of the dissolved CO_2 contained in the hypolimnium. The volume of the water column below the saturation level is about 0.08 km^3 and the CO_2 dissolved in it could be as much as 28.8 gmol (0.66 km^3 at STP) and if a significant fraction of this is released, it would constitute a huge reservoir of gas that would flow out of the lake. Because it is denser than air, the released CO_2 would flow like a river down the valleys surrounding the lake, asphyxiating every living creature: people, livestock and even insects in its path, until the concentration is significantly reduced by mixing with the surrounding air (1 gmol of CO_2 ~ 44,000 mt ~ 0.0227 km^3 at STP).

7.3 Outgassing

On 21 August 1986, around 21:30, when people living in the villages below the lake had finished their supper and either gone to bed or preparing to do so, some disturbance in the lake triggered an explosive release of the dissolved CO_2 in the lake. A rumbling was heard coming from the lake, immediately followed by a huge jet of water leading to a white cloud over the lake as much a 100 m thick. The cloud contained CO_2 in high concentrations, which spilled over the northern wall of the crater, flowed down the valley like a river 50 m thick at speeds of 20–50 km h^{-1}, asphyxiating 1,700 people, many in their sleep, and 3,500 livestock, as well as most living creatures in its path up to 25 km from the lake. The worst affected were villages of

Nyos, Cha and Subum. There were just a few dazed survivors. The hypothesised CO_2 concentrations before (no measurements exist) and the measured ones after the outgassing are shown in Fig. 7.1, with possible sequence of events (Kusakabe, 2015).

The exact amount of gas released by this limnic eruption is unknown since no measurements of dissolved gases in the lake exist prior to the event. Estimates vary from 5 gmol to as much as 45 gmol (1 km³), but Kusakabe (2015) estimates it to be about 14 gmol (0.32 km³ STP or 616,000 m). The eruption caused a seiche 25 m high wiping out vegetation on the south side of the crater wall and stewing vegetative debris into the lake. The lake surface waters turned red due to the oxidation of anoxic ferrous iron-rich hypolimnic water, when it came into contact with air. Figure 7.2 shows the lake surface in its normal blue pristine state and the deep

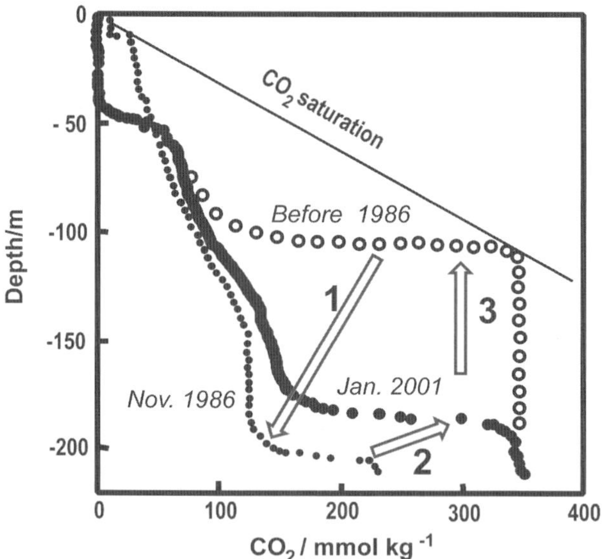

Fig. 7.1. Hypothetical CO_2 concentration before the limnic eruption (arrow 1) leading to the concentrations observed in the aftermath (November 1986) and the subsequent buildup to January 2011 (arrow 2). Arrow 3 suggests reoccurrence when the CO_2-laden waters reach the saturation depth of 110 m if the lake is not degassed (from Kusakabe, 2015).

red after the eruption. The water escaping the lake during the seiche lowered its level by about a metre.

In the following two days, villagers from nearby villages discovered what had happened to their neighbours. Many of the dead had lesions on the skin and the survivors had respiratory problems. Because of the tropical heat quickly decomposing the bodies, the dead had to be buried in a hurry in mass graves. Three days later, the world learned of the disaster that had befallen the people living near lake Nyos. While the death toll was relatively small compared to other well-known disasters such as earthquakes and tsunamis, the mysterious event, with a then unknown cause and trigger, sparked worldwide interest in the media, as well as among volcanologists and limnologists. Scientists from France, UK, USA and Japan arrived to determine what happened and find means to prevent it from happening again. Meanwhile, the Cameroonian government, not knowing the trigger and whether such an event could be repeated, ordered the evacuation of the surrounding area, disrupting the lives of thousands living there. Many thousands had fled the area in the immediate aftermath and could not return to their homes until a definite cause and trigger could be identified.

There was some initial confusion among scientists as to whether the trigger was limnic or phreatic (volcanic in origin). Only people, and air-breathing animals and creatures had been affected, and no

Fig. 7.2.   Lake Nyos (left) in its normal pristine state and (right) for a few days after the eruption (from Rouwet *et al.*, 2015).

damage to plants and trees, and non-living objects in the valleys was evident. There were no telltale signs of a volcanic eruption such as sulphur or damage to inanimate objects in the valley. The event appeared to be quite similar to one, that occurred on 15 August 1984 at 23:30 to the nearby Lake Monoun which had killed 37 people. There a truck travelling along the road near the lake the next morning had stalled. But only people who got off the truck to investigate the cause had collapsed, lost consciousness and died, while people riding above the truck had survived, pointing to asphyxiation by a dense gas that had pooled over the road. Further investigation of the lake confirmed its CO_2 content and pointed to it as the cause of the deaths at Monoun. Scientists investigating that incident (Sigurdsson *et al.*, 1987) had arrived at the conclusion that the event was most likely the outgassing of dissolved CO_2 in the lake water.

Regardless of the trigger which precipitated the disaster at Nyos, the culprit was therefore quickly identified as outgassed CO_2 (e.g. Freeth 1987; Kling *et al.*, 1987; Tietze 1987; Kusakabe *et al.*, 1989), but it took years of patient study and measurements to put together the exact sequence of events leading to the disaster, so that measures could be taken to prevent it from happening again (see Kusakabe, 2015).

7.4 Aftermath

Volcanic lakes suddenly became an intense topic of study, culminating in the recent publication of a comprehensive book on the subject (Rouet *et al.*, 2015), which presents the knowledge accumulated by scientific studies, many as the direct result of the Nyos tragedy. Until the limnic eruption of Lake Nyos, not much attention had been given to the topic. But the tragedy led to intense examination of volcanic lakes around the world, especially meromictic crater lakes around the world to see if a similar event could happen elsewhere. Concerns were raised about the much larger and much deeper Lake Kivu. Even though Lake Kivu is a rift lake along the

Albertine Rift, not volcanic, it is next to two active volcanoes and contains enormous amounts of dissolved CO_2 that dwarfs the amounts in Lake Nyos.

Immediately after the event, both Lake Nyos and Lake Monoun were periodically monitored (e.g. Kusakabe *et al.*, 2008) and the measurements indicated an alarming build up of CO_2 in the hypolimnium. Figure 7.3 shows the evolution of temperature (top), conductivity (middle) indicative of total dissolved solids (TDS) and CO_2 concentration (bottom) in Lake Monoun from October 1986 to January 2003 (left panels) and in Lake Nyos from November 1986 to January 2011. Note the build-up of CO_2-laden water in the hypolimnium in both cases. Measurements indicated a continuous recharge of CO_2 by seepage of CO_2-laden water at the lakebed, with concentrations of 160 mmol kg^{-1} in Lake Monoun and 360 mmol kg^{-1} in Lake Nyos. In Lake Monoun, this seepage had raised the CO_2-laden waters in 2003 to 58 m from the surface, close to the saturation level of 50 m. At a rate of increase of its thickness of 1 m a^{-1}, recurrence of outgassing was considered highly probable after 2011. Urgent degassing was needed to avert a disaster.

Artificial degassing had been proposed earlier by some. The idea was to insert a pipe to the CO_2-laden waters near the lakebed and start the flow of water up toward the surface by a pump. Once the flow starts, the release of dissolved gases creates a buoyant two-phase flow in the pipe that is self-sustaining. The CO_2 in the water shooting out of the pipe is released to the atmosphere and CO_2-depleted water falls back to the lake. Thus, the thickness of the CO_2-rich water column in the hypolimnium can be gradually reduced to harmless levels.

The Nyos-Monoun Degassing Project (NMDP) was set up in 1996. In the face of warnings by some that degassing could actually destabilise the lake and precipitate another limnic eruption, a French team had demonstrated feasibility of safe degassing first in Monoun and later in Nyos (Halbwachs *et al.*, 2004) using small diameter pipes. Permanent degassing had been started at Lake Nyos in 2001 with a pipe of 140 mm inside diameter with its intake at

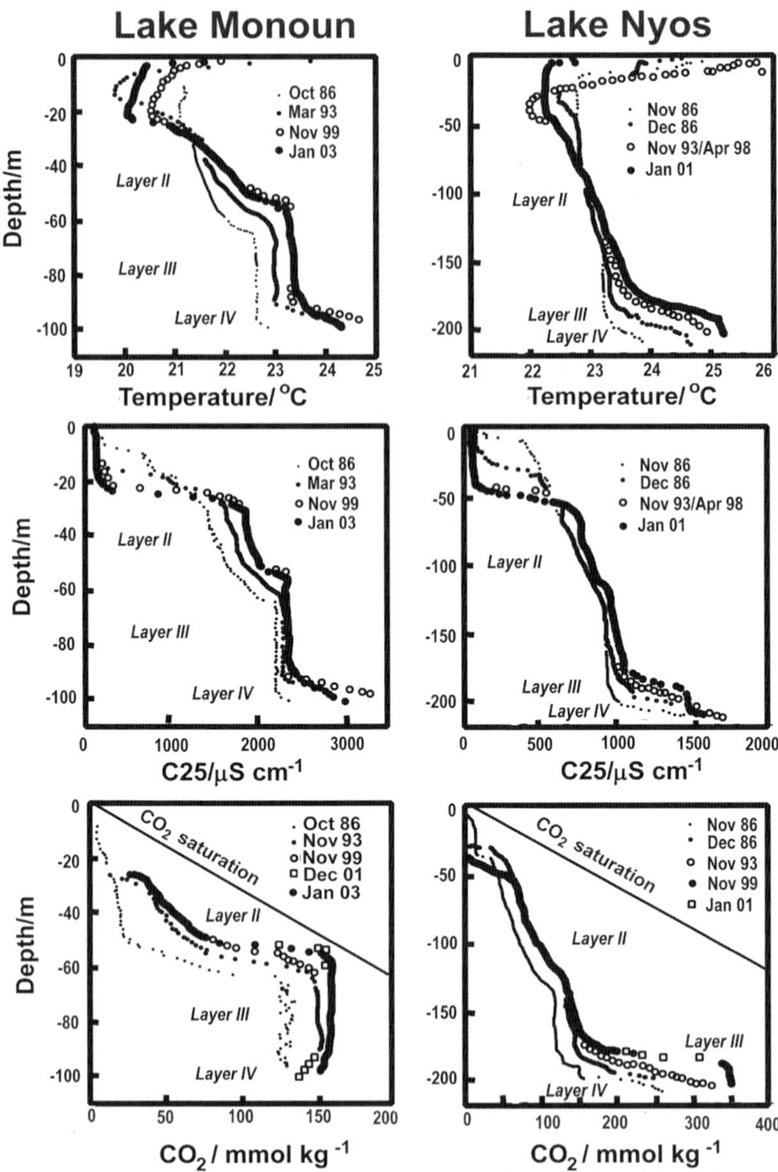

Fig. 7.3. The evolution of Temperature (top), Conductivity (middle) indicative of total dissolved solids (TDS) and CO_2 concentration (bottom) in Lake Monoun from October 1986 to January 2003 (left panels) and in Lake Nyos from November 1986 to January 2011. Note the build-up of CO_2-laden water in the hypolimnium (from Kusakabe, 2015).

203 m depth. Once the alarm was raised, it was also started in Lake Monoun in 2003 with a 102 mm pipe with intake at 73 m depth.

The projects were carried out under NMDP, funded by French, US and Cameroonian governments. Two more pipes were installed in Lake Monoun in 2006 with the intake lowered to 93 m depth. The three pipes operating together removed most of the gas in Lake Monoun rendering it harmless. The gas content decreased from 0.61 gmol in 2003 to less than 0.036 gmol in 2011. The degassing actually removed the self-lift capability (resulting in 8 m tall water fountains) itself (Kusakabe, 2015).

The degassing of Lake Nyos was comparatively slow and hence two more larger 257 mm diameter pipes were installed in 2011 and 2012 using EU funding. The CO_2 content was reduced from 14+ gmol in 2001 to 9+ gmol in 2012 (Kusakabe, 2015). It appears that the three pipes in operation will remove most of the gas in a few years. The result of degassing operations on CO_2 concentrations in the lakes up to year 2011 is shown in Fig. 7.4 (Kusakabe, 2015).

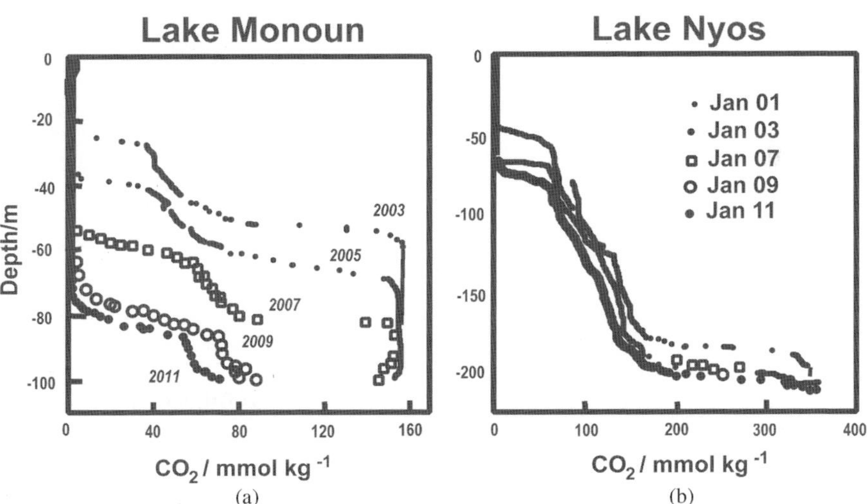

Fig. 7.4.   The change in CO_2 concentrations in Lake Monoun (left) and Lake Nyos (right) as a result of degassing operations under NMDP (from Kusakabe, 2015).

Degassing operation resulted in a spectacular "soda water" fountain reaching to heights of as much as 30 m, depending on the CO_2 concentration in the water flowing through the intake. Figure 7.5 shows the fountain (left panel) during the experimental degassing in 2011, and the three fountains and the intensely reddish surface waters resulting from the operation of all three degassing pipes in 2012 (right panel) in Lake Nyos. The initial fears that the degassed, cooled water emerging from the fountain would sink to the bottom destabilising the water column and causing inadvertent limnic eruption did not materialise. Successful degassing operations in Lake Monoun and Lake Nyos has suggested that similar operations at Lake Kivu could be undertaken if needed, although the additional complexities involving combustible methane have to be considered. Potential recovery of abundant methane, a cleaner-burning fuel compared to petroleum products is an added attraction. However,

Fig. 7.5. (Left) A 45 m tall water fountain emerging from the degassing pipe in Lake Nyos in 2011. (Right) The three degassing pipes in operation in Lake Nyos in 2012 resulting in red colouration of the surface waters due to oxidation of Ferrous iron-rich hypolimnium water on contact with air (from Kusakabe, 2015).

massive involvement and funding by the international community would be needed to accomplish the task, but it is feasible to avert a future catastrophe.

However, because of continuous recharging, degassing operations will have to be resumed in the future. Even after successful degassing, danger still persists at Lake Nyos, due to potential for a dam break, which would cause severe flooding and fatalities in the valley below, all the way to the border with Nigeria 200 km to the north. It may even release the pressure on the lakebed thereby causing any CO_2 gas in the sediments to be released. The 40 m high pyroclastic dam holding the lake waters appears to be weak. However, degassing allows the lake water level to be lowered to relieve the load on the dam, without the danger of the decrease in pressure releasing dissolved gases in the hypolimnium. The dam can also be reinforced to prevent its failure and efforts to do so are underway. However, continuous monitoring and future degassing of the lake will still be necessary. Numerical models capable of simulating and predicting the water column structure (Kantha and Freeth, 1996, McCord and Schladow, 1998; Schmid *et al.*, 2006) can also assist in the effort.

7.5 Other Volcanic Lakes

Lake Monoun, 100 km SE of Lake Nyos, is half the depth of Lake Nyos and therefore the saturation depth of 50 m can easily be reached by the CO_2-laden waters seeping through the lakebed. It has a surface area of 0.526 km^2 and a total water volume of 0.01393 km^3 compared to Nyos (area ~1.58 km^2 and volume 0.1794 km^3). More important from the point of view of CO_2 content is the volume below the saturation level, which is 0.01 km^3 and 0.08 km^3, respectively (Kling *et al.*, 2005). With concentrations of 150 mmol kg^{-1} and 360 mmol kg^{-1} in the water seeping in through the lakebed, this means that the maximum possible CO_2 content in the lakes is 1.5 gmol and 28.8 gmol, respectively. The recharge rates in Lake Monoun and Lake Nyos appear to be 8.2 mmol a^{-1} and 0.12 gmol a^{-1}, respectively, so that the recharge time to lethality is 183 years at Lake

Monoun and 240 years at Lake Nyos, assuming the lakes have been completely degassed.

Observations since 1986 (Fig. 7.3) show that atmospheric forcing can deepen the upper mixed layer to a maximum depth of only about 20 m in Lake Monoun, and 50 m in Nyos, perhaps 60 m during rare cooling events. The smaller mixed layer depth in Lake Monoun is because of the relatively significant amounts of fresh river water flowing into the lake. Therefore, it is most unlikely for the overturning to have been initiated by mixed layer deepening in either lake. However, the fact that both limnic eruptions took place in the month of August, with a difference of only a week (but 2 years apart) is curious and may not be just happenstance. August, in Cameroon is a month of cold rains and it is possible that with the lakes fully charged to almost the level of saturation (50 m and 113 m, respectively), cold water from rain and/or streams flowing into the lakes, sinking down and displacing the CO_2-laden waters upward could have prepared the lakes for outgassing. All that needed was a small trigger in the form of a rock slide or wind-induced seiche or some other perturbation. It is estimated that water colder than 20°C would be needed to penetrate the chemocline (Kling *et al.*, 2015). The 1980's were an unusually cool period and it is likely that the rainwater was as cold as 18.5°C that August in 1986. But this remains speculative, since there were no atmospheric measurements around Nyos.

Lake Kivu (located around 2°S, 29°E) has a mean depth of 240 m and is 485 m deep at its deepest point. It has a surface area of 2,400 km^2 (1,600 times that of Lake Nyos) and the volume of fresh water it holds is an enormous 560 km^3. The chemocline there is at a depth of 250 m and the waters below contain not only 300 km^3 of dissolved CO_2, but also 55–60 km^3 of bacterially-generated methane, CH_4 (Kling *et al.*, 2006). The amount of CH_4 is so huge that Lake Kivu is one of the largest natural gas deposits in the world. Nearly 2.5 million people live around the lake. This, combined with the fact that the amount of dissolved gases dwarfs the amount of dissolved gas in Lake Nyos, makes an outgassing event at Lake Kivu truly a disaster of epic proportions.

Unfortunately, Lake Kivu, a rift (not crater) lake, is located close to two active volcanoes. The fear is that lava from a volcanic eruption can flow into the lake and in sufficient quantities, reach below the chemocline depth and displace dissolved gases-laden hypolimnic waters up toward the saturation level. The 2002 eruption of the volcano Mount Nyiragongo caused its lava to flow into the lake and reach depths of 70–100 m. The eruption stopped well before the lava got to the chemocline depth. Next time around, luck may not prevail. There are also volcanic vents underneath the lakebed. There is an evidence in sediments of massive extinctions around the lake roughly every 1000 years, suggesting limnic eruptions, trigger unknown. A limnic or phreatic eruption would lead to the release of asphyxiating CO_2 as well as highly combustible CH_4, with catastrophic results to the millions living around the lake and unimaginable horror.[2] However, Vaselli *et al.* (2015) conclude that in spite of increasing gas concentrations in recent years, the lake is very stable and a limnic eruption is unlikely for the next 80–200 years. But this does not rule out the possibility of outgassing by volcanic eruptions.

Panoramic Lake Quilotoa in Equador is a 3 km wide, 250 m deep caldera lake in the Equatorial Andes at an altitude of 3,900 m, formed by an eruption 800 years ago. It gets its greenish colour from dissolved minerals. The lakebed has fumeroles and hot springs exist on its eastern flank. It appears that there have been four recorded eruptions, all in the 18th century, with the one in 1797 involving a limnic eruption of CO_2 that killed all the cattle on its shores. With a volume of 0.25 km^3 the lake contains enormous quantities of dissolved CO_2 according to a survey in November 1993. However, Aguilera *et al.* (2000) conclude that the risk of a dangerous eruption is negligible, but recommend periodic surveys to monitor any more buildup of CO_2 in the deep lake waters.

Finally, CO_2 emissions from volcanic lakes are very small compared to the 40 Gta^{-1} of anthropogenic CO_2 emitted by fossil fuel

---

[2]However, extraction of methane for power generation holds the promise of cheap energy for the surrounding countries. A 35-MW power plant project is underway.

burning. Even if entire Kivu gas content were to be released by outgassing, only 0.6 Gt would be added. Therefore, from global warming point of view, gas bearing volcanic lakes are a non-issue.

7.6 Concluding Remarks

The Nyos outgassing, as tragic as it was, resulted in many positive outcomes. It brought worldwide attention to the possibility of limnic eruptions. It brought an international team of volcanologists and limnologists together to study volcanic lakes, assess their potential hazard and come up with ways to avert future disasters. It led to the establishment of the International Working Group on Crater Lakes in 1990's, which was succeeded by IAVCEI Commission on Volcanic Lakes (CVL). Rich, developed nations around the world, France, USA, Japan and EU pitched in to defuse the danger of "killer lakes" in Cameroon. Urgent studies were conducted on Lake Kivu, following the eruption of Mount Nyiragongo, to assess the possibility of an even more catastrophic degassing event and explore possible ways to avert the disaster. Studies also ruled out limnic eruptions in some other crater lakes such as the Ruapehu crater lake in New Zealand and Lake Quilotoa. The communities in Africa, and the world as a whole, are better off as a result of this focus by world community on limnic eruptions, not just volcanic eruptions. We also know much more about volcanic lakes than before the Nyos disaster (e.g. Rouwet *et al.*, 2015). The reader is urged to peruse this treatise, especially Christenson and Tassi (2015), Christenson *et al.* (2015), Costa and Chiodini (2015), Kling *et al.* (2015), Kusakabe (2015), Masot and Bernard (2015) and Vaselli *et al.* (2015) for specific articles related to Lakes Nyos, Monoun and Kivu.

References

Aguilera, E., Chiodini, G., Cioni, C., Guidi, M., Marini, L. and Raco, B. (2000). Water chemistry of Lake Quilotoa (Equador) and assessment of natural hazards, *Journal of Volcanology Geothermal Research*, **97**, 271–285.

Aka, F. T. (2000). Noble gas systematics and K–Ar chronology: Implications on the geotectonic evolution of the Cameroon Volcanic Line, West Africa. Ph.D. Thesis, University Okayama, Japan, p. 175.

Aka, F. T., Kusakabe, M., Nagao, K. and Tanyileke, G. (2001a). Noble gas isotopic compositions and water/gas chemistry of soda springs from the islands of Bioko, *Applied Geochemistry*, **16**, 323–338.

Aka, F. T., Kusakabe, M. and Nagao, K. (2001b). New K–Ar ages for Lake Nyos maar, Cameroon: Implications for hazard evaluation, *Journal of Geoscience Society of Cambridge*, **1**, 25–26.

Aka, F. T., Nagao, K., Kusakabe, M., Sumino, H., Tanyileke, G., Ateba, B. and Hell, J. (2004). Symmetrical helium isotope distribution on the Cameroon Volcanic Line, *Chemical Geology*, **203**, 205–223.

Aka, F. T., Yokoyama, T. Kusakabe, M., Nakamura, E., Tanyileke, G., Ateba, B., Ngako, V., Nnange, J. M. and Hell, J. V. (2008). U-series dating of Lake Nyos maar basalts, Cameroon (West Africa). Implications for potential hazards on the Lake Nyos dam, *Journal of Volcanology Geothermal Research*, **176**, 212–224.

Aka, F. T. and Yokoyama, T. (2013). Current status of the debate about the age of Lake Nyos dam (Cameroon) and its bearing on potential flood hazards, *Natural Hazards*, **65**(1), 875–885.

Christenson, B. and Tassi, F. (2015). Gases in volcanic lake environments, in Rouwet, D., Christenson, B., Tassi, F. and Vandemeulebrouck, J. (eds.), *Volcanic Lakes, Advances in Volcanology* (Springer-Verlag, Berlin), pp. 125–153.

Christenson, B., Nemeth, K., Rouwet, D., Tassi, F., Vandemeulebrouck, J. and Varekamp, J. (2015). Volcanic lakes, in Rouwet, D., Christenson, B., Tassi, F. and Vandemeulebrouck, J. (eds.), *Volcanic Lakes, Advances in Volcanology* (Springer-Verlag, Berlin), pp. 1–20.

Clarke, T. (2001). Taming Africa's killer lake, *Nature*, **409**, 554–555.

Costa, A. and Chiodini, G. (2015). Modeling air dispersion of CO_2 from limnitic eruptions, Cameroon, in Rouwet, D., Christenson, B., Tassi, F. and Vandemeulebrouck, J. (eds.), *Volcanic Lakes, Advances in Volcanology* (Springer-Verlag, Berlin), pp. 451–466.

Freeth, S. J. (1987) The Lake Nyos gas disaster. *Nature*, **325**, 104–105.

Halbwachs, M., Sabroux, J. C., Grangeon, J., Kayser, G., Tochon-Danguy, J. C., Alain, F., Beard, J. C., Villevieille, A., Vitter, G., Richon, P., Wüest, A., and Hell, J. (2004) Degassing the 'Killer Lakes' Nyos and Monoun, Cameroon, *EOS Transactions of American Geophysical Union*, **85**(30), 281–288.

Kantha, L. H. and Freeth, S. J. (1996). A numerical simulation of the evolution of temperature and stratification in Lake Nyos since the 1986 disaster, *Journal of Geophysical Research*, **101**(B4), 8187–8203.

Kling, G. W., Clark, M. A., Compton, H. R., Devine, J. D., Evans, W. C., Humphrey, A. M., Koenigsberg, E. J., Lockwood, J. P., Tuttle, M. L. and Wagner, G. N. (1987). The 1986 Lake Nyos gas disaster in Cameroon, West Africa, *Science*, **236**, 169–175.

Kling, G. W., Evans, W. C., Tuttle, M. L. and Tanyileke, G. (1994). Degassing of Lake Nyos. *Nature*, **368**, 405–406.

Kling, G. W., Evans, W. C., Tanyileke, G., Kusakabe, M., Ohba, T., Yoshida, Y. and Hell, J. V. (2005). Degassing Lakes Nyos and Monoun: Defusing certain disaster, *Proceedings of National Academy of Sciences*, **102**(40), 14185–14190.

Kling, G. W., MacIntyre, S., Steenfelt, J. and Hirslund, F. (2006). Lake Kivu gas extraction: Report on lake stability. Report produced for the World Bank and the Government of Rwanda, p. 104.

Kling, G. W., Evans, W. C. and Tanyileke, G. Z. (2015). The comparative limnology of lakes Nyos and Monoun, Cameroon, in Rouwet, D., Christenson, B., Tassi, F. and Vandemeulebrouck, J. (eds.), *Volcanic Lakes, Advances in Volcanology* (Springer-Verlag, Berlin), pp. 401–426.

Kusakabe, M., Ohsumi, T. and Aramaki, S. (1989). The Lake Nyos disaster: Chemical and isotopic evidence in waters and dissolved gases from three Cameroonian crater lakes, Nyos, Monoun and Wum, *Journal of Volcanology Geothermal Research*, **39**, 167–185.

Kusakabe, M., Ohba, T., Issa, Yoshida, Y., Satake, H., Ohizumi, T., Evans, W. C., Tanyileke, G. and Kling, G. W. (2008). Evolution of CO_2 in Lakes Monoun and Nyos, Cameroon, before and during controlled degassing, *Geochemical Journal*, **42**, 93–118.

Kusakabe, M. (2015). Evolution of CO_2 content in Lake Nyos and Monoun, and sub-lacustrine CO_2-recharge system at Lake Nyos as envisaged from $CO_2/^3He$ ratios and noble gas signatures, in Rouwet, D., Christenson, B., Tassi, F., Vandemeulebrouck, J. (eds.), *Volcanic Lakes, Advances in Volcanology* (Springer-Verlag, Berlin), pp. 427–450.

McCord, S. A. and Schladow, S. G. (1998). Numerical simulations of degassing scenarios for CO_2-rich Lake Nyos, Cameroon, *Journal Geophysical Research*, **103**, 12355–12364.

Rouwet, D., Christenson, B., Tassi, F. and Vandemeulebrouck, J. (2015). (eds.) *Volcanic Lakes, Advances in Volcanology*, (Springer-Verlag, Berlin), pp. 427–450.

Schmid, M., Halbwachs, M. and Wüest, A. (2006). Simulation of CO_2 concentrations, temperature, and stratification in Lake Nyos for different degassing scenarios, Geochemistry Geophysics Geosystsiens, 7: Q06019. doi:10.1029/2005GC001164.

Sigurdsson, H., Devine, I. D., Tchoua, F. M., Presser, T. S., Pringle, M. K. and Evans, W. C. (1987). Origin of the lethal gas burst from Lake Monoun, Cameroon, *Journal of Volcanology Geothermal Research*, **31**, 1–16.

Tietze, K. (1987). Results of the German-Cameroon research expedition to Lake Nyos (Cameroon) October/November 1986. Interim-Report, Bundesanstalt für Geowissenschaften und Rohstoffe Archive no. 100470, p. 84.

Vaselli, O., Tedesco, D., Cuoco, E. and Tassi, F. (2015). Are limnic eruptions in the CO_2-CH_4-rich gas reservoir of Lake Kivu (Democratic Republic of Congo and Rwanda) possible? Insights from physico-chemical and isotope data, in Rouwet, D., Christenson, B., Tassi, F. and Vandemeulebrouck, J. (eds.), *Volcanic Lakes, Advances in Volcanology* (Springer-Verlag, Berlin), pp. 489–506.

Chapter 8

Kuwait Oil Fires (1991): A Deliberate Environmental Disaster During Wartime

Michelle L. Bell

School of Forestry and Environmental Studies,
Yale University, USA

8.1 Introduction

During the first Gulf War, or the Persian War, a coalition of forces from the US, Britain, Germany, Japan, Egypt, Syria and other countries joined to remove Iraqi military forces from Kuwait. The Iraqi military, led by Saddam Hussein, had invaded Kuwait on 2 August 1990. Massive air attacks from coalition forces in early 1991 crippled Iraqi infrastructure, weapons manufacturing, communications and ground troops, resulting in most Iraqi forces exiting the area by the end of February 1991. This conflict, for which the US-led air operations were named Operation Desert Storm, resulted in thousands of deaths, as well as health, emotional and social impacts on survivors (Barth *et al.*, 2009; Bullman *et al.*, 2005; Haley, 2003; Writer *et al.*, 1996; Rose, 2003; Kang and Bullman, 2001). However, the war also resulted in an environmental disaster as a result of Iraq's "scorched earth" strategy, as Iraqi forces intentionally set fire to over 600 of Kuwait's

1000 oil wells. These oil fires exceeded in number all previous oil fires combined (Arkin, 1996).

The deliberate destruction of the environment during wartime, sometimes regarded an environmental war crime is not new, and such events, including the burning of structures and crops, have been recorded in earlier wars and in ancient times (Al-Damkhi, 2007; Leebaw, 2014). For example, the US explored the use of fire to destroy forests in southern regions of Vietnam in the mid-1960s (Shapley, 1972) (see Box 8.1). As technology has advanced, the scope and damage of environmental warfare have grown. The scale of the intentional burning of oil wells in Kuwait produced high levels of air pollution that likely led to substantial public health burdens for

---

Box. 8.1.   Fire as a Weapon in the Vietnam War

In 1965, the US Department of Defence commissioned research on the feasibility of igniting forest fires as a military tactic for the Vietnam War. The originally classified report, titled "Forest Fire as a Military Weapon", notes that forest fires can substantially damage enemy troops and facilities (U.S. Department of Agriculture Forest Service, 1970). The U.S. intentionally burned forests in the Mekong Terrace in South Vietnam in the mid-1960s in several attempts to increase visibility for air scouting by removing the top layer of jungle canopy and to aid ground-level warfare by eliminating vegetation that could be used for cover. The overall aim was to develop technology and methods to incite a "fire storm", a large and self-sustaining fire that could be used under field conditions. The process included spraying herbicides, drying the area and igniting the forest such as through an incendiary bomb. One senator described this tactic as "unprecedented environmental warfare" (Shapley, 1972). These burning efforts, under code names, such as Sherwood Forest, Hot Top and Operation Pink Rose, were not successful given the dense jungle canopy and high humidity (U.S. Department of Agricultural Forest Service, 1970; Wilmington Morning Star, 1972).

those who were exposed. While the impact of the human health consequences from the fires is not fully known, many returning veterans suffered from a wide range of symptoms that have been loosely defined as Gulf War syndrome. Exposure to environmental contaminants including the smoke and other aspects of the burning oil well fires may have been a contributor to such illness, along with many other physical and psychological factors. This chapter describes the burning of the oil wells in Kuwait and subsequent studies on the resulting air pollution's effect on human health concerns, especially for veterans as this was the most studied population, along with a summary of the challenges and opportunities of this research.

8.2 The Burning of Kuwaiti Oil Wells

The fires began in mid-January 1991, continuing into February 1991, as Iraqi forces strategically set fire to oil wells. Charges of about 30–40 pounds of C-4 plastic explosives were detonated at over 600–700 wells, of which approximately 650 burned, although the estimate varies. The wells emitted crude oil and plumes of black smoke, with some white plumes from wells with a high salt water content. In addition to oil wells, Iraqi forces destroyed oil storage tanks, oil transfer pipelines, and oil rigs to weaken Kuwait's capability to produce and process oil by targeting related infrastructures. Approximately 71% of Kuwaiti oil fields were burned, 13% damaged, 5% gushed oil and 11% remained intact (Rostker, 2000). Figure 8.1 shows the plume of a burning well. Figure 8.2 provides a timeline of some of the events of the oil fires and efforts to control them.

The burning of oil wells and associated structures was a series of carefully planned and coordinated attacks with detailed preparation and wiring of oil fields well before the wells were ignited. Prior to the full sabotage of oil wells, Iraqi forces performed trial explosions of six oil wells in December 1990 (Arkin, 1996). The burning likely had multiple motivations (Al-Damkhi, 2007). This systematic destruction simultaneously weakened Kuwait's economy and environment. Smoke from the fires also provided cover that obscured movement of Iraqi forces and hindered movement of the Coalition's ground troops. Based on intercepted Iraqi documents, some initially

Fig. 8.1.   Burning oil well and smoke plume (Taschen, 2016).

hypothesised that the oil well fires were also intended to disperse harmful chemical agents, although later translations found no evidence to support such claims (Rostker, 2000). In spring of 1991, the burning wells emitted an average of about 5 million barrels of oil each day (Hobbs and Radke, 1992) and 70–100 m^3 of natural gas each day (Rostker, 2000). The released crude oil combined with smoke to create rain-like droplets that produced a black tar-like coating of "oil rain".

8.2.1 *Non-burning wells*

Some oil well heads that were severely damaged, but were not burning gushed crude oil, which is flammable and toxic. Further, some burning wells gushed crude oil due to incomplete burning. The oil from these wells formed some 400 lakes with approximately 156 million barrels of oil (Al-Damkhi, 2007), with other sources reporting over 100 oil lakes with 25–40 million barrels of oil (Husain, 1995; Al-Besharah, 1991). The oil lakes varied greatly in depth, size and oil viscosity, with one lake covering 72 km^2. The lakes greatly hampered

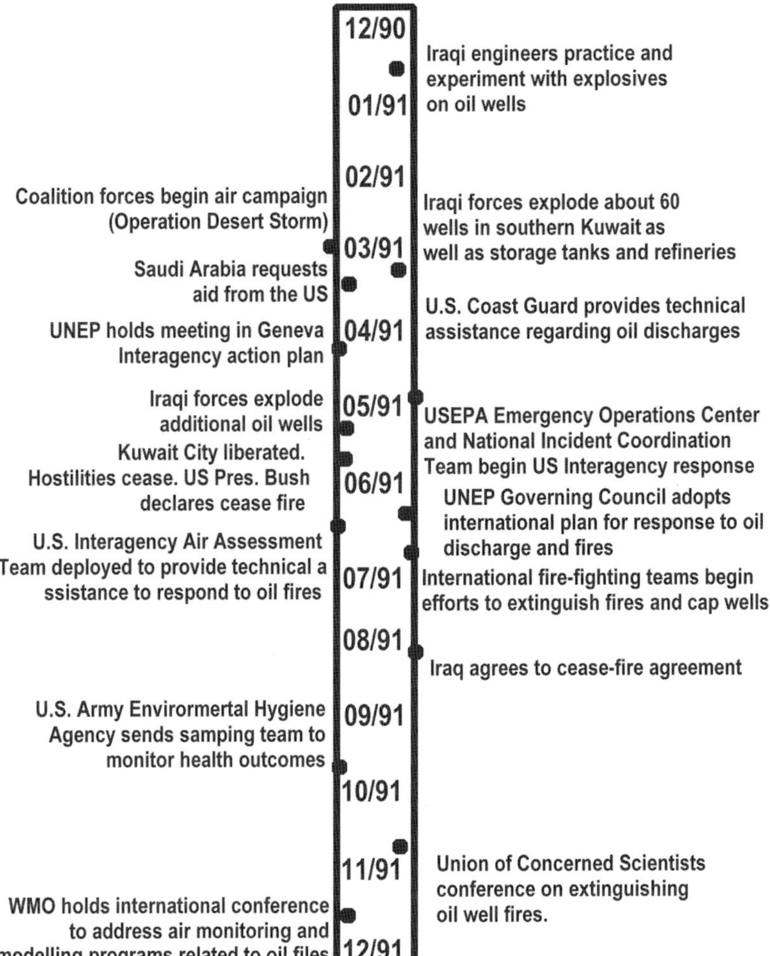

Fig. 8.2.   Timeline of key events.

Notes: 8/90: Iraqi military forces invaded Kuwait. 11/91: Last oil well capped. WMO = World Meteorological Organization. UNEP = United Nations Environment Program.

fire-fighting efforts by blocking access to the fires, contaminating groundwater, and harming wildlife and vegetation. Many other water bodies were contaminated as water from lakes containing oil were used to combat burning wells. Efforts to slow the spread of oil lakes included construction of sand barriers and pumping oil out of the lakes. After the war, a large portion of the split oil remained as oil lakes, contaminating soil and prohibiting economic development of these areas (Elgibaly, 1999; Barker and Bufarsan, 2001; Bader and Clarkson, 1995).

Oil wells that were not burning also presented hazards if emitting flammable gases, such as methane, benzene, or hydrogen sulphide (H_2S). H_2S, a colourless, ignitable gas, occurs naturally in crude oil and is produced by petroleum drilling, with the level of hazard depending on the H_2S gas content. Exposure to H_2S in other conditions has been observed to cause nausea; irritation to the eyes, nose and throat; headaches, memory loss, coughing, altered breathing, and at high concentrations death (OSHA, 2005).

8.2.2 *Extinguishing the flames*

By the end of March 1991, international fire-fighting teams had begun developing plans to extinguish fires and cap wells. Efforts to stop the fires proved challenging, due to the high levels of smoke that lowered visibility and presented risk to fire fighters. Oil lakes hindered access to fires. Workers also faced land mines, unexploded bombs, lack of infrastructure and equipment and little access to water (Husain, 1995). Many of the methods used to extinguish the burning oil wells were designed to permit future use of oil wells, and thereby did not always employ the most efficient methods, as the decision makers balanced multiple objectives. The international effort of about 10,000 fire fighters, primarily from the US but representing about 40 countries, extinguished over 600 oil fires and capped about 150 additional impaired oil wells. This included many professional, experienced firefighters as well as volunteers. The enormous plumes of smoke from the burning oil wells exposed troops and citizens to extremely high levels of harmful air pollutants over a period of about 7–9 months before the fires were extinguished (Rostker, 2000). The fire fighters themselves were exposed to high levels of air pollution and extreme heat, and in many cases they were covered in spewing oil as they worked.

Several firefighters and related workers died during the efforts to extinguish the oil fires (Wald, 1991). The last oil well was capped in early November 1991, when a previously controlled well was relit and then ceremoniously extinguished by Sheik Jaber al-Ahmed al-Sabah, the Emir of Kuwait, in a red carpet ceremony with ceremonial sword dancers.

8.3 Impacts on Air Quality

When the fires began and during the early periods of burning, most areas had few if any air monitors, yet personal accounts give some indication of the very high levels of air pollution. Troops and workers described how the smoke turned day into the appearance of night: "the smoke . . . turned the sunlit, bright day into a dark night"; there was "no capability of distinguishing the sun from the moon for the first 6 weeks after the liberation of Kuwait" (Stead, 1991). "It was darker than night due to smoke cover. The sun couldn't pierce the smoke" (Klingbeil, 2012). "Oil at times flowed like a river, 16 inches deep across the road. Fires blazed relentless on all sides"; "Even at [10 am], the sky was pitch-black with smoke from the burning oil fires. . . If you could picture Dante's Inferno, this had to be it." (Vriesenga, 1994).

The resulting plumes from reached 3–5 km into the air and hundreds of km across (Limaye *et al.*, 1991). The pollution travelled far distances, up to 2000 km east of Kuwait. Smoke from the fires was observed by astronauts aboard the US National Aeronautics and Space Administration's (NASA's) space shuttles and could be seen in satellite imagery, as seen in Fig. 8.3. Much of the pollution failed to reach the upper atmosphere, so it did not spread globally through long-range transport. However, the smoke had regional effects on climate, lowering surface temperature by as much as 10°C near the source of smoke in 1991 (Arkin, 1996; Bakan *et al.*, 1991).

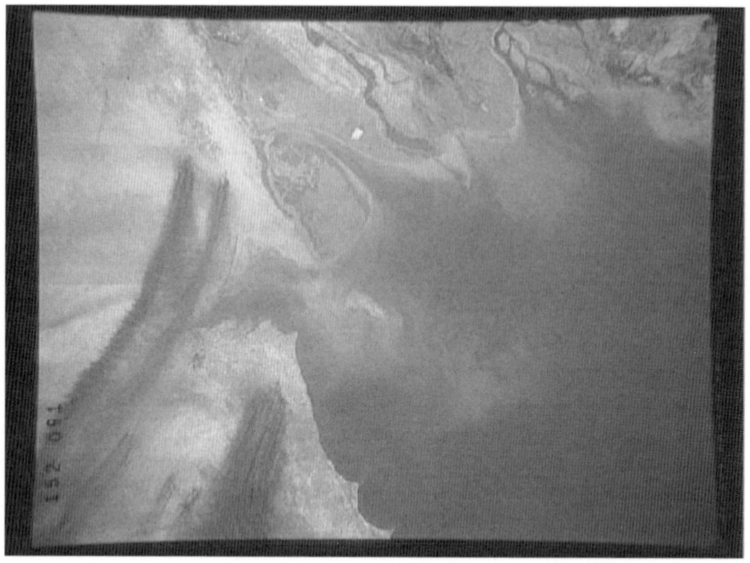

Fig. 8.3.   Satellite image of black smoke plumes in northwestern part of the Persian Gulf. The plumes are blowing to the South.

Source: NASA, STS037-152-091 Oil Well Fires and Smoke, Kuwait, April 1991.

Research efforts to estimate the level and location of air pollution from the burning wells included air quality modelling, such as Lagrangian models that estimate how parcels of air and thereby pollution are transported and dispersed through the atmosphere (Draxler *et al.*, 1994). Output from such models can be compared to surface and aircraft measurements for evaluation. Dispersion modelling was used to estimate how US military were exposed to air pollution from burning oil wells by combining information on troop movements, measurements from monitoring stations and satellite imagery (Heller, 2011). Multiple US agencies including each branch of the military participated.

Multiple monitoring studies were conducted to measure air pollution levels resulting from the burning wells. These programs involved several US agencies including the US Army, Environmental Protection Agency, Centers for Disease Control and Prevention, National Oceanic and Atmospheric Administration, Department of

Defence, Coast Guard, and Department of Energy, among others. Many other countries and agencies also played critical roles. Central efforts were conducted by the United Nations Environment Program, International Maritime Organization, Japan Environment Agency, Kuwait Environmental Protection Department, World Meteorological Organization, Saudi Arabia Meteorology and Environmental Protection Agency and British Meteorological Office.

The largest monitoring campaign was performed by the US Army Environmental Hygiene Agency (AEHA), later renamed the Center for Health Promotion and Preventive Medicine. No comprehensive systematic monitoring program was instituted until May 1991, well after the oil fires were initiated (Spektor, 1998), which limits understanding of the levels of air pollution for the first few months of the burning wells, although about 92% of the fires were still burning at the start of this campaign. The US AEHA monitoring was performed 5 May 1991, through 3 December 1991, with all fires extinguished around early November 1991. This campaign collected over 5,000 environmental samples including sand, soil and air (Heller, 2011).

Air pollution from the burning oil wells peaked towards the end of February 1991. Air pollution directly from the production of oil through drilling and processing was lower during and after the burning of oil wells as the oil industry was disrupted, but after operations were reinstated oil production increased resulting in an increase in this source of air pollution.

Smoke from the burning oil wells and lakes contained numerous harmful air pollutants, including carbon monoxide (CO), volatile organic compounds (VOCs), H_2S, precursors of ozone (O_3), polycyclic aromatic hydrocarbons (PAHs), sulphur dioxide (SO_2) and heavy metals. A key pollutant in the smoke from burning oil wells is particulate matter (PM), which itself is a complex chemical mixture (Shapley, 1972) (see Box 8.2). These pollutants are known to be harmful to human health when emitted from other sources under conditions with much lower levels than those created during the Kuwait oil fires. Table 8.1 provides a brief description of some of the key air pollutants produced during the burning of the oil wells, as well as their common sources and some of the health

---

Box. 8.2. Particulate Matter Air Pollution

Particulate matter is a common air pollutant that comes from many sources including vehicles and industry. This pollutant was also produced in large quantities by the burning oil wells in Kuwait (Brown *et al.*, 2008). Airborne particles are a complex mixture of gases and solids that include a wide array of chemical structures that can differ based on the original sources of the pollution, their chemical and physical transformation in the atmosphere, and season (Bell *et al.*, 2007). Particulate matter is often measured and analysed with respect to its size, such as particles ≤10 μm in aerodynamic diameter (PM_{10}), ≤2.5 μm ($PM_{2.5}$), or ≤0.1 μm (ultrafine PM); or particles with aerodynamic diameter <10 μm and >2.5 μm (coarse PM). While smaller particles, which penetrate deeper into the respiratory system, appear to be more harmful, all particle sizes have been associated with detrimental human health effects.

Table 8.1. Pollutants from burning oil wells, their traditional sources and health effects.

Pollutant	Description	Examples of traditional sources	Example health effects
Carbon dioxide (CO_2) ~133 million tonnes from oil well fires	Primary greenhouse gas	Vehicles, industry, power production	Contributor to climate change and associated health impacts
Nitrogen oxides (NO_x) ~80,000 tonnes from oil well fires	Highly reactive gas	Vehicles, power production. Contributes to formation of particulate matter. Precursor to ozone	Emergency department visits, hospital admissions, respiratory health

(*Continued*)

Table 8.1.   (*Continued*)

Pollutant	Description	Examples of traditional sources	Example health effects
Particulate matter (e.g. PM_{10}, $PM_{2.5}$)	Complex chemical mixture of airborne particles of different sizes	Dust, combustion, construction, power production	Mortality, hospital admissions, exacerbation of asthma, cardiovascular and respiratory health
Carbon monoxide (CO) ~0.8 million tonnes from oil well fires	Colourless, odourless gas	Combustion processes including vehicles	Chest pain, hospital admissions, mortality at very high levels
Hydrogen sulphide (H_2S)	Ignitable, colourless gas	Oil and natural gas production	Effects on central nervous system, eye irritation, respiratory health, dizziness
Sulphur dioxide (SO_2) ~2.56 million tonnes from oil well fires	Highly reactive gas	Fossil fuel combustion, industrial processes	Asthma symptoms, emergency department visits, hospital admissions, respiratory health
Ozone (O_3)	Colourless gas formed from the precursors of NOx and VOCs	Precursor sources: industry, power generation, vehicles, chemical solvents	Mortality, respiratory symptoms, lung function, doctor visits

Note. The pollutants, sources, and health effects listed here are examples and do not represent complete lists (Sadiq and McCain, 1993; Pope and Dockery, 2006; Lippmann, 1989; Bell *et al.*, 2004, 2014; Hesterberg *et al.*, 2009; Atkinson *et al.*, 2016; Goldsmith and Aronow, 1975; Sykes and Walker, 2016).

effects that have been observed in the scientific literature under typical conditions.

Measurement studies indicated that SO_2 levels in Kuwait, Saudi Arabia, Bahrain and Iraq were below Meteorology and Environmental Protection Agency air quality standards, although values exceeded historical levels, likely due to the influence of burning oil wells (Sadiq and McCain, 1993; Al Ajmi, 1994). Authors of measurement studies found that levels of NOx and O_3 did not appear to be greatly affected by the oil well fires, relative to other pollutants (Amin and Husain, 1994). PM_{10} content of nickel, vanadium and chromium were high. Multiple measurement campaigns identified very high levels of PM_{10} reaching 326, 530, 600 and 636 μg m^{-3} in Kuwait (Mulholland *et al.*, 1991; Sadiq and McCain, 1993). The World Health Organization (WHO) recommends that in order to protect human health daily levels of $PM_{2.5}$ not exceed 25 μg m^{-3} (World Health Organization, 2006).

The pollution from the oil wells reached far beyond Kuwait. For instance, in Bahrain, 320 km downwind of the oil wells, measurements of air pollution were taken 31 July 1991–4 August 1991, during which fires were still burning (Madany and Raveendran, 1992). PM_{10} levels ranged from 139 μg m^{-3} to 673 μg m^{-3}, averaging 199 μg m^{-3}, with high nickel and vanadium content.

Any evaluation of air quality due to the burning of oil wells in the Gulf War should consider that levels of some air pollutants can be high for other reasons, such as dust storms. This is demonstrated by a study that included measurements taken at three monitoring sites in Kuwait in 2004–2005, well after the oil fires (Brown *et al.*, 2008). The site locations included a residential and commercial area with local sources such as traffic, a northern site in a remote desert region, and a southern site with urban and industrial sources. Measurements of daily (24-h values) began February 2004 and were taken through October 2005, February 2005, and July 2005 for the central, northern and southern sites, respectively. The authors identified different levels of particulate matter chemical components at the three sites, indicating variation in the contributions of traffic and other pollution sources by region. Coarse particles, for which dust storms are a common source, compromised a considerable fraction (50–60%) of PM_{10}. Levels of $PM_{2.5}$ averaged 37.6, 30.8 and 36.5 μg m^{-3} for the central, northern, and southern sites, respectively. Although these levels do not represent annual averages, they are several times higher than the WHO's air quality guidelines for $PM_{2.5}$, which include an annual mean of 10 μg m^{-3} for $PM_{2.5}$ (World Health Organization, 2006). Similarly, the overall PM_{10} levels at the three sites measured 92.8, 65.8 and 90.4 μg m^{-3} for the central, northern and southern sites, respectively, in contrast to the WHO's annual PM_{10} guideline of 20 μg m^{-3}. Maximum daily (24-h) levels reached 94.5, 76.7 and 82.5 μg m^{-3} for $PM_{2.5}$ and 197.9, 178.4 and 224.1 μg m^{-3} for PM_{10}, exceedingly higher than the WHO daily guidelines of 25 μg m^{-3} for $PM_{2.5}$ and 50 μg m^{-3} for PM_{10} (World Health Organization, 2006), indicating high exposure levels in this region from sources other than the burning oil wells.

8.4 Human Health Effects from Air Pollution

8.4.1 *Health warnings and protection*

Several governments issued warnings and distributed protective equipment in attempts to reduce exposure to smoke and oil from burning wells for military and civilian personnel. These included guidance to reduce exposure through behaviour, such as to stay a specified distance (e.g. 200 m) from fires, reduce physical activity, stay upwind of fires, and keep vehicle windows closed. Protective equipment and related guidance included use of surgical or dust masks, goggles, scarves and hoods.

8.4.2 *Research on air pollution and human health*

Many of the studies of the human health consequences of the burning oil wells have focused on air pollution, especially for military personnel. US agencies conducted many of the studies using a similar methodology to estimate exposure to air pollution for military forces. Exposure methods for those studies included self-assessment through questionnaires and modelled estimates of air pollution from burning oil wells.

Research performed to estimate the health outcomes of exposure to air pollution from the burning oil wells has inconsistent results. Some studies found an indication of increased risk of adverse health outcomes whereas others did not find such evidence. In general, the studies suffer from limitations relating to both exposure information or health information, which will be described in Sec. 8.4.2.3.

8.4.2.1 *Research finding evidence of an association between burning oil wells and health*

Six to eight months after potential exposure to oil wells, smoke, US Army veterans were evaluated at Department of Defence military facilities by military physicians or civilian physicians employed by the military (Cowan *et al.*, 2002). There were no definitive clinical tests for asthma; for this study "cases" of asthma were identified based on physician-diagnosed asthma as determined by the physical examination and patient responses. It did not include information from previous medical records. These study participants were compared to "controls", selected as veterans without asthma. Exposure to air pollution from burning oil wells was estimated in two ways: air quality modelling and questionnaires. Exposure from air quality modelling was estimated at 15 km and 24-h average resolution using an atmospheric advection and diffusion model and applied to the coordinates of each company, removing study participants with uncertain exposure estimates. During examination by physicians, participants were asked if they were exposed to smoke from burning oil wells (yes/no). These exposure methods and variations were used in many US military studies of health effects of air pollution from the burning wells. Results from the two measures of exposure (i.e. modelled and self-reported) were not very similar. However, in both cases, risk of asthma was higher for those exposed to pollution than for those unexposed, with evidence of increasing risk with increasing exposure, indicating a dose-response relationship, using the modelled exposure. The risk of asthma for those exposed to 6–30 days of smoke with levels ≥65 μg m^{-3} was 48% higher (95% confidence interval 12–77%) than those with no days of exposure.

In another study of US veterans, a self-administered questionnaire was sent to 1599 veterans in October 1991, approximately 4 weeks after they returned from Kuwait (Petruccelli, 1999). The survey asked about health issues prior to deployment in Kuwait, during their time in Kuwait, and since returning, with a focus on eye, nose and throat irritation; respiratory problems, allergies, rashes and fatigue. This study used proximity to estimate exposure on the assumption that those closer to oil fires would have higher exposure. The study participants self-reported their distance from oil fires during their time in Kuwait, although the surveys did not inquire as to the time spent in such environments. Prevalence of some symptoms (burning eyes, shortness of breath, cough and productive cough) was higher for those who reported being closer to oil fires. For example, 25% of those who reported being <10 miles from an oil fire reported cough compared to 16.1% and 8.3% of those who had been 11–100 miles and >100 miles, respectively. The authors concluded that some health symptoms were associated with recalled distance to oil fires.

Surveys of patients in Kuwaiti clinics and emergency rooms were conducted by multiple agencies from Kuwait, the US and France (Osman, 1997). The authors compared the patterns and trends of air pollution levels to those of medical records. Patterns of upper respiratory irritation were similar to those of air sampling for airborne particles. Patient visits for some health outcomes (psychiatric, chronic bronchitis, emphysema and gastrointestinal complaints) were higher after the oil wells began to burn.

8.4.2.2 *Research not finding evidence of an association between burning oil wells and health*

Several studies and reports have been published finding no evidence of a link between exposure to smoke from oil fires and health risks, most of which were conducted by the US agencies to assess the health of veterans. One study conducted by the US Army Environmental Hygiene Agency, renamed the US Army Center for Health Promotion and Preventive included three key aims: (1) environmental monitoring of air pollution and soil sampling to characterise the concentrations of pollutants to which Department of Defence personnel were exposed during deployment, (2) industrial hygiene air monitoring to characterise occupational exposure of DoD personnel with potentially high levels of exposure to pollution from burning oil wells, with an emphasis on outdoor workers and persons working near oil fields next to Kuwait City, and (3) biological measurements of exposure and effect in real time to attempt to observe biological responses or markers of exposure to pollutants from burning oil wells in a cohort of US soldiers (U.S. Army Environmental Hygiene Agency, 1994; Deeter, 2011). Risk assessment methodology was used to estimate the long-term health risk to cancer and non-cancer health outcomes. The Army concluded that health risks were low with an estimated upper limit of lifetime cancer risk no higher than three excess cancers per a million persons exposed, and a low risk of non-cancer health outcomes (U.S. Army Environmental Hygiene Agency, 1994; Technology Assessment Board of the 103rd Congress, 1994).

A later study used a similar approach to estimate health risk, but estimated risk for each individual troop unit by incorporating information on actual troop movement within the theatre and modelled estimates of the location of fire smoke (US Army Center for Health Promotion and Preventative Medicine, 1998). This study used environmental monitoring data, backwards trajectory modelling from the National Oceanic and Atmospheric Administration Hybrid Single Particle Lagrangian Integrated Trajectory (HYSPLIT) model to estimate the location and concentration of air pollutants, and data on the location of troop units, with similar methods and modelling results applied to other US military studies. This study also reported minimal risk for long-term health consequences with cancer risk no higher than two in a million. Non-carcinogenic risk, primarily from air pollutants of VOCs from burning oil wells, was also estimated to be low.

US veterans were surveyed by telephone, 5 years after the War, regarding respiratory symptoms (Lange *et al.*, 2002). Exposure to air pollution from oil fires was estimated through the survey (yes/no/don't know) as well as modelled at 15-km resolution. Study subjects who responded "yes" regarding exposure were asked about how many days they had been exposed (<5, 6–30, or >31 days). For 1,560 veterans who served at some point between 2 August, 1990 and 31 July, 1991, estimated modelled exposure was matched to troop locations from global position systems records. Self-reported and modelled exposures were moderately correlated. Pre-deployment health was assessed by asking military personnel, 5 years after deployment, about their health in August 1990 and during the period August 1990–July 1991. Health at the time of the survey was assessed by asking about respiratory symptoms within the past month, symptoms of depression within the past 2 weeks, and injuries within the past 3 months. For the modelled estimates of exposure, results did not indicate a relationship between exposure and risk of respiratory illness; although, for asthma and bronchitis higher rates of illness were associated with higher self-reported exposures.

Another analysis considered Department of Defence hospitals from August 1991 to July 1999 for 405,142 US Gulf War veterans. Particulate matter from oil fire smoke was estimated using HYSPLIT modelling and information on locations of troops (Smith, 2004). The study did not identify evidence that smoke increased risk of hospitalisations.

Hospitalisation records during the Gulf War were analysed for US military on active duty, Reserve or National Guard status and deployed in the Gulf War for one or more days within the period August 1990 through July 1991, including 18,631 personnel who were hospitalised (Smith *et al.*, 2004). Exposure to smoke from oil well fires was estimated using the US Army Center for Health

Promotion and Preventive Medicine modelling approach that also was used in other health analyses (Smith *et al.*, 2002). Authors considered whether military personnel exposed to oil fire smoke ("under plume") at the time of hospitalisation had higher rates of hospitalisation than those unexposed ("not under plume"), and considered adjustment for key confounders. They identified no key differences in hospitalisations based on exposure to smoke from the oil fires.

A literature review conducted by RAND's National Defence Research Institute and commissioned by the US Office of Secretary of Defence explored the scientific literature on the health effects of the oil fires on military personnel (Spektor, 1998). The review focused on several studies performed during and shortly after the War on US personnel such as measurements in blood, soil and air particulates, as well as biological samples from Kuwaiti feral cats and self-reported health surveys of veterans. The authors of the review concluded that levels of many air pollutants were lower than those in urban environments, with the exception of particulate matter. For this pollutant, the review concluded that the high concentration of particles resulted from the naturally high levels in the region rather than the burning oil wells.

A survey of Australian veterans considered 1,456 who served in the Gulf War at some point from 2 August 1990 to 4 August 1991, and 1,588 controls as other military personnel who served during the same time period but not in the Gulf War (Kelsall *et al.*, 2004). Health was assessed by postal questionnaire and through medical examinations including spirometric tests. The survey and health assessment were conducted several years after the War, August 2000–April 2002. As the study participants include few women, analysis was restricted to men. The Gulf War veterans had higher prevalence of some respiratory symptoms than did the controls. Veterans who self-reported exposure to smoke from burning oil wells had slightly poorer forced vital capacity than those who did not. The authors concluded that results indicated no evidence of major impact on lung function from self-reported exposure to smoke from burning oil wells or dust storms.

8.4.2.3 *Limitations for the existing research on air pollution and human health for the Kuwaiti oil fires*

As described above, the research on the health impacts of air pollution from the burning of oil wells report inconsistent results, with some studies indicating an association between exposure and adverse health outcomes, whereas other studies did not. When considered collectively, little research has been conducted on the Kuwaiti oil fires given that many of the studies have duplicative methods such that any limitations (or strengths) of the study design are repeated in other studies, and similarly many of the studies have the same or overlapping teams of authors or sponsoring agencies.

The scientific study of the unusual conditions of wartime environmental exposures faces many challenges. This type of research has limitations because in many of the studies, health effects and exposure are assessed well after they occur. Self-assessment of exposure is prone to "recall bias", in which those with health problems can be more likely to remember exposure to environmental contaminants than those without health concerns. Exposure assessments were often simplistic (e.g. binary indicators, proximity to oil wells). Estimated exposures using Lagrangian modelling is difficult for these conditions given the lack of validated input variables typical in more common air quality modelling exercises (e.g. emissions databases) and the lack of surface measurements during the early stages of the fires for model validation. Poor to moderate correlation between different exposure methods demonstrates the difficulties of estimating exposure. Further, some studies assumed that exposure was a homogenous condition whereas under real conditions the level of pollutants varied even for those in high pollution areas.

While researchers made efforts to adjust for factors that may vary across individuals, such as gender or smoking status, there may be unmeasured confounders, other differences between the exposed and unexposed populations that were not addressed. Many of the studies did not consider baseline health status before the War, although for military personnel the population is likely to be quite healthy. Finally, the studies were primarily on troops, with very few examining the health impacts on civilians. Potentially susceptible subpopulations were excluded, such as children, the elderly and those with pre-existing conditions.

These limitations do not diminish the importance and value of such studies, which provide critical evidence on the exposure and health response from air pollution from the burning oil wells. In fact, the studies show the ingenuity of the scientific community to conduct research under challenging conditions. They do, however, indicate that the existing studies may not provide a full understanding of the health impacts of air pollution from the burning wells.

Photos from the War and measurements of air pollution, especially for particulate matter, clearly indicate exceptionally poor air quality. Numerous studies have shown that PM_{10} and $PM_{2.5}$ are associated with harmful human health outcomes including premature death (Pope and Dockery, 2006). Measurements of PM_{10} in Kuwait reached levels over 25 times that of the WHO air quality guidelines. Therefore, those in the presence of these plumes quite likely suffered damage to their health.

8.5 Other Impacts on Human Health

While much of the human health research on this environmental disaster has focused on air pollution from the burning oil wells, many other health outcomes could have resulted. These include those resulting from exposure to "oil rain", spilt crude oil, extreme heat from the fires, contaminated water, lowered agricultural production, and associated debris from destroyed equipment and infrastructure from the sabotage of the oil industry infrastructure. Those working to extinguish the fires also were exposed to pools of burning oil and unexploded mines.

8.6 Other Environmental Damage

In addition to the burning of oil wells, the War resulted in other environmental degradation such as from the movement and encampments of troops, land mines and production of waste. This stemmed from both deliberate sabotage and the consequences of military activities. Damage included soil erosion, soil contaminated by oil, severely damaged ecosystems from deposition from the air pollution, and impacts on vegetation (Bejarano and Michel, 2010; Hussain and Gondal, 2008; Omar *et al.*, 2009).

Iraqi forces intentionally released about 11 million barrels of oil into the Arabian Gulf from January to May 1991. The resulting oil spill was over 20 times larger than that of the Exxon Valdez of March 1989. One oil slick was about 2,000 km². This discharged oil had dramatic effects on bird species, as bird populations declined and some became covered in oil. Impacts on local ecosystems included lower temperature of seawater, lowered density of intertidal infauna in very contaminated areas, and loss of animal life. Production of fish eggs and larvae declined in areas with oil slicks or sheens. Fisheries had greatly reduced production, smaller shrimp and diseased fish. Marine turtles lost nesting areas on island beaches and some developed lesions. Dolphins and other marine mammals also perished (Sadiq and McCain, 1993). Surveys of the intertidal Gulf coast line of Saudi Arabia in 2002–2003 found that the majority of mangroves, tidal flats, salt marshes, rock shores and sand flats were still affected with lower diversity than areas without oil contamination; the recovery time for marshes without tidal flushing was estimated at over a 100 years (Jones *et al.*, 2008).

8.7 Conclusion

The burning of the oil wells during the Gulf War created one of the world's worst environmental disasters. An unusual aspect of this event, compared to many others such as the London Fog of 1952 (Chapter 10), is that the extreme conditions were generated intentionally as Iraqi Forces attempted to maximise damage. The destruction created severe air pollution and oil lakes, which led to contamination of soil and water, as well as harm to vegetation and terrestrial and marine wildlife.

The effects of the air pollution on human health are not well understood, especially the long-term consequences, due to the lack of exposure information for the early portion of the time during which the fires were burning and due to the challenges inherent in this type of research of wartime conditions. The limited research that has been performed provides inconsistent evidence, with many studies finding no indication of health effects. However, the severe levels of air pollution were well above health-based regulatory standards and guidelines, especially for particulate matter, giving credibility to the hypothesis that exposed persons, both military and civilian populations, suffered substantial health impacts.

References

Al Ajmi, D. N. (1994). *The Air Quality Impact of Kuwait Oil Field Fires*, in *Transactions on Ecology and the Environment* (WIT Press, Southampton, U.K.).

Al-Besharah, J. (1991). *The Kuwait Oil Fires and Oil Lakes — Facts and Numbers*, in *Environmental and Health Impact of the Kuwaiti Oil Fires: Proceedings of an International Symposium, held at the University of Birmingham 17 October 1991* (The University of Birmingham: Institute of Occupational Health, Birmingham).

Al-Damkhi, A. M. (2007). Kuwait's oil well fires, 1991: Environmental crime and war, *International Journal of Environmental Studies*, **64**, 31–44.

Amin, M. B. and Husain, T. (1994). Kuwaiti oil fires — air quality monitoring, *Atmospheric Environment*, **28**, 2261–2276.

Arkin, M. W. (1996). The environmental threat of military operations, *International Law Studies*, **69**, 116–135.

Atkinson, R. W., Butland, B. K., Dimitroulopoulou, C., Heal, M. R., Stedman, J. R., Carslaw, N., Jarvis, D., Heaviside, C., Vardoulakis, S., Walton, H. and Anderson, H. R. (2016). Long-term exposure to ambient ozone and mortality: A quantitative systematic review and meta-analysis of evidence from cohort studies, *BMJ Open*, **6**, e009493.

Bader, M. S. H. and Clarkson, W. W. (1995). An overview of the oil lakes in Kuwait — the problem and remedial methods, *Journal of Environmental Sciences & Health, Part A*, **30**, 1039–1057.

Bakan, S., Chlond, A., Cubasch, U., Feichter, J., Graf, H., Grassl, H., Hasselmann, K., Kirchner, I., Latif, M., Roeckner, E., Sausen, R., Schlese, U., Schriever, D., Schult, I., Schumann, U., Sielmann, F. and Welke, W. (1991). Climate response to smoke from the burning oil-wells in Kuwait, *Nature*, **351**, 367–371.

Barker, C. and Bufarsan, A. (2001). Evaporative losses from the surface "oil lakes" of Southern Kuwait, *Environmental Geoscience*, **8**, 123–129.

Barth, S. K., Kang, H. K., Bullman, T. A. and Wallin, M. T. (2009). Neurological mortality among US veterans of the Persian Gulf War: 13-year follow-up, *American Journal of Industrial Medicine*, **52**, 663–670.

Bejarano, A. C. and Michel, J. (2010). Large-scale risk assessment of polycyclic aromatic hydrocarbons in shoreline sediments from Saudi Arabia: environmental legacy after twelve years of the Gulf war oil spill, *Environmental Pollution*, **158**, 1561–1569.

Bell, M. L., Dominici, F., Ebisu, K., Zeger, S. L. and Samet, J. M. (2007). Spatial and temporal variation in $PM_{2.5}$ chemical composition in the United States for health effects studies, *Environmental Health Perspectives*, **115**, 989–995.

Bell, M. L., McDermott, A., Zeger, S. L., Samet, J. M. and Dominici, F. (2004). Ozone and short-term mortality in 95 US urban communities, 1987–2000, *Journal of the American Medical Association*, **292**, 2372–2378.

Bell, M. L., Zanobetti, A. and Dominici, F. (2014). Who is more affected by ozone pollution? A systematic review and meta-analysis, *American Journal of Epidemiology*, **180**, 15–28.

Brown, K. W., Bouhamra, W., Lamoureux, D. P., Evans, J. S. and Koutrakis, P. (2008). Characterization of particulate matter for three sites in Kuwait, *Journal of the Air & Waste Management Association*, **58**, 994–1003.

Bullman, T. A., Mahan, C. M., Kang, H. K. and Page, W. F. (2005). Mortality in US Army Gulf War veterans exposed to 1991 Khamisiyah chemical munitions destruction, *American Journal of Public Health*, **95**, 1382–1388.

Cowan, D. N., Lange, J. L., Heller, J., Kirkpatrick, J. and DeBakey, S. (2002). A case-control study of asthma among US Army Gulf War

Veterans and modeled exposure to oil well fire smoke, *Military Medicine*, **167**, 777–782.

Deeter, D. P. (2011). The Kuwait Oil Fire Health Risk Assessment Biological Surveillance Initiative, *Military Medicine*, **176**(Suppl 7), 52–55.

Draxler, R. R., McQueen, J. T. and Stunder, B. J. B. (1994). An evaluation of air pollutant exposures due to the 1991 Kuwait oil fires using a Lagrangian model, *Atmospheric Environment*, **28**, 2197–2210.

Elgibaly, A. A. M. (1999). Cleanup of oil-contaminated soils of Kuwaiti oil lakes by retorting, *Energy Sources*, **21**, 547–565.

Goldsmith, J. R. and Aronow, W. S. (1975). Carbon monoxide and coronary heart disease: A review, *Environmental Research*, **10**, 236–248.

Haley, R. W. (2003). Excess incidence of ALS in young Gulf War veterans, *Neurology*, **61**, 750–756.

Heller, J. M. (2011). Oil well fires of Operation Desert Storm — defining troop exposures and determining health risks, *Military Medicine*, **176**, 46–51.

Hesterberg, T. W., Bunn, W. B., McClellan, R. O., Hamade, A. K., Long, C. M. and Valberg, P. A. (2009). Critical review of the human data on short-term nitrogen dioxide (NO_2) exposures: Evidence for NO_2 no-effect levels, *Critical Reviews in Toxicology*, **39**, 743–781.

Hobbs, P. V. and Radke, L. F. (1992). Airborne studies of the smoke from the Kuwait oil fires, *Science*, **256**, 987–991.

Husain, T. (1995). *Kuwaiti Oil Fires: Regional Environmental Perspectives*, (Elsevier Science Ltd, Oxford).

Hussain, T. and Gondal, M. A. (2008). Monitoring and assessment of toxic metals in Gulf War oil spill contaminated soil using laser-induced breakdown spectroscopy, *Environmental Monitoring and Assessment*, **136**, 391–399.

Jones, D. A., Hayes, M., Krupp, F., Sabatini, G., Watt, I. and Weishar, L. (2008). *The Impact of the Gulf War (1990–91) Oil Release Upon the Intertidal Gulf Coast Line of Saudi Arabia and Subsequent Recovery.* In *Protecting the Gulf's Marine Ecosystems from Pollution*, Abuzinada, A. H. *et al.* (eds.), (Birkhäuser, Basel), pp. 237–254.

Kang, H. K. and Bullman, T. A. (2001). Mortality among US veterans of the Persian Gulf War: 7-year follow-up, *American Journal of Epidemiology*, **154**, 399–405.

Kelsall, H. L., Sim, M. R., Forbes, A. B., McKenzie, D. P., Glass, D. C., Ikin, J. F., Ittak, P. and Abramson, M. J. (2004). Respiratory health status of Australian veterans of the 1991 Gulf War and the effects of exposure to oil fire smoke and dust storms, *Thorax*, **59**, 897–903.

Klingbeil, A. (2012). Three Safety Boss veterans recall the Kuwait oil fires, *Alberta Oil*, **8**. Available at: https://www.albertaoilmagazine.com/ 2012/08/three-oilfield-veterans-recall-the-kuwait-oil-fires/.

Lange, J. L., Schwartz, D. A., Doebbeling, B. N., Heller, J. M. and Thorne, P. S. (2002). Exposures to the Kuwait oil fires and their association with asthma and bronchitis among Gulf War veterans, *Environmental Health Perspectives*, **110**, 1141–1146.

Leebaw, B. (2014). Scorched earth: Environmental war crimes and international justice, *Perspectives on Politics*, **12**, 770–788.

Limaye, S. S., Suomi, V. E., Velden, C. and Tripoli, G. (1991). Satellite observations of smoke from oil fires in Kuwait, *Science* **252**, 1536–1539.

Lippmann, M. (1989). Health effects of ozone. A critical review, *Journal of the Air Pollution Control Association*, **39**, 672–695.

Madany, I. M. and Raveendran, E. (1992). Polycyclic aromatic hydrocarbons, nickel and vanadium in air particulate matter in Bahrain during the burning of oil fields in Kuwait, *Science of the Total Environment*, **116**, 281–289.

Mulholland, G. W., Benner, B. A., Fletcher, R. A., Steel, E., Wise, S. A., May, W. E., Madrzykowski, D. and Evans, D. A. (1991). *Analysis of Smoke Samples from Oil Well Fires in Kuwait* (National Institute of Standards and Technology, Gaithersburg, MD).

Omar, S. A., Bhat, N. R., and Asem, A. (2009). Critical assessment of the environmental consequences of the invasion of Kuwait, the Gulf War, and the aftermath. In *Handbook of Environmental Chemistry Volume 3U* Environmental consequences of war and aftermath (pp. 141–170). Springer Berlin Heidelberg.

Osman, Y. (1997). Environmental surveys conducted in the Gulf region following the Gulf War to identify possible neurobehavioral consequences, *Environmental Research*, **73**, 207–210.

Petruccelli, B. P., Goldenbaum, M., Scott, B., Lachiver, R., Kanjarpane, D., Elliott, E., Francis, M., McDiarmid, M. A. and Deeter, D. (1999). Health effects of the 1991 Kuwait oil fires: A survey of US Army troops, *Journal of Occupational & Environmental Medicine*, **41**, 433–439.

Pope III, C. A. and Dockery, D. W. (2006). Health effects of fine particulate air pollution: Lines that connect, *Journal of the Air & Waste Management Association*, **56**, 709–742.

Rose, M. R. (2003). Gulf War service is an uncertain trigger for ALS, *Neurology*, **61**, 730–731.

Rostker, B. (2000). *Environmental Exposure Report: Oil Well Fires — Interim Report*, (U.S. Department of Defense, Washington, D.C.).

Sadiq, M. and McCain, J. C. (1993). *The Gulf War Aftermath: An Environmental Tragedy*, (Springer Science & Business Media, New York).

Shapley, D. (1972). Technology in Vietnam: Fire storm project fizzled out, *Science*, **177**, 239–241.

Smith, T. C., Corbeil, T. E., Ryan, M. A., Heller, J. M. and Gray, G. C. (2004). In-theater hospitalizations of US and allied personnel during the 1991 Gulf War, *American Journal of Epidemiology*, **159**, 1064–1076.

Smith, T. C., Heller, J. M., Hooper, T. I., Gackstetter, G. D. and Gray, G. C. (2002). Are Gulf War veterans experiencing illness due to exposure to smoke from Kuwaiti oil well fires? Examination of Department of Defense hospitalization data, *American Journal of Epidemiology*, **155**, 908–917.

Spektor, D. M. (1998). *A Review of the Scientific Literature as it Pertains to Gulf War Illnesses: Oil Well Fires* (National Defense Research Institute, RAND, Washington, D.C.).

Stead, C. F. (1991). *Oil Fires, Petroleum, and Gulf War Illnesses. CDC Conference on the Health Impact of Chemical Exposures During the Gulf War* (Centers for Disease Control, Atlanta).

Sykes, O. T. and Walker, E. (2016). The neurotoxicology of carbon monoxide — historical perspective and review, *Cortex*, **74**, 440–448.

Technology Assessment Board of the 103rd Congress. (1994). *The Department of Defense Kuwait Oil Health Fire Risk Assessment (The "Persian Gulf Veterans' Registry")* (Health Program, Office of Technology Assessment, U.S. Congress, Washington, D.C.).

U.S. Army Center for Health Promotion and Preventative Medicine. (1998) *Environmental Surveillance Health Risk Assessment, Kuwait Oil Fires, Draft Final Report*, 47-EM-7121-98 (U.S. Army, Washington, D.C.).

U.S. Army Environmental Hygiene Agency. (1994). *Kuwait Oil Fire Health Risk Assessment, 5 May – 3 December 1991, Final Report*, No. 39-26-L192-91, (U.S. Army, Washington, D.C.).

U.S. Department of Agriculture Forest Service. (1970). *Forest Fire as a Military Weapon* (U.S. Department of Defense, Washington, D.C.).

U.S. Occupational Safety and Health Administration. (OSHA). (2005). *OSHA Fact Sheet: Hydrogen Sulfide (H_2S)* (U.S. Department of Labor, Washington, D.C.).

Vriesenga, M. P. (1994). *From the Line in the Sand: Accounts of the USAF Company Grade Officers in Support of Desert Sheild/Desert Storm* (Air University Press, Maxwell AFB, Alabama).

Wald, M. L. (1991). *Amid Ceremony and Ingenuity, Kuwait's Oil-Well Fires are Declared Out* (The New York Times, New York, NY).

Wilmington Morning Star. (1972). Brush Fires a Weapon in South Vietnam War, *Wilmington Morning Star* — Marine edition (Wilmington, NC).

World Health Organization (WHO). (2006). *WHO Air Quality Guidelines for Particulate Matter, Ozone, Nitrogen Dioxide, and Sulfur Dioxide: Global Update 2005, Summary of Risk Assessment* (WHO, Geneva).

Writer, J. V., DeFraites, R. F. and Brundage, J. F. (1996). Comparative mortality among US military personnel in the Persian Gulf region and worldwide during Operations Desert Shield and Desert Storm, *Journal of the American Medical Association*, **275**, 118–121.

Chapter 9

Kuwait Oil Fires (1991): Searching for Long-range Deposits on Heritage

Alessandra Bonazza

Institute of Atmospheric Sciences and Climate,
National Research Council (ISAC-CNR), Italy

9.1 Introduction

During the Gulf War, the environment in Kuwait and neighbouring countries suffered damage from oil well fires that burned for months, releasing airborne contaminants. Typically, burning crude oil produces smoke containing a large number of particulate and gaseous species such as carbon dioxide, carbon monoxide, sulphur dioxide, nitrogen oxides, volatile organic compounds, polycyclic aromatic hydrocarbons, hydrogen sulphide, acidic aerosols and soot (Hobbs and Radke, 1992). The effects of air pollution on cultural heritage are well-known, being soiling and blackening, alongside surface recession, the most conspicuous deterioration processes (Bonazza *et al.*, 2005). Iranian authorities claimed that in some provinces the cultural heritage was severely damaged as a consequence of the smoke produced by the fires. The United Nations Compensation Commission issued its final report on awards of compensation for environmental and public health damage resulting from Iraq's 1990–1991 invasion and occupation of Kuwait on 30 June, 2005. No compensation for Iran's claim for damage to cultural heritage was awarded due to insufficient evidence regarding nature and extent of damage and the contribution of other factors, such as local sources of pollution from motor vehicle emissions, regional oil refining and human occupation of historic sites. This chapter investigates in detail the cause of soiling and blackening in archaeological sites and monuments of Khuzestan and Fars provinces (Iran), which were claimed to be affected by oil well fires. The potential damage to heritage of this episode is also discussed in relation to the chemical composition of emissions and the coverage area of the smoke plume.

9.2 Studies of Smoke from the 1991 Kuwait Oil Fires

Of more than 740 oil wells sabotaged by the Iraqi troops in February 1991 during the Persian Gulf War, approximately 520 were still afire in June 1991. The first well fires were extinguished in early April 1991, with the last well capped on 6 November, 1991. In May 1991, several months after their ignition, the fires were still emitting soot into the atmosphere at a rate equivalent to that from three million heavy-duty diesel trucks, particles (<3.5 μm diameter) at a rate of 10% of worldwide anthropogenic emissions, and SO_2 at about 5% of global anthropogenic emissions, causing one of the largest man-made environmental disasters in history with severe consequences on air quality and on the weather of the Persian Gulf region (Cofer *et al.*, 1992; Ferek *et al.*, 1992; Hobbs, 1992).

Two main groups of fires were caused: one located north of Kuwait City (the "north field") and the other in the Greater Burgan oil field south of Kuwait City (the "south field"). The individual fires in these fields generally produced distinct, isolated plumes over short distances, after which they combined to form a "supercomposite" plume, which was about 40 km (Ferek *et al.*, 1992; Hobbs and Radke, 1992). The smoke plumes were often channelled in relatively narrow rivulets towards southeast due to the prevailing warm and dry winds that generally blow northwesterly during late spring and summer in the Persian Gulf region (the "Shamal" wind). Meteosat visible and infrared satellite observations conducted from May to July 1991 indicated that the smoke, after crossing the southern coast of the Arabian Peninsula, travelled over the northern Arabian Sea, circulated over the central Arabian desert and finally drifted directly eastward toward Iran reaching its southern part (Limaye *et al.*, 1992).

About 25% of the plumes from individual fires were observed to emit white or light grey smoke, while others emitted black or a mixture of black and white smoke. A few of the fires emitted no visible plume at all. The differences in plume characteristics were probably due to variations in the oil/gas/water mixture ejected from the uncontrolled wells, as well as compositional differences between the different reservoirs and the condition of the damaged well heads (Ferek *et al.*, 1992). Generally, the Kuwait oil fires burned very efficiently, with 95–98% of the total carbon (TC) emitted as CO_2 and only 0.6%–1.4% emitted as CO for the composite plumes, and according to Ferek *et al.* (1992) the emitted SO_2 and NO_x were relatively rapidly removed in the smoke from fires through reactions with coarse-mode (>2.5 μm diameter) soil dust, reducing in this way their impact on the Persian Gulf region. Following the studies of Laursen *et al.* (1992) on the emission factors for particles, elemental carbon and trace gases from the Kuwait oil fires, it has been concluded that the overwhelming majority of the burned carbon was present as CO_2. The contribution of soot (elemental carbon) to the TC was low, varying from 0.3% to 0.6% for the composite plumes, while the total organic carbon was found to range between 0.5% and 0.7%. CH_4 also encountered for only a small fraction of the burned carbon (0.2–0.4% for the composite plumes).

Compositional investigations on source oil-fire smoke conducted by Cofer *et al.* (1992) confirmed that combustion was surprisingly efficient and categorised the smoke plumes into three classes based on the type of oil fire and the resulting smoke colour. Black smoke plumes (about 60–65% of the oil fires) were found to emit the highest concentration of elemental (4,470 μg m^{-3}) and organic carbon (1,050 μg m^{-3}), while white plumes (25–30% of the oil fires) were characterised by 107 μg m^{-3} of organic carbon and almost no elemental carbon. In summarising the airborne studies of the smoke from the Kuwait oil fires, Hobbs and Radke (1992) concluded that the smoke had marked effects on air quality and some aspects of weather in the Persian Gulf, but only small effects beyond the Gulf region and insignificant effects on a global scale, due to the low particle emissions, its short atmospheric residence time and the low height of its transport in the atmosphere.

9.3 Soiling and Blackening of Buildings and Archaeological Sites

The driving role of air pollution in surface deterioration of monuments facades and archaeological remains is worldwide recognised. Several studies have been conducted in order to increase the knowledge about the damaging phenomena occurring particularly on marble and limestone exposed to gaseous pollutants and aerosols emitted by anthropogenic activities and evidence has been given to how changes in atmospheric concentration of pollutants are reflected on the chemistry of damage layers forming on architectural surfaces (Bonazza and Sabbioni, 2016). Additionally, recent studies argued on the possible future tuning of colour of facades of monuments located in urban areas, because of the higher contribution in modern atmosphere of organic compounds from vehicular traffic, nowadays the major cause of pollution in city centres (Grossi, 2016).

The numerous studies reported in the literature on damage affecting monuments located in urban areas have demonstrated that the main deterioration process occurring is black crust formation. These dark layers are observed on stone surfaces wetted by rainwater but sheltered from intensive runoff and wash out. They are due to the dry and wet deposition of SO_2 and black particles, and their subsequent reaction with the carbonate (CC) material leading to the formation of gypsum (i.e. calcium sulphate dihydrate), thought the widely studied sulphation process.

It has been demonstrated that sulphur and carbon are the main chemical elements present in the dark damage layers that result from atmospheric deposition (Bonazza and Sabbioni, 2016). In addition, it has been highlighted that the TC content in black layers is composed of two main fractions: CC and non-carbonate carbon (NCC):

$$TC = CC + NCC.$$

While the CC fraction is linked to the stone substrate, the NCC fraction on monuments has three different origins (Ghedini *et al.*, 2006 and included references):

(1) deposition of atmospheric particles containing elemental and organic carbon, as primary and secondary pollutants;
(2) biological weathering due to the action of organisms: among the products of their metabolism, formic, acetic and oxalic acids react with the underlying materials, leading to the formation of calcium formate, acetate and oxalate;
(3) surface treatments, such as oils, waxes, proteins, etc., frequently used in the past to protect monuments and historic buildings.

NCC is composed of elemental (EC) and organic carbon (OC). While EC is a distinctive tracer for combustion, being entirely due to particles emitted by combustion processes into the atmosphere, OC may have a number of different and often simultaneous origins, including pollutants deposition, biological weathering and the decay of protective organic treatments.

The approach used to identify pollutant impact and the causes of blackening of architectural surfaces is discussed here; reporting as a paradigmatic case study work done in the Southern provinces in Iran aimed at identifying the potential role of Kuwait oil fires in damaging monuments and archaeological sites affected by the smoke following the 1991 Gulf War.

9.4 Impact of Kuwait Oil Fires on Cultural Heritage

9.4.1 *Iranian case studies*

During a sampling campaign of September 2003, valuable archaeological sites of Khuzestan and Fars provinces, in Southern Iran, have been examined in detail with the aim of understanding the causes of blackening and soiling of Iranian Cultural Heritage; claimed by Iranian authorities to be affected by the oil well fires in the 1991 Gulf War. The two provinces contain a large number of archaeological sites and a wide range of stone remains, in the form of buildings, relics, bas-reliefs and carvings. Specifically, samples of surface deposits and damage layers were collected from Persepolis, Pasargadae Complex, Sassanide Grand Fire Temple and Qal'eh (Castle) Dokhtar at Firooz Abad, as well as Shapour bas-reliefs (Figs. 9.1 and 9.2) and the City ruins at Bishapour, all in Fars province. In Khuzestan, specimens were collected from Izeh reliefs, Qal'eh Shush (Susa Castle) and Apadana (Palace) of Darius in Susa. Generally, on visual inspection, most of the archaeological sites appeared to be in a good state of overall conservation. However, there were clear signs of biological weathering, while soiling and blackening were observed, particularly at the Shapour bas-reliefs.

Fig. 9.1.   Shapour relief, Fars province. Grey-black areas are distinguishable.

Fig. 9.2.   Shapour relief, Fars province. Grey-black areas are clearly observable on unsheltered areas not protected to rain washing out.

All the archaeological sites investigated are located in rural areas, with exception of the complexes in Susa, which show the features typical of an urban location. Specimens were taken from sheltered areas that had not undergone cleaning or previous restoration works aiming at understanding the major causes of surface soiling and blackening, including Kuwait oil well fires. Table 9.1 reports a list of the archaeological sites selected for sampling, a brief description of the samples collected, with details on the sampling point.

9.4.2 *Analysis*

After collection, the samples were dried and stored at 20°C in a nitrogen environment, until the time of analysis. First, the minero-petrographic characterisation of damage layers, surface deposit and underlying stone material were executed by optical microscopy

Table 9.1.   Samples collected from archaeological sites and monuments in Iran (see Bonazza *et al.*, 2007).

Sample	Site	Sampling point	Descriptions
PER 2	Persepolis Complex	Inside a niche, 2.00 mm from the ground, located right of the west door of the southern wall of the Hundred Columns Hall	Black deposit
PER 3	Persepolis Complex	Lefthand corner of a niche, 2.50 m from the ground, in the western wall of the Hundred Columns Hall	Black deposit
PER 4	Persepolis Complex	Base of column located in the western area of the Apadana Palace	Black-grey material
PAS 2	Cyrus Tomb, Pasargadae Complex	South side of the base, about 3.00 m from the ground	Black surface deposit and fragment, with co-presence of deposit and stone substrate
FIR 7	Sassanian Fire Temple, Firooz Abad	West-facing wall about 1.60 m from the ground	Stone fragment with black layer
FIR 9	Dokhtar Castle, Firooz Abad	Mound of collapsed building materials, collected at about 7.00 m from the base	Fragment of material with black deposit
FIR 10	Dokhtar Castle, Firooz Abad	Natural outcrop halfway up the hill slope to the fortress	Fragment of incoherent material with dark deposit
BIS 1	Shapour reliefs, Bishapour	First bas-relief, 1.30 m from the base	Black deposit

(*Continued*)

Table 9.1.   (*Continued*)

Sample	Site	Sampling point	Descriptions
BIS 2	Shapour reliefs, Bishapour	First bas-relief, 0.50 m from the base below sample BIS 1	Black deposit
BIS 3	Shapour reliefs, Bishapour	First bas-relief, 0.50 cm from the base, on the right of sample BIS 2	Fragment of black crust and undamaged material
EE 4B	Izeh reliefs	Bas-relief n°1	Black deposit
SUS 2	Palace of Darius, Susa	Column base	Surface fragment

(OM) observations of thin transversal sections, using an Olympus BX51. Identification of the crystalline phases was integrated by applying X-ray diffraction (XRD) analysis with a Philips PW 1730. Any atmospheric carbonaceous particles deposited on the surface were identified by observations using a scanning electron microscope (Philips XL 20) equipped with a dispersive energy analyser (SEM-EDX), while the thermal behaviour of the samples was evaluated by means of differential and gravimetric thermal analysis (DTA–TGA) using a Mettler Toledo TGA/SDTA 851. Ion chromatographic analyses were performed using a Dionex Chromatograph (4500i model) for the measurement of soluble salt concentration. Subsequently, to discriminate and quantify the carbon species in the damage layers, the powder specimens underwent the chemical–thermal quantitative methodology for carbon speciation and measurement set up by Ghedini *et al.* (2006). In this case, for NCC quantification, samples were placed in silver capsules, subjected to two repeated steps of acidification with HCl concentrate solution to remove CC, and then fully oxidised to quantify the carbon content by flash combustion/gas chromatographic analysis using a commercial carbon–hydrogen–nitrogen–sulphur–oxygen (CHNSO EA 1108 FISONS Instruments) analyser. Finally, thermally assisted hydrolysis and methylation (thermochemolysis), gas chromatography-mass spectrometry and molecular analyses were performed following the procedure described by Bonazza *et al.* (2007).

9.4.3 *Results and discussion*

The results obtained using the different analytical techniques applied to the collected samples revealed a number of different damage typologies occurring at the surface of the monuments and archaeological sites exposed to atmospheric deposition:

(a) inter-crystalline decohesion, due to wind erosion produced by the impact of airborne soil dust, mainly composed of quartz and calcite, at the stone surface (Fig. 9.3);

(b) biological weathering linked to colonisation of airborne biological particles at the stone surface (Fig. 9.4);

(c) formation of red-orange patinas due to dedolomitisation processes occurring in oxidising conditions, during which

Fig. 9.3. SEM showing surface decohesion on metamorphic CC stone from Shapour reliefs (Bishapour, Fars).

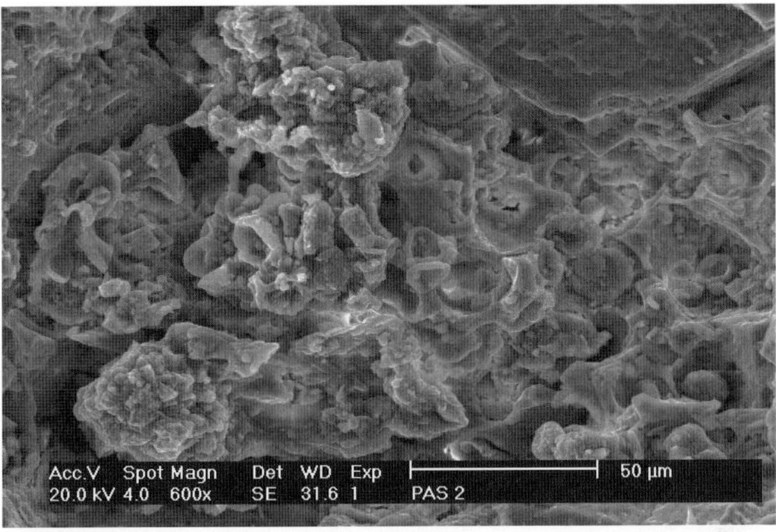

Fig. 9.4. SEM showing biological colonisation on samples from the Cyrus Tomb (Pasargadae Complex, Fars).

any ferrous iron in the precursor dolomite is oxidised to produce iron oxides (in Fig. 9.5);

(d) surface blackening and soiling.

The results from the diverse analytical techniques are discussed in detail in Bonazza *et al.* (2007).

Petrographical characterisation performed by OM indicated that the original building materials collected at the archaeological sites investigated are metamorphic and sedimentary CC stones, mainly composed of calcite (calcium carbonate) and/or dolomite (calcium–magnesium carbonates). Almost every sample revealed the presence of a red-brown coloured surface layer mainly consisting of secondary calcite and dolomite, quartz, sheet silicates, and oxides and hydroxides of iron. Observations by OM, failed to reveal any evidence of black particles, embedded in the damage layers, linked to combustion processes. XRD analysis confirmed and supported the mineralogical composition of the surface layer revealed

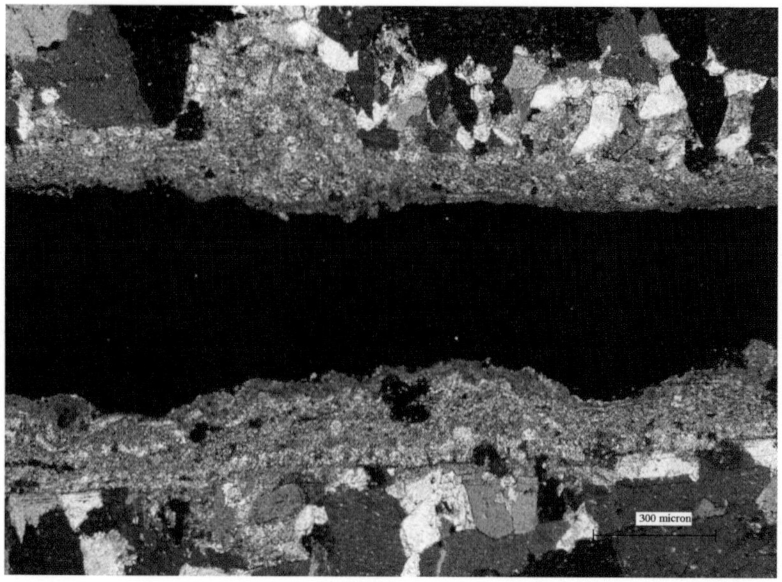

Fig. 9.5.   Optical micrograph (crossed polars) showing red-brown coloured surface layer on metamorphic CC stone from Shapour reliefs (Bishapour, Fars).

by petrographical analysis and reaffirming the dedolomitisation process by detecting ankerite in sample PAS 2.

As introduced in Sec. 9.3, TC content in black sulphated layers is composed of two main fractions: CC and NCC. The NCC fraction represents the carbon component not linked to the CC substrate but originating from anthropogenic activities. It is caused by the deposition of atmospheric particles containing elemental and organic carbon, as primary and secondary pollutants. Also, biological compounds account for some of the NCC. Elemental carbon is considered a marker for combustion, being linked to carbonaceous particles emitted by mobile and stationary sources. Bonazza *et al.* (2007) stated that in the Iranian samples SEM-EDX analyses, a search of aluminosilicate and carbonaceous particles, indicating coal and oil combustion processes was unsuccessful. The NCC fraction ranged from 0.20% in Izeh reliefs, 2.78% in Pasargadae Complex, 10.07% in Bishapour reliefs to 11.56% in Persepolis. NCC contents in black crusts from different European monuments range between 0.75% and 5.13% (Bonazza *et al.*, 2005, Bonazza and Sabbioni, 2016).

Sulphates were the most abundant anions present in the samples (Fig. 9.6), but in concentrations much lower than those normally found in European black crusts (52–$542 \cdot 10^3$ $\mu g/g$). Gypsiferous soils usually develop in Iranian arid and semi-arid areas, and the country is one of the world's largest producer of gypsum, with 345 mines; 11 in Khuzestan and Eight in Fars province (two in Firooz Abad), therefore, wind dispersion of gypsum is a likely origin for the sulphates.

Short-chain organic anions, such as oxalate, formate and acetate were also quantified (Fig. 9.6). Their biological origin is unquestioned, in particular concerning oxalates.

Smaller and variable quantities of NO_3^- and NO_2^-, were also found (Fig. 9.6). In general, nitrogen acids form very soluble salts (nitrate and nitrite), with the calcium of the stone, and the relatively high amount detected in this study compared to the quantity generally measured in black crusts present in European urban sites

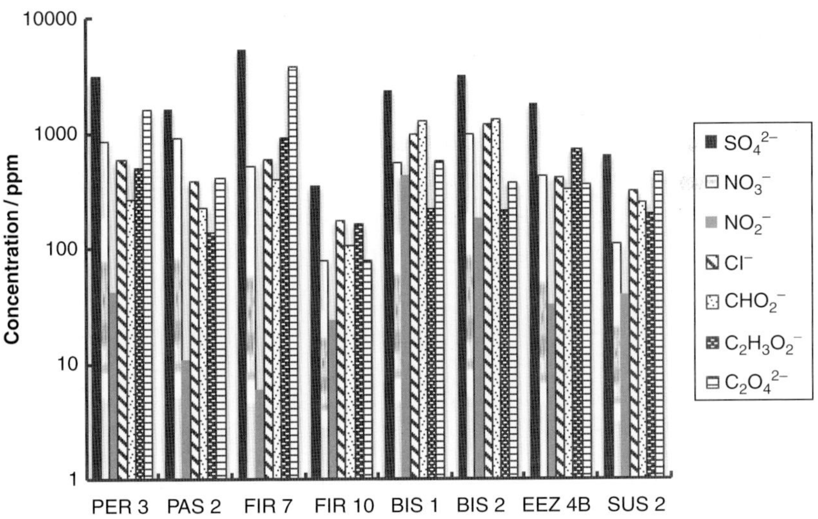

Fig. 9.6.   Anion concentration in samples of black deposit and crusts collected at the Iranian archeological sites.

(Bonazza *et al.*, 2005), can be related to infrequent episodes of seasonal run off.

In the samples, the low concentrations of sulphates, the presence of oxalates and of typical *n*-fatty acids patterns in the organic fraction of all the Iranian black stains studied were highly suggestive of a cyanobacterial contribution (Bonazza *et al.*, 2007). Additional evidence supported a biological origin for the extended black areas covering the Iranian bas-reliefs and monuments investigated. Bonazza *et al.* (2007) explained that DNA extraction of samples, with cyanobacterial 16S rDNA fragments as the specific target of PCR amplifications, was positive for the Persepolis Complex, the Cyrus Tomb (Pasargadae Complex), the Sassanide Grand Fire Temple and the Dokhtar Castle at Firooz Abad (Fars province).

9.5 Concluding Remarks

Results by analysing samples of black deposit and layers from the archaeological sites under investigation suggested that the oil well fires were not a source of the soiling and blackening found on Iranian bas-reliefs engraved on rocks, but rather it arose as a biological contribution from phototrophic microorganisms; a natural phenomenon in cliff face areas with water percolation through the rock and leakage. Evidence of black cover due to microbial growth was also found in other Iranian monuments investigated.

Even though it is evident from the studies conducted on the 1991 Kuwait oil fires that the smoke reached the southern Iran, its effect on the blackening and soiling of Iranian cultural heritage was negligible, largely due to the high efficiency of combustion, the low soot concentration and the rapid removal of SO_2 in smoke from the fires, as discussed in Sec. 9.2.

Acknowledgements

I wish to thank Cristina Sabbioni, Nadia Ghedini, Cesareo Saiz-Jimenez and Alessandro Sardella for the stimulating and fruitful discussions.

References

Bonazza, A. and Sabbioni, C. (2016). Composition and chemistry of crusts on stone. Urban Pollution and changes to materials and building surfaces, in Brimblecombe, P. (ed.), *Air Pollution Reviews*, Vol. 5, (Imperial College Press, Singapore), pp. 103–126.

Bonazza, A., Sabbioni, C. and Ghedini, N. (2005). Quantitative data on carbon fractions in interpretation of black crusts and soiling on European built heritage, *Atmospheric Environment*, **39**, 2606–2618.

Bonazza, A., Sabbioni, C., Ghedini, N., Hermosin, H., Jurado, V., Gonzalez, J. M. and Saiz-Jimenez, C. (2007). Did smoke from the Kuwait oil well fires affected archaeological sites and monuments of Iranian cultural heritage?, *Environmental Science Technology*, **41**, 2378–2386.

Cofer, W. R., III, Stevens, R. K., Winstead, E. L., Pinto, J. I., Sebacher, D. I., Abdulraheem, M. Y., Al-Sahafi, M., Mazurek, M. A., Rasmussen, R. A., Cahoon, D. R. and Levine, J. S. (1992). Kuwaiti oil fires: Composition of source smoke, *Journal of Geophysical Research*, **97**, 14521–14525.

Ferek, R. J., Hobbs, P. V., Herring, J. A. and Laursen K. K. (1992). Chemical composition of emissions from the Kuwait oil fires, *Journal of Geophysical Research*, **97**, 14483–14489.

Ghedini, N., Sabbioni, C., Bonazza, A. and Gobbi, G. (2006). Chemical-thermal quantitative methodology for carbon speciation in damage layers on building surfaces, *Environmental Science Technology*, **40**, 939–944.

Grossi, M. G. (2016). Soiling and Discolouration, in Brimblecombe, P. (ed.), *Air Pollution Reviews*, Vol. 5, (Imperial College Press, London), pp. 127–142.

Hobbs, P. V. (1992). Studies of smoke from the 1991 Kuwait oil fires. Introduction, *Journal of Geophysical Research*, **97**, 14481.

Hobbs, P. V. and Radke, F. L. (1992). Airborne studies of the smoke from the Kuwait oil fires, *Science*, **256**, 987–991.

Laursen, K. K., Ferek, R. J. and Hobbs, P. V. (1992). Emission factors for particles, elemental carbon and trace gases from the Kuwait oil fires, *Journal of Geophysical Research*, **97**, 14991–14497.

Limaye, S. S., Ackerman, S. A., Fry, P. M., Isa, M., Ali, H., Ali, G., Wright, A. and Rangno, A. (1992). Satellite monitoring of smoke from the Kuwait oil fires, *Journal of Geophysical Research*, **97**, 14551–14563.

Chapter 10

Southeast Asian Forest Fires (1997/1998): El Niño as a Driver of Regional Impacts

Mohd Talib Latif[*,†], Murnira Othman[†],
Ahmad Makmom Abdullah[‡], Md Firoz Khan[§],
Fatimah Ahamad[§] and Liew Juneng[*]

*School of Environmental and Natural Resource Sciences,
Faculty of Science and Technology, Universiti Kebangsaan Malaysia,
Malaysia
†Institute for Environment and Development (LESTARI),
Universiti Kebangsaan Malaysia,
Malaysia
‡Faculty of Environmental Studies, Universiti Putra Malaysia,
Malaysia
§Centre for Tropical Climate Change System (IKLIM),
Institute of Climate Change, Universiti Kebangsaan Malaysia,
Malaysia

10.1 Background

Biomass burning and haze episode are major problems in Southeast Asia. Agricultural activities especially related to "slash and burn" of tropical forest particularly from the expansion of plantations such as oil palm has led to biomass burning emission from large areas. Combustion processes in peat soil areas create a smouldering type of combustion which is hard to contain and can last several months. Dry weather conditions, especially related to El Niño phenomenon, increase the occurrence of forest fires in large areas i.e. in Sumatra and Kalimantan, Indonesia as well as several other parts in Peninsular Malaysia, Sabah and Sarawak. In 1997, the El Niño phenomenon contributed to the worst air pollution episode in Southeast Asia. With the persistence of the El Niño that started in 1997 and an abnormally short wet season (Levine, 1999), forest fires raged again in 1998 and resulted in another more localised haze event (Radojevic and Hassan, 1999).

According to Duncan *et al.* (2003), the influence of the 1997 fires in Kalimantan and Sumatra on ambient air quality started in July, peaked in September/October and weakened by November, when the delayed monsoonal rain extinguished the fires and scavenged the atmosphere. Satellite imagery from United States National Aeronautics and Space Administration (NASA) showed a smoke haze layer which expanded over an area of more than 3 million km^2, covering large parts of Sumatra and Kalimantan. Smoke from these two areas moved via southwest monsoon to other countries such as Malaysia, Singapore, Brunei as well as parts of Thailand and the Philippines (Brauer and Hisham-Hashim, 1998). During this period, particulate matter (PM) concentrations frequently exceeded the ambient air quality guidelines and standards suggested by each affected countries. Monthly mean horizontal visibility at most locations in Sumatra and Kalimantan in September 1997 was below 1 km and daily maximum visibility was frequently below 100 m. The area most affected by transboundary emission in 1997 was Sarawak, Malaysia. In the city of Kuching, Sarawak, the ambient particle concentration recorded at Malaysian Department of Environment monitoring station increased significantly 5–20 times above background levels. In total, 32 days were recorded in the 'unhealthy' to 'hazardous' range based on Malaysian Air Pollutant Index (API). Visibility decreased from above 15 km to below 0.5 km during this period (Heil and Goldammer, 2001). Prolonged El Niño episode caused a continued dry period in early 1998. In contrast to the situation in 1997, the fire-related air pollution episode in the first half of 1998 was essentially restricted to Borneo. The weakened southerly monsoonal flow by that time reduced the effect of transboundary emission from local biomass burning episodes to other areas. However, the population in Kalimantan and Borneo-Malaysia was exposed to distinctively elevated air pollution for a period of months (Heil and Goldammer, 2001).

10.2 Peat Forest and Agricultural Activities

In the Southeast Asian region, particularly Malaysia and Indonesia, the natural vegetation is predominantly lush tropical rainforests. However, a large portion of the land is covered by peat and swamp forests (Fig. 10.1). During the 1997 fires, more than 90% of the area that burned in central Kalimantan was peatlands (Page *et al.*, 2002). It was estimated that over 1,000,000 ha of peat and swamp forest was burnt in the 1997/1998 fires in Kalimantan (Tacconi, 2003). In tropical areas, peat forest is characterised by wet soil which is high in organic content. The anaerobic decomposition of organic matter in water logged soil results in the formation of acidic peat soil. Peat soil can reach depths up to 20 m with the accumulation of this partially decomposed organic matter. Monsoonal rains and the location of these forests in lowlands allow the soil to retain moisture. However, deforestation and agricultural activities typically result in drainage of water from the soil. The drier peat soil is more susceptible to burning. Besides agricultural activity, natural causes such as drought caused by strong El Niño events are also linked to increased risks of peat forest and peatland fires.

Slash and burn method for agriculture has been traditionally practised in Southeast Asia. The fires in 1997 were partially attributed to nomadic cultivators and smallholders who use this method for clearing land. However, large-scale deforestation in Borneo and Sumatra for logging as well as to open lands for cash crops such as oil palm has also been indicated as a cause. Fire has also been used

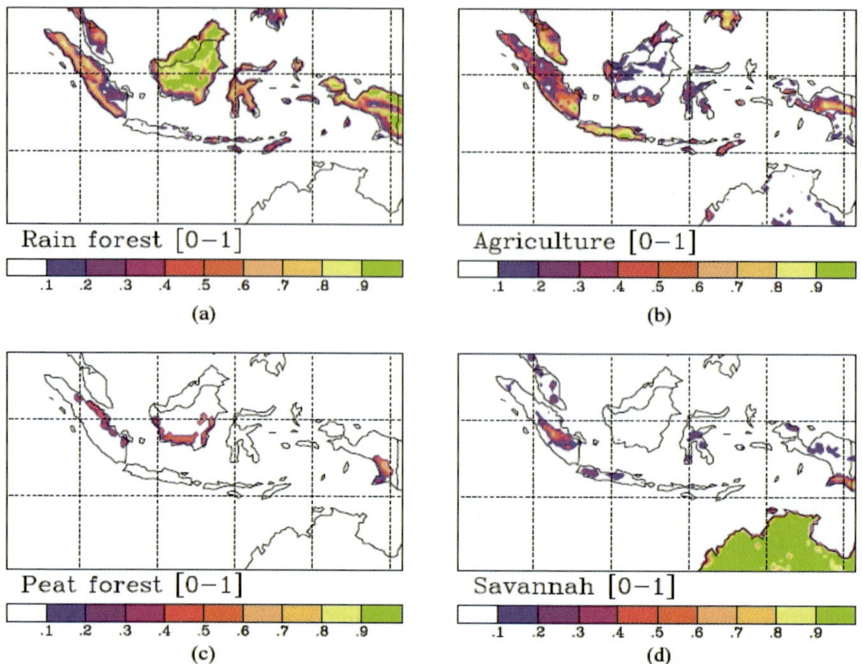

Fig. 10.1.   Distribution of the major vegetation classes as used in the REMO model domain in fraction (0–1) per model grid cell (0.5° × 0.5°) for (a) rain forest, (b) agriculture, (c) peat forest and (d) savannah.

Source: Langmann and Heil (2004).

by large plantation companies to claim land from local cultivators and *vice versa.* Center for International Forest Research (CIFOR) final report on *The Underlying Causes and Impacts of Fires in Southeast Asia* identified four direct causes of fires which were: (1) fire being used to assist with land clearing, (2) fire used as a weapon in land tenure or land-use disputes, (3) accidental or escaped fires, and (4) fire connected with resource extraction. The drainage and burning activities due to agricultural practices coupled with strong El Niño event caused large-scale fires which were hard to contain (Tacconi, 2003; Applegate, 2001; Suganto *et al.*, 2004).

A consistent correlation was found between the strong 1997–1998 El Niño driven rainfall deficit and the number of active fire counts (Wooster *et al.*, 2012). The study also showed a difference in the fire diurnal cycle between El Niño and non-El Niño periods using active fire data from Giglio (2007). During non-El Niño periods, the fires typically occur around noon and last for a few hours. During El Niño periods, the fire diurnal cycle is damped and the peak shifts to around 9.00 p.m. The deficit in rainfall and the subsequent drought increased the risk of fires. Figure 10.2 indicates the distribution of hotspot in September 1997. The land mass covers natural forests, agricultural areas and peatlands. Recently logged forests were much more susceptible to fires than primary forests and forests logged long ago (Siegert *et al.*, 2001). Besides large-scale logging and plantation of cash crops particularly oil palm, the one million hectare rice project in Central Kalimantan was also been cited as the cause of the large scale burning occurring in Indonesia (Tacconi, 2003; Siegert *et al.*, 2001). The drought also increased incidences of uncontrolled fire caused by local village activities such

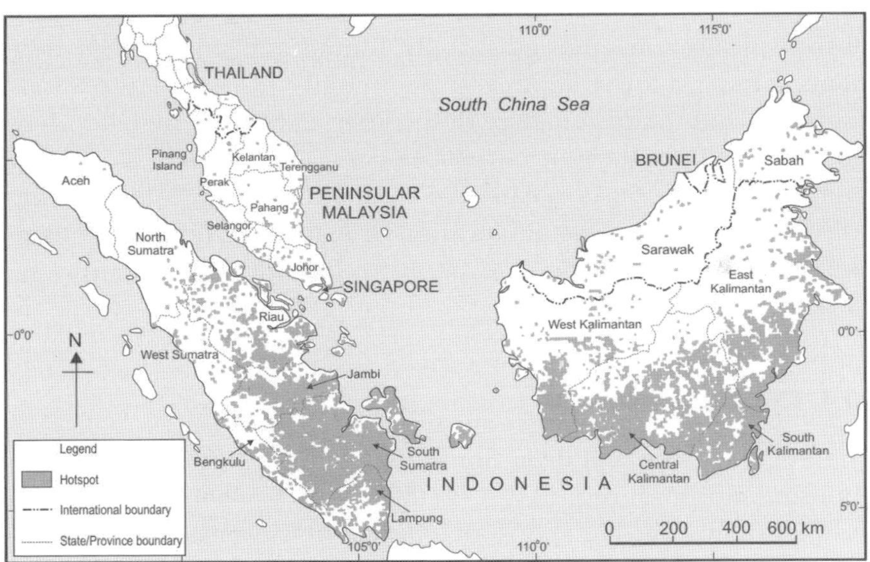

Fig. 10.2.   Distribution of hotspots in September 1997 monitored by NOAA satellite.

Source: Mahmud (2009).

as the use of fire to clear pathways to be used as short-cuts during the dry season which increases risks of fire spreading and burning swamp forests. Campfires that are not properly extinguished by migrating communities seeking out smalls pools and shallow channels concentrated with fish during the dry seasons have also been cited as causes of fire (Dennis *et al.*, 2000).

10.3 Burning Type — Smouldering versus Flaming

The burning of peat soil is typically characterised by smouldering instead of flaming type of combustion. These fires propagate slowly and can last from a few days to a few months compared to a few hours during flaming combustion. Smouldering combustion can reach peak temperatures of 400–700°C while temperatures during flaming combustion can reach up to 1,800°C (Usup *et al.*, 2004; Rein *et al.*, 2008). One reason for the difficulty in managing peatland fires arises because burning not only occurs on the surface, but also in the subsurface layer of the soil. Peat soil burning at depths up to 20 cm typically consumes grass-roots, humus and small woody fragments as fuel. At lower depths, the main fuel consists of large woody fragments and the peat matrix (Usup *et al.*, 2004). Average depth of burning in Indonesia is around 30 ± 8 cm (Van der werf *et al.*, 2010). Smouldering combustion can spread laterally underground, giving little indication of the actual location or extent the fire unlike flaming combustion which provides more visual indicators (Usup *et al.*, 2004; Rein *et al.*, 2008).

10.4 Climate Factors

The regional climate over Southeast Asia is controlled by two opposing monsoon regimes — (i) the winter monsoon and (ii) the summer monsoon. The monsoon circulations reversal and the associated meridional migration of the inter-tropical convergence zone (ITCZ) drives the wet and dry seasons over the Southeast Asia. During the winter monsoon season (DJF), the low-level winds near the equator is predominantly northeasterly and the ITCZ with the associated monsoon trough is located close to the equator, which brings abundant of moisture over the maritime continent. During the summer monsoon season (JJA) when the surface wind is predominantly southwesterly, the ITCZ is relocated northward leaving the maritime continent climatologically dry.

In addition to the annual cycle, the El Niño Southern Oscillation (ENSO), a coupled atmosphere–ocean phenomenon over the Pacific Ocean, modulates the interannual variability of the regional climate over Southeast Asia. The warm phase of ENSO is called El Niño while the cold phase is called La Niña. During ENSO events, the global atmospheric circulation is altered. Located over the rising branch of the Walker Circulation, Southeast Asia is directly influenced by the phenomenon (Tanggang *et al.*, 2010). During non-ENSO years, the prevailing trade-winds converge leading to abundant moisture over the maritime continent sustaining convective activities. This results in generally low-pressure centre over this region. However, during El Niño events, the anomalous warming of the tropical Pacific sea surface temperature shifts the low-pressure centre from the maritime continent to the central of Pacific. This shifts the rising branch of Walker cell and establishes an anomalously high-pressure centre; hence, modifying the regional low-level circulation anomalies. The resulting low-level divergence causes lower precipitation during an El Niño event. Regionally, the impact of El Niño shows considerable spatial and seasonal variations (Juneng and Tangang, 2005). During the summer monsoon, the impact is most felt over the maritime continent when the region is climatologically dry. The anomalously high surface pressure further exacerbates the drought. It is during this time that the number of detected fires hot-spots and incidence of trans-boundary smoke episodes elevates. The drought during the El Niño years can be extended into September–October–November (SON) and December–January–February and further deteriorate the regional air quality and environment.

The 1997/1998 El Niño event is the strongest El Niño ever recorded during the 20th century. Its extensive impact on the global as well as regional climate and environment has been substantially reported in many studies. McPhaden (1999) provided an overview of the evolution of the 1997/1998 El Niño event. The initiation of the 1997 El Niño is modulated by the intraseasonal variability, called Madden–Julian Oscillation (MJO) which has a periodicity of 30–60 days. The intraseasonal signal manifested itself as a series of westerly wind events which generally increased in intensity along the equator and eventually weakened the trade winds. The weakened and reversed trade wind in early 1997 weakened the eastern Pacific coastal upwelling and rapidly increased the central and eastern Pacific equatorial sea surface temperature. The 1997 event developed so rapidly from June to December 1997. At the peak of the event, the equatorial pacific sea surface temperature (SST) reached ~29°C, about 4°C above its climatological mean. After the peak in December 1997, the SST remained warm due to weaker local winds. This persisted until the trade winds returned to their normal strength in mid-1998 and the eastern pacific upwelling system recovered to bring back the cold subsurface cold water.

The impact of 1997 El Niño events was largely felt over the maritime continent during the fall due to the extended period of drought season. Figure 10.3 shows the seasonal 950 hPa wind anomalies during the evolution of the 1997/1998 El Niño event over the maritime continent. The direction of the anomalies suggested increasing summer monsoon wind speeds over the region with a predominantly easterly component. Also noted is the large negative rainfall anomaly, which extended into the SON season. This greatly influenced the cross-equatorial flow (Koe *et al.*, 2001). The extended drought enhanced the incidence of a large number of forest fires and elevated aerosol emissions, while the stronger easterly winds and cross equatorial flow allowed more extensive transport of smoke (Koe *et al.*, 2001). The forest fires were mostly detected over Sumatra and Kalimantan where the rainfall anomalies were largely negative due to the anomalously high regional pressure modulated by the El Niño (Radojevic and Hassan, 1999). In addition, the high subsidence and atmospheric stability during the El Niño event confined the aerosol to lower part of the atmosphere and prohibit wet deposition (Chandra *et al.*, 1998). Hence, advection was the dominant mechanism of haze transport.

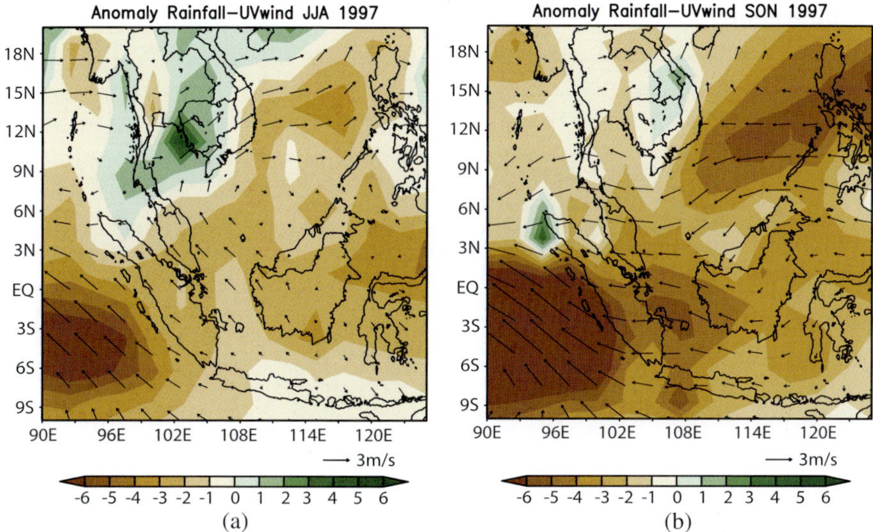

Fig. 10.3. The seasonal rainfall anomalies (shaded) and the 925 hPa wind anomalies (vectors) during (a) JJA 1997 and (b) SON 1997. The 925 hPa wind values are from NCEP/NCAR Reanalysis product Kalnay *et al.* (1996) whilst the rainfall values were obtained from Xie and Arkin (1997).

10.5 Major Air Pollutants

10.5.1 *Particulate matter and its composition*

The optical properties of the aerosols found during haze episode in 1997/1998 point predominantly to smouldering rather than flaming combustion sources (Gras *et al.*, 1999; Nakajima *et al.*, 1999). In addition, the aerosols exhibited a strong hygroscopic growth in scattering comparable to that of peat smoke. Aerosols from smouldering also consist of small amount of elemental carbon content (Vokelson *et al.*, 1997). These findings indicate that much of the emission was produced by smouldering combustion (Natukuwa *et al.*, 1999). This information is also supported by the high amount of sulphate found, which reflects its presence in peat soil (Gras *et al.*, 1999; Balasubramanian *et al.*, 1999).

The concentration of PM_{10} and total suspended particle (TSP) recorded in selected stations during haze episode in 1997 far exceeded the limit set by the United States Environmental Protection Agency (USEPA) which are 150 μg m^{-3} and 260 μg m^{-3}, respectively. In the areas where the biomass burning took place in Sumatra and Kalimantan, the concentration of TSP exceeded the concentration of 2000 μg m^{-3}. The amount of particles in the atmosphere will affect the visibility recorded at the stations due to light scattering processes (Fig. 10.4). Monthly mean of horizontal visibility at the monitoring sites such as Jambi and Pekan Baru, Sumatra and Pontianak, Kalimantan (Indonesia) reduced below 1 km during the peak of the haze episode between July and September 1997 (Heil and Goldammer, 2001). Simultaneous measurement of TSP and PM_{10} at Malaysian Meteorological Department in Petaling Jaya showed that both TSP and PM_{10} during haze episode exceeded the limit of concentration as suggested by Malaysian Department of Environment for TSP (260 μg m^{-3}) and PM_{10} concentration (150 μg m^{-3}). Table 10.1 shows that on average PM_{10} represents about 66% of total TSP recorded during haze episode between July and mid-November 1997. When the PM_{10} concentrations exceeded 150 μg m^{-3}, the PM_{10}/TSP ratio ranged between 70% and 93% (average 78%).

Characteristically for emissions from vegetation burning, the additional atmospheric particle loading during the smoke-haze episode was predominantly due to an increase of the fraction below or equal to 2.5 μm in diameter ($PM_{2.5}$) (Heil and Goldammer, 2001). A later study conducted by See *et al.* (2000) revealed a comprehensive characteristics of aerosol particles. Comparison of the physical, chemical and optical properties of aerosol during hazy and clear sky showed that the average concentration of $PM_{2.5}$ and most chemical components increased approximately by a factor of 2 on hazy days.

A study of $PM_{2.5}$ and its composition during haze episode in November 1997, a USEPA project was conducted by Pinto *et al.* (1998) at Petaling Jaya, Malaysia and two sites in Indonesia (Palembang

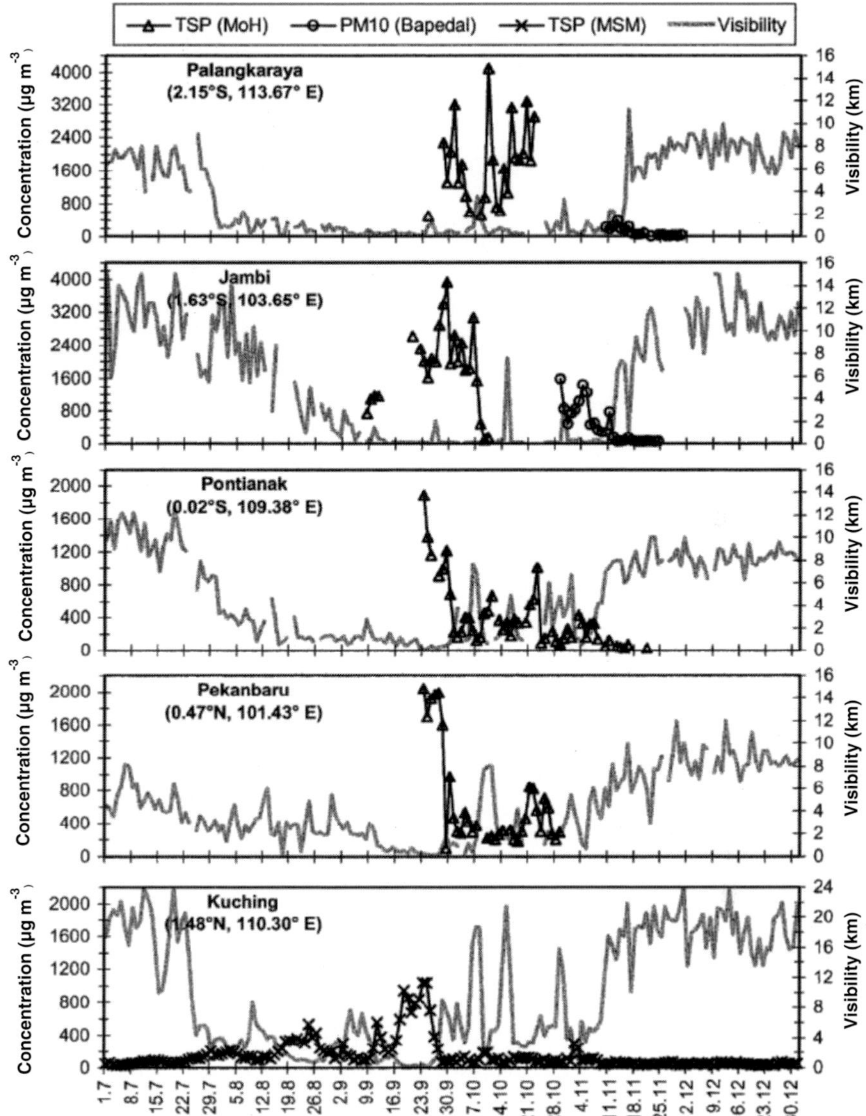

Fig. 10.4. These daily mean concentration of PM$_{10}$ and TSP recorded at selected station in Southeast Asia during haze episode in 1997.

Source: Heil and Goldammer (2001).

Table 10.1.   The concentration of PM_{10} and TSP during and after the smoke haze episode at Petaling Jaya, 1997.

	Haze (1 Jul–15 Nov)		Post-haze (16 Nov–31 Dec)
	Total	$PM_{10} > 150$	Total
PM_{10}	107	247	38
	(28–424)	(153–424)	(22–56)
TSP	155	314	83
	(52–525)	(204–525)	(51–117)
PM_{10}/TSP (%)	66	78	46
	(26–93)	(70–93)	(33–70)

Note: Mean (and range) concentration in $\mu g\ m^{-3}$.
Source: Heil and Goldammer (2001).

and Sriwijaya University). The results on $PM_{2.5}$ concentration and its composition recorded at these three areas are listed in Table 10.2. Higher $PM_{2.5}$ concentration was observed in Palembang and Sriwijaya University compared to Petaling Jaya station and the primary contribution was from biomass burning in Indonesia. During this sampling campaign, high concentrations of trace elements such as S, K and Si were recorded at all stations, and the possible contributors were biomass burning and motor vehicles. The total polycyclic aromatic hydrocarbons (PAHs) concentration was recorded at its highest at Palembang, followed by Sriwijaya University and Petaling Jaya with concentrations of 318 ng m^{-3}, 222 ng m^{-3} and 135 ng m^{-3}, respectively, with phenanthrene showing the highest concentration at all locations.

A study by Abas *et al.* (2004) showed that aerosol samples collected during the haze episode in Kuala Lumpur contain *n*-alkanes, *n*-alkan-2-ones, *n*-alkanols, methyl *n*-alkanoates, *n*-alkylnitriles, *n*-alkanals, *n*-alkanoic acids, levoglucosan, PAHs, and unresolved complex mixture as the dominant components, with minor amounts of terpenoids, glyceryl esters and sterols. During the haze episode, the organic components were transported from areas outside the region assuming all smoke components are external to the city, they amount to about 30% of the total organic particle burden. Rainwater analyses suggest

Table 10.2.   Mass concentration of $PM_{2.5}$ and its chemical composition in Malaysia and Indonesia in November 1997.

Elements	Malaysia (Petaling Jaya)	Indonesia (Palembang)	Indonesia (Sriwijaya University)
$PM_{2.5}$ μg m^{-3}			
$PM_{2.5}$	62.1	341	264
Organic carbon	26.1	282	200
Elemental carbon	1.9	5.4	3.2
Metal oxides	10.0	15	13
Sulfate	10.0	44	29
Trace elements ng m^{-3}			
Al	bd	bd	bd
Si	160	200	115
S	2400	11000	6900
Cl	70	4500	4600
K	280	1400	1500
Ca	98	79	47
Ti	27	11	6.5
V	9.3	bd	bd
Cr	0.2	bd	bd
Mn	4.5	1.7	bd
Fe	120	83	71
Ni	2.2	3.8	<0.1
Cu	9.3	2.1	3.9
Zn	34.3	13	6.4
As	2.3	1.2	1.3
Se	0.7	3.9	1.4
Br	9.8	95	72
Pb	39	64	7.7
PAHs ng m^{-3}			
Naphthalene	<4.9	12.1	<4.9
Acenaphthylene	6.0	11.8	<4.0

(*Continued*)

Table 10.2. *(Continued)*

Elements	Malaysia (Petaling Jaya)	Indonesia (Palembang)	Indonesia (Sriwijaya University)
Acenaphthene	3.9	<2.8	<2.8
Fluorine	12.5	47.8	26.9
Phenanthrene	50.9	187.8	108.6
Anthracene	12.1	32.8	17.1
Fluoranthene	10.9	38.5	21.0
Pyrene	11.9	42.8	18.1
benz[a]anthracene	<4.1	8.7	<4.1
Chrysene	<4.1	11.7	<4.9
benzo[*b*]fluoranthene	<6.5	10.4	<6.5
benzo[*k*]fluoranthene	<4.0	<4.0	<4.0
benzo[*e*]pyrene	<4.0	6.3	<4.0
benzp[*a*]pyrene	<3.8	7.1	<4.0
indeno[1,2,3-*cd*]pyrene	<3.6	4.8	<3.6
dibenzo[*a,h*]anthracene	<4.4	<4.4	<4.4
benzo[*g,h,i*]perylene	<4.1	5.8	<4.1
Coronene	<4.7	<4.7	<4.7
Perylene	<4.0	<4.0	<4.0
Total PAHs	135.0	318.0	222.0

Note: bd, beneath detection limits.
Source: Pinto *et al.* (1998).

that aerosols from forest fires do not have a major acidifying effect because dissolved acidic gases (e.g. SO_2) are neutralised by alkaline substances (e.g. Ca, Mg, K) that are also emitted by forest fires (Balasubramanian *et al.*, 1999).

10.5.2 *Gases*

According to Heil and Goldammer (2001), biomass burning released gaseous and other compounds such as carbon dioxide (CO_2), ammonia (NH_3), hydrogen (H_2), several hydrocarbons including methane (CH_4), formaldehyde and methyl chloride as well as carbonaceous materials. Levine (1999) estimated the emission of CO_2, CO, CH_4, NO_x, NH_3, O_3 and total PM from agricultural/plantation emission, forest fires emission and peat fires emission. The author found peat fires produced the highest emission of CO_2 and total PM that accounts for 89% and 94% of the total. Agricultural and forest fire emissions does not seem to produce significant amount of pollutants compared to peat fires.

The study by Levine (1999) estimated the gaseous emissions resulting from the 1997 fires: on a daily basis, the calculated emissions of CO_2, CO, CH_4, and NO_X from the Kalimantan and Sumatra fires of 1997 significantly exceeded the emissions from the Kuwait oil fires of 1991. Davies and Unam (1999) indicated that the smoke haze can elevate directly atmospheric concentrations of particulate matter, SO_2, CO, CH_4 and CO_2. A study by Kunii *et al.* (2002) on CO, SO_2, NO_2 and O_x concentration during haze event in the year 1997 in three areas in Jambi, Indonesia found high concentration of CO, with an average concentration of 20 ppm followed by NO_2 and SO_2, both with an average concentration of 0.01 ppm, while O_x recorded an average concentration of 0.04 ppm. Radojevic (2003), who studied atmospheric pollutant concentrations in Brunei during the 1998 haze showed a result with the highest concentration of O_3 at 63.2 μg m^{-3} followed by NO_2 with a concentration of 41.5 μg m^{-3}, while CO recorded the lowest concentration at 4.2 mg m^{-3}. Sawa *et al.* (1999) recorded high CO mixing ratios of 3–9 ppm from aircraft measurements in Kalimantan, during haze episode in October 1997, which suggested that CO concentration was heavily accumulated in haze layer.

Koe *et al.* (2001) identified large amounts of greenhouse gases and the concentration of pyrolysed products from the forest fires in the 1997 episode. Fujiwara *et al.* (1999) observed pronounced enhancements of total and tropospheric O_3 with the Brewer spectrophotometer and ozonesondes at Watukosek (7.5°S, 112.6°E), Indonesia in 1994 and in 1997 when extensive forest fires were reported in Indonesia. The integrated tropospheric O_3 increased from 20 DU to 40 DU in October 1994 and to 55 DU in October 1997. On 13 October 1994, most O_3 mixing ratios were more than

50 ppbv throughout the troposphere and exceeded 80 ppbv at some altitudes. On 22 October 1997, the concentrations of O_3 were more than 50 ppbv throughout the troposphere and exceeded 100 ppbv at several altitudes. The coincidence of O_3 enhancements with the forest fires suggest the photochemical production of tropospheric O_3 resulting from precursors emitted during the fires for both cases. The years of 1994 and 1997 correspond to El Niño events when convective activity is weaker in Indonesia. Thus, in this region, it is likely that pronounced enhancement of tropospheric O_3 associated with extensive forest fires due to sparse precipitation may take place with a period of a few years coinciding with El Niño events.

10.6 Impact of Biomass Burning

10.6.1 *Physical environment-visibility*

The biomass burning activities in Southeast Asia cause notable degradation of air quality and hazy conditions, where haze usually refers to reduced visibility. Visibility depends on several factors such as relative humidity and amount of sunlight. Haze is a phenomenon that occurs when there are too much of certain types of atmospheric pollutants in the ambient air. Trace compound originating from biomass burning act as reactants in atmospheric chemistry, with PM which then absorb and scatter the incident radiation and decreases visibility (Abas *et al.*, 2004). Referring to Odman *et al.* (2009), $PM_{2.5}$ is one of the main contributors to visibility impairment and Wang *et al.* (2004) made the most reliable and sensitive measurement of haze is visibility, while Field *et al.* (2004) commented that visibility value is important in helping to provide excellent source of *in situ* data for air quality measurement. For example, in Korea visibility status plays significant role in determining the existence of a haze episode, where less than 5 km visibility by eye was considered as hazy day (Kang *et al.*, 2004). Other measurement of visibility is the mean extinction coefficient (B_{ext}) that measure the degree of visible light that is absorbed and scattered by aerosol at a certain distance (Brauer and Saksena, 2002; Field *et al.*, 2009).

Visibility is related to the composition of fine PM itself; water-soluble inorganic ions for example SO_4^{2-}, NO_3^- and NH_4^+ and carbonaceous components (Kang *et al.*, 2004; Tan *et al.*, 2009; Xiao *et al.*, 2011). Odman *et al.* (2009) studied the source of regional haze in the Southern US and found that the summertime $PM_{2.5}$ can decrease with the reduction of SO_2 emissions, which can also improve visibility. Another study by Cheng *et al.* (2011) in Jinan, China had found that the water content in particles with diameters ranging from 1.0–1.8 μm as well as SO_4^{2-} concentrations play a significant role in visibility impairment.

High smoke concentrations from biomass burning, which leads to impaired visibility has affected a number of Southeast Asian countries. According to Odihi (2003), areas located closed to biomass burning will receive much more soot, pungent burning smells and low visibility. Lower visibility was reported in Padang City, Indonesia where the value was less than 100 m from December 1997 to the early months of 1998 (Yonneda *et al.*, 2000). Okada *et al.* (2001) also reported that the average visibility measured in Banjarmasin and Pangkalanbun on 23 October 1997 were 2.2 km and 0.16 km, respectively, and in 25 October 1997, reduction of visibility was observed with value of 1.4 km and 0.0 km, respectively. Long-range analysis of B_{ext} and other parameters such as total particulate emitted matter and precipitation in Sumatra and Kalimantan from year 1960 to 2006 was determined by Field *et al.* (2000).The study concluded that during the 1997 haze event, the B_{ext} value was 16.6 km^{-1} which showed the severity of the haze event in Kalimantan, while in Sumatra the highest B_{ext} value was recorded in the year 1991 followed by the year 1997 with a range of 4–6 km^{-1}. Field *et al.* (2004) studied the regional visibility for Sumatra and Kalimantan using the year 1995 values as reference due to the absence of severe haze event in that year (Fig. 10.5). The results showed that there were three haze events which were from September to October 1994, July to September 1997 and February to March 1998. The lowest visibility was recorded only in the year 1997 (15.1%) showing the severity of the haze event in 1997. Radojevic and Tan (2000) also reported that low visibility was experienced by Brunei, with 200–700 m visibility

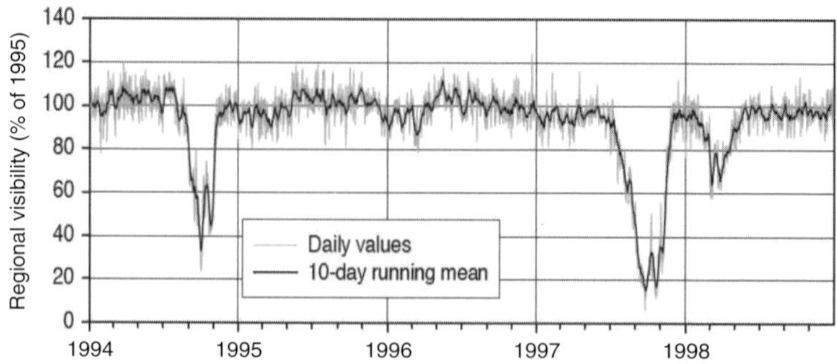

Fig. 10.5.   Regional visibility (Sumatra and Kalimantan) and its 10 days running mean.

Source: Field *et al.* (2004).

recorded for one week during the 1997 haze episode and the severe haze episode in early 1998 caused visibility to drop to 100 m.

10.6.2 *Health impact*

The haze episode of 1997/1998 creates significant impact to human health. High pollutant loading in ambient air can enter human body easily during breathing which can cause lung, heart and circulatory problems. According to Othman *et al.* (2014), high pollutant concentrations in air can cause alveolar inflammation, worsen existing lung disease and increase risk of cardiovascular events. Radojevic (2003) also added that the symptoms of health effect during haze episode were upper respiratory tract infection, asthma, conjunctivitis, bronchitis, eye and throat irritation, coughing, breathlessness, blocked and runny nose, and skin rashes. The elevated amount of air pollutant, especially PM can significantly affect human health. Fine particulates can easily enter human body via inhalation. It may then penetrate into the lower respiratory tract and remain there for a long period while larger particulates may be deposited in the upper respiratory system (Brauer and Hisham-Hashim, 1998). Referring to Odihi (2003), one of the highest recorded disease during haze episode particularly during the haze episode 1998 was chronic obstructive pulmonary disease (COPD).

During the 1997 haze episode, approximately 300 million people in Southeast Asia were exposed to elevated particle concentrations with about five-fold increase in respiratory diseases were observed in Malaysia compared to normal days (Awang *et al.*, 2000; Heil *et al.*, 2007). As reported by Sastry (2009) and Sahani *et al.* (2014), there was an increment of daily natural mortality among those aged between 65 and 74 years residing in Kuala Lumpur during the 1997 haze episode. A study by Odihi (2001) in Brunei Darussalam found that the haze episode in 1997 significantly affected young people (age 1–5 years) and older people (> 60 years) with urban population more adversely affected compared to those dwelling in rural areas. This result showed that children and older population were most vulnerable to health problems during haze and needed attention. Moreover, construction industry workers were also among the vulnerable groups impacted by the haze event. Radojevic (2003) also studied the impact of 1998 haze to human health at Brunei Darussalam and found that PM_{10} concentration was 60% higher than WHO guideline for PM_{10}. In Singapore, during the specific haze event, there was about 30% increment of hospital attendance, where an increase of PM_{10} concentration from 50 μg m^{-3} to 150 μg m^{-3} resulted in an increase of 12% upper respiratory tract illness, 19% of asthma and 26% rhinitis (Heil and Goldammer, 2001).

For Indonesia, the number of bronchial asthma, bronchitis and acute respiratory infection cases was observed to be approximately 1,802,340 cases among 12,360,000 people who were affected during 1997 haze episode in eight provinces (Aditama, 2000). Acute respiratory infection recorded higher number of cases compared to asthma and bronchitis. West Sumatra had the highest number of cases due to the large population. The details of the provinces in Indonesia and number of cases are shown in Table 10.3. Kunii *et al.* (2002) performed a health effects investigation using 534 respondents in Indonesia to determine the impact of haze episode in year 1997 and found that more than 90% of the respondents reported respiratory problems with elderly individuals with age more than 60 years old were most affected.

Table 10.3.   Health impact in eight provinces in Indonesia from September to November 1997.

Province	No. of population risk	No. of cases		
		Asthma	Bronchitis	Acute respiratory infection (ARI)
Riau	17,01,000	410,280	7,995	199,107
West Sumatra	24,11,000	58,164	11,332	282,087
Jambi	14,78,000	35,650	6,947	172,926
South Sumatra	23,55,000	56,803	11,069	275,535
West Kalimantan	14,78,000	44,574	8,686	216,216
Central Kalimantan	716,000	17,574	3,366	83,772
South Kalimantan	17,33,000	41,800	8,145	202,716
East Kalimantan	118,000	2,846	555	13,806
Total	12,36,0000	298,125	58,095	1,446,120

Source: Aditama (2000).

The toxicity effect of fine particles during haze episode in 1997/1998 was not well understood due to the lack of specific studies on particle compositions and toxicity effect. Nevertheless, Tan *et al.* (2000) had performed a study on the association between acute air pollution during the 1997 haze episode in Singapore and peripheral white blood cells (WBC) in human. The result showed that high air pollution concentrations during haze cause elevated band neutrophil counts (percentage of total polymorphonuclear leukocytes). This suggested that acute air pollution during haze episode lead to pathogenesis of cardiorespiratory morbidity. A later study evaluating cytotoxic effect of PAHs in $PM_{2.5}$ during 2010 haze and non-haze period in Singapore showed statistically significant decrease of metabolically active (human lung epithelial cell line A549) cells following direct exposure to samples collected during haze event. The percentage of metabolically active cells decreased significantly following a direct exposure to PM samples collected during the haze period (Pavagadhi *et al.*, 2013). Another study by Huang *et al.* (2016) on particulate-bound trace elements (TrElems) in $PM_{2.5}$ during haze episode in 2014 indicated the cumulative cancer risks exceeded the acceptable level (1 in a million i.e. 1×10^{-6}).

The haze episode of the year 1997/1998 was also suspected of contributing to the emergence of Nipah virus (NiV) in Malaysia and Singapore which produced significant health impact to human and pigs. According to Chua *et al.* (2002), an outbreak of fatal febrile encephalitis was discovered in Perak and Northern Peninsular Malaysia, where the symptoms were respiratory illness and encephalitis in pigs while 105 human deaths were recorded. There were 246 cases reported on febrile encephalitis in human that occurred between September 1998 and April 1999 in Peninsular Malaysia and Singapore. Studies by Chua *et al.* (2002) had suggested that the haze episode and El Niño related climatic condition drove the migration of fruit bats/flying fox into cultivated orchards in order to gain an improved food supply and transmitted the NiV to pigs and human populations. However, Pulliam *et al.* (2012) refutes this cause for the emergence of NiV. The author suggested that the driver of NiV emergence was due to the agriculture intensification of mango production and livestock farming which produced a virus pathway in bat/flying fox and infected the pig population.

10.6.3 *Economic impact*

The smoke blanket that covered Southeast Asia affected economic activities in the region. Restricted visibility caused land, air and water traffic to be delayed or cancelled, which in turn affected tourism, industrial and fishing activities (Heil and Goldammer, 2001). There was also an incident of aircraft crash in September 1997 in northern Sumatra where 234 deaths were recorded and ship collisions occurring in the Strait of Melaka, where dozens of people were killed (Heil and Goldammer, 2001). Moreover, the haze episodes that occurred in recent years have reduced the interest of tourists in visiting countries in Southeast Asia. Referring to Narayanan (2002), the tourist industry was most affected during haze episode where tourist tends to cancel or shorten the duration of their stay providing bad publicity abroad.

The economic losses during the haze episode in1997 were evaluated for Malaysia, Indonesia, Singapore and other Southeast Asian countries. Mohd Shahwahid and Othman (1999) estimated that the total cost of damage for Malaysia was US$321 million (RM801.90 million), which includes cost of illness, productivity lost during the state of emergency, decline in tourist arrivals, flight cancellation, decline in fish landings, fire-fighting cost, cloud seeding activity cost and mask distribution cost (Table 10.4). Nasir *et al.* (2000) and Sahani *et al.* (2014) also reported that during the 1997 haze episode, the economic loss in Malaysia due to restricted activities cost about 79.3% of total health damage and 10.7% of asthma attack. For the tourism industry in Malaysia, it was estimated that the economic loss due to the decline of tourist arrivals was RM318.6 million with RM6.5 million sales loss for flight cancellations. For Indonesia, as reported by Ruitenbeek (1999), the economic cost was estimated at US$4085.25 million for several impacts such as health impact cost, tourism and airline/airport cost, and fire impact on timber and agriculture. The tourism industry alone recorded around US$46 million loss with airport authority managed by PT Angkasa Pura

Table 10.4.   Aggregate value of haze damage in Malaysia due to haze in 1997.

Type of damage	RM million	USD million	Percentage (%)
Adjusted cost of illness	21.02	8.41	2.62
Productivity loss during the state of emergency	393.51	157.40	49.07
Decline in tourist arrivals	318.55	127.42	39.72
Flight cancellations	0.45	0.18	0.06
Decline in fish landings	40.58	16.23	5.00
Cost of fire-fighting	25.00	10.00	3.12
Cloud seeding	2.08	0.83	0.26
Expenditure on masks	0.71	0.28	0.09
Total damage cost	801.90	321.00	100

Source: Mohd Shahwahid and Othman (1999).

showing losses of approximately US\$10 million (Narayanan, 2002). It was estimated that, the severe haze episode in year 1997 caused 1037 cancelled flights in Sumatra and 2027 cancelled flights in Kalimantan, which highlights the large economic losses during that event.

Other researchers also reported the direct and indirect cost of haze, which include transportation, agriculture, timber and non-timber forest resources, and human health cost during haze episode in year 1997/1998 for Indonesia was estimated between US\$8.7 and US\$9.2 billion (Nichol, 1998; Heil and Goldammer, 2001; Sastry, 2002; Field *et al.*, 2004). The impact of haze on Singapore economy was studied by Hon (1999) where the total economic losses were estimated between US\$78.8 million with US\$69.3 million contributed from health effect, tourism and airlines industries losses. The government of Singapore also spent around US\$1.5 million to help extinguish the fire in Sumatra (Forsyth, 2014). The economic loss for tourism industry alone was estimated to be US\$58.4 where this sector was highly impacted compared to other industries (Narayanan, 2002). For Brunei, the severe 1997 and 1998 haze episodes damaged economic activities and caused 3.75% reduction in tourist number to the country (Anaman and Lool, 2000). The authors also reported that the economic loss for tourism industry was B\$1 million and the reduction in the number of tourists was noticeable even two months after the event ended.

Hours of attendance at school were also affected by the high concentration of pollutants during haze episode, which exceeded the limit that was allowed by governmental regulation, e.g. standards such as the PSI (Pollutants Standard Index) in Singapore and Indonesia, and API in Malaysia. Several schools were closed during haze episode in order to reduce the exposure of children to haze. For example, in Malaysia, when the API exceeded 500 during the 1997 haze episode in Kuching, Sarawak, all outdoor activities were prohibited and restricted to essential economic activities involving food retailing, electricity and water provision, and law enforcement (Mohd Shahwahid and Othman, 1999).

10.6.4 *Other impacts*

Severe biomass burning coupled with haze adversely affects biological species. For example, Risk *et al.* (2003) studied the impact to coral (*Porites lobate*) during the 1997 haze episode and found decreased $\delta 13$ C value in the coral skeleton suggesting an effect from decreased photosynthesis due to reduced sunlight. A study by Radojevic and Tan (2000) looked at the impact of biomass burning on the pH water in Brunei in the years 1994, 1997 and 1998, but results showed no significant acidifying effect of biomass burning in water compared to Crutzen and Andreae (1980) who reported an acidifying impact on rainwater connected to biomass burning in the tropics. However, Balasubramanian *et al.* (1999) who conducted a research on rainwater acidity from July to December 1997 in Singapore had found that the mean pH values of the rainwater samples were between 3.79 and 6.20 and the volume weighted mean was 4.35. These results suggested the influence of anthropogenic derived pollutants on rainwater during haze episode.

Agricultural sector was also affected during haze episode reflected in low yields. One of the influencing factors for plant metabolism is ratio of red and infra-red radiation (Nichol, 1997). However, during haze event, high particulate materials in the air will scatter and absorb the amount of solar radiation which in turn decreases the ratio of red and infra-red radiation. Another important factor is the photosynthetic photon flux density (PFD). According to Yanhong *et al.* (1998), in tropical forests, PFD is usually a limiting resource. Yanhong *et al.* (1996) found that there was 50% reduction of PFD readings and lower value of photosynthetic carbon gain during haze day in year 1994 compared to non-haze day in Pasuh Forest Research, Negeri Sembilan Malaysia and concluded that the growth of plant can be affected by haze episodes. Davies and Unam (1999) have studied the impact of haze on photosynthetically active radiation (PAR), an important parameter to evaluate the photosynthetic rates of plant species. The results showed that during the haze episode of 1997, the PAR value in Kuching, Malaysia was severely affected. The values were 42–92% lower compared to clear days and this had an effect on forest level carbon fixation. Moreover, during haze episode, two varieties of hybrid rice in Malaysia suffered about 50% reduction in growth rate and showed abnormal ripening (Nichol, 1997). There was severe reduction in the output of fresh fruit bunches from a 10 acre oil palm plantation in Melaka during 1997/1998 due to the haze episode and drought, such that output was delayed for 6 months. Yoneda *et al.* (2000) reported that, during June 1997, the rice yield from paddy field and bananas yield decreased 10% and 25%, respectively, in Padang, Indonesia.

10.6.5 *Climate change*

The deforestation process and biomass burning activities released large amounts of carbon to the environment. Crutzen and Andreae (1990) reported that rapid removal of tropical forest releases 1.1–3.6 Pg C a^{-1} while emission of biomass burning produced 0.5–1.4 Pg C a^{-1}. Moreover, the emission of CO_2 from biomass burning was estimated to contribute 40% to the global anthropogenic annual gross release of CO_2 (Andreae *et al.* 1996; Heil and Goldammer, 2001). In Southeast Asia, the paddy rice plantations involve large areas which then produce agricultural waste such as rice straw. Rice straw was estimated to be 31% of agricultural waste in Southeast Asia where it is usually burned and produces significant amount of carbon (Corutzen and Andreae, 1990).

The main contributor of carbon release in Southeast Asia is peatland exploitation and peat fires. Peatlands area in Southeast Asia is mostly found in Indonesia and Malaysia. According to Miettinen and Liew (2010) and Jaenicke *et al.* (2008), around 10% of Indonesia land surface consists of peatland where it covers approximately 13 Mha. Hooijer *et al.* (2010) reported that there was a much larger area of peatland in Indonesia where over 22.5 Mha was documented. However, this rich carbon reservoir experienced exploitation due to human activities such as logging, drainage and conversion of forests to plantation. The activities causing the lowering of the water table makes peatland areas more susceptible to fire (Miettinen and Liew, 2010).

Peat fires released significant amounts of carbon to the atmosphere. Page *et al.* (2002) studied the emission of carbon during the 1997 peatland fires where the estimated results showed that the carbon released during the peatland fires in Indonesia in the year 1997 was between 0.81 and 2.57 Gt which is about 13–40% of current global emission from fossil fuel burning. Moreover, Heil *et al.* (2007) reported that about 480–2570 Tg of carbon and 55 Tg of PM_{10} were produced during 1997 biomass burning and peatland fires. Another study by Duncan *et al.* (2003) had found that during the 1997 fires in Indonesia, about 700 Tg C were emitted with an estimated 3–4% of Indonesia's total carbon stored in peatland. Increasing carbon emission was also associated with CO_2 emission. According to Jaenicke *et al.* (2008), the emission of CO_2 is associated with peat oxidation after drainage in order to transform to plantation area and peat combustion. It was estimated by Heil *et al.* (2007) that the emission from peat fires in 1997 contributed around 3,470 Tg CO_2 emission and 345 Tg CO emission while a study by Duncan *et al.* (2003) showed 230 Tg CO_2 and 130 Tg CO emissions from September to November 1997. Another interesting finding by Matsueda *et al.* (2004) had found that the CO_2 cycle in the upper troposphere was changed during the 1997 haze and biomass burning event and it was estimated that the increasing mean concentration of CO_2 during that event impacted the Southern Hemisphere and Northern Hemisphere concentration with value 0.3–0.4 ppm and 0.1 ppm, respectively. Thus, the strategy to mitigate emission from tropical forest and peatland is crucial in order to maintain the carbon store.

The devastating haze episode in the year 1997 was also described by the production greenhouse gas (GHG), where GHG was also known to contribute to global warming and climate change. The GHG production from Indonesia during 1997/1998 haze event only was estimated to be 30% of the annual total global average GHG emission from land use change calculated for the period from year 1989 to 1995 (Page *et al.*, 2002; Tacconi, 2003; Tacconi *et al.*, 2007). The estimation of GHG emission from peat fires in Southeast Asia for year 1997 accounts for 13–40% of global man-made emission for that particular year. This result precisely showed the significant emission of GHG from peat fires which act as one of the main contributors of future climate change.

10.7 Conclusion

Haze episode in 1997/1998 significantly affected human health and had a large economic impact on countries in Southeast Asia. The prolonged dry period that resulted from El Niño lasted until the first quarter of 1998 and contributed to increased burning episodes in the region, particularly Indonesia. The amount of smoke significantly contributed to low visibility in affected countries due to long-range transport from Sumatra and Kalimantan, Indonesia. Monsoonal winds influenced the transport of fine PM into parts of Malaysia, Singapore, Thailand, Brunei and Philippines.

Ground based and airborne investigations of the haze 1997 indicated that fires on peat swamp vegetation made a substantial contribution to the development of haze. Smouldering type of burning within peat soil produced high emission of fine PM in ambient air. Based on the relevance of peat swamp fires to transboundary smoke haze, particularly during dry periods, emission reduction and control strategies will have to focus on the prevention of fires in peat swamp as a matter of priority. The 1997/1998 episode has made it evident that in addition to a sound fire management, a fundamental revision of the land conversion and fire use policies is required to prevent the reoccurrence of similar episodes. Sustainable agricultural activities within these areas will prevent yearly smoke haze episode in Southeast Asian region.

The haze episode in 1997/1998 resulted in the intensification of regional measures towards cooperation in fire and smoke management, especially between Association of Southeast Asian Nations (ASEAN) countries. These measures include the establishment of ASEAN Haze Technical Task Force and the implementation of Regional and National Haze Action Plans. These plans define the ASEAN's countries contribution to fire prevention, monitoring, fighting and other mitigation measures. Among others, it is also targeted to upgrade the national air quality and meteorological monitoring networks in order to strengthen the region's early warning and monitoring system in respect to smoke haze. In 2002, 10 ASEAN Member Countries signed the ASEAN Agreement on Transboundary Haze Pollution in Kuala Lumpur, Malaysia. The Agreement is the first regional arrangement in the world that binds a group of contiguous states to tackle transboundary haze pollution resulting from land and forest fires. However, haze episodes have been recurring since the 1997/1998 event. Clearly, there is still much room for improvement in terms of regional cooperation and fire management.

References

Abas, M. R., Oros, D. R. and Simoneit, B. R. T. (2004). Biomass burning as the main source of organic aerosol particulate matter in Malaysia during haze episodes, *Chemosphere*, **55**, 1089–1095.

Aditama, T. Y. (2000). Impact of haze from forest fire to respiratory health: Indonesian experience, *Respirology*, **5**, 169–174.

Anaman, K. A. and Looi, C. N. (2000). Economic impact of haze-related air pollution on the tourism industry in Brunei Darussalam, *Economic Analysis and Policy*, **30**(2), 133–143.

Andreae, M. O., Atlas, A., Cachier, H., Cofer III, W. R., Geoffrey, W. H., Helas, G., Koppmann, R., Lacaux, J.-P. and Ward, D. E. (1996) Trace gas and aerosol emissions from savanna fires., in Levine, J. S. (ed.), *Biomass Burning and Global Change* (MIT Press: Cambridge, Massachusetts) pp. 278–295.

Applegate, G., Chokkalingam, U. and Suyanto, S. (2001). The underlying causes and impacts of fires in south-east Asia. Final Report. (CIFOR, ICRAF, USAID, USFS, Bogor).

Awang, M. B., Jaafar, A. B., Abdullah, A. M., Ismail, M., Hassan, M. N., Abdullah, R., Johan, S. and Noor, H. (2000). Air quality in Malaysia: Impacts, management issues and future challenges, *Respirology*, **5**, 183–196.

Balasubramanian, R., Victor, T. and Begum, R. (1999). Impact of biomass burning on rainwater acidity and composition in Singapore, *Journal of Geophysical Research*, **104**(D21), 26881–26890.

Brauer, M. and Hisham-Hashim, J. (1998). Fires in Indonesia: Crisis and reaction, *Environmental Science and Technology*, **32**(17), 404A–407A.

Brauer, M. and Saksena, S. (2002). Accessible tools for classification of exposure to particles, *Chemosphere*, **49**, 1151–1162.

Chandra, S., Ziemke, J., Min, W. and Read, W. (1998). Effects of 1997/1998 El Niño on tropospheric ozone and water vapor, *Geophysical Research Letters*, **25**, 3867–3870.

Cheng, S.-H., Yang, L.-X., Zhou, X.-H., Xue, L.-K. and Gao, X.-M. (2011). Size-fractionated water-soluble ions, situ pH and water content in aerosol on hazy days and the influences on visibility impairment in Jinan, China, *Atmospheric Environment*, **45**, 4631–4640.

Chua, K. B., Chua, B. H. and Wang, C. W. (2002). Anthropogenic deforestation, El Niño and the emergence of Nipah virus in Malaysia, *The Malaysian Journal of Pathology*, **24**(1), 15–21.

Crutzen, P. J. and Andreae, M. (1990). Biomass burning in the tropics: impact on atmospheric chemistry and biogeochemical cycle, *Science*, **250**(4988), 1669–1678.

Davies, S. J. and Unam, L. (1999). Smoke-haze from the 1997 Indonesian forest fires: Effects on pollution levels, local climate, atmospheric CO_2 concentrations, and tree photosynthesis, *Forest Ecology and Management*, **124**, 137–144.

Dennis, R., Erman, A., Stolle, F. and Applegate, G. (2000). *The Underlying Causes and Impacts of Fires in South-east Asia: Site 5* (Danau Sentarum, West Kalimantan Province, Indonesia).

Duncan, B. N., Bey, I., Chin, M., Mickley, L. J., Fairlie, T. D., Martin, R. V. and Matsueda, H. (2003). Indonesian wildfires of 1997: Impact on tropospheric chemistry, *Journal of Geophysical Research D: Atmospheres*, **108**(15), ACH 7-1–ACH 7-25.

Field, R. D., van der Werf, G. R. and Shen, S. S. P. (2009). Human amplification of drought-induced biomass burning in Indonesia since 1960, *Nature Geoscience*, **2**, 185–188.

Field, R. D., Wang, Y., Roswintiarti, O. and Guswanto (2004). A drought-based predictor of recent haze events in western Indonesia, *Atmospheric Environment*, **38**, 1869–1878.

Forsyth, T. (2014). Public concerns about transboundary haze: A comparison of Indonesia, Singapore, and Malaysia, *Global Environmental Change*, **25**, 76–86.

Fujiwara, M., Kita, K., Kawakami, S., Ogawa, T., Komala, N., Saraspriya, S. and Suripto, A. (1999). Tropospheric ozone enhancements during the Indonesian Forest Fire Events in 1994 and in 1997 as revealed by ground-based observations, *Geophysical Research Letters*, **26**, 2417–2420.

Giglio, L. (2007). Characterization of the tropical diurnal fire cycle using VIRS and MODIS observations, *Remote Sensing of Environment*, **108**(4), 407–421.

Gras, J. L., Jensen, J. B., Okada, K., Ikegami, M., Zaizen, Y. and Makino, Y. (1999). Some optical properties of smoke aerosol in Indonesia and tropical Australia, *Geophysical Research Letters*, **26**, 1393–1396.

Heil, A. and Goldammer, J. (2001). Smoke-haze pollution: A review of the 1997 episode in Southeast Asia, *Regional Environmental Change*, **2**(1), 24–37.

Heil, A., Langmann, B. and Aldrian, E. (2007). Indonesian peat and vegetation fire emissions: Study on factors influencing large-scale smoke haze pollution using a regional atmospheric chemistry model, *Mitigation and Adaptation Strategies for Global Change*, **12**, 113–133.

Hon, P. M. L., *Singapore*, in Glover, D. and Jessup, T. (eds.), *Indonesia's Fires and Haze: The Cost of Catastrophe* (Institute of Southeast Asian Studies, Singapore.

Hooijer, A., Page, S., Canadell, J. G., Silvius, M., Kwadijk, J., Wösten, H. and Jauhiainen, J. (2010). Current and future CO_2 emissions from drained peatlands in Southeast Asia, *Biogeosciences*, **7**, 1505–1514.

Huang, X., Betha, R., Tan, L. Y. and Balasubramanian, R. (2016). Risk assessment of bioaccessible trace elements in smoke haze aerosols versus urban aerosols using simulated lung fluids, *Atmospheric Environment, Part B*, **125**, 505–511.

Jaenicke, J., Rieley, J. O., Mott, C., Kimman, P. and Siegert, F. (2008). Determination of the amount of carbon stored in Indonesian peatlands, *Geoderma*, **147**, 151–158.

Juneng, L. and Tangang, F. T. (2005). Evolution of ENSO-related rainfall anomalies in Southeast Asia region and its relationship with atmosphere-ocean variations in Indo-Pacific sector, *Climate Dynamics*, **25**, 337–350.

Kalnay, E., Kanamitsu, M., Kistler, R., Collins, W., Deaven, L., Gandin, M., Iredell, M., Saha, S., White, G., Woollen, J., Zhu, Y., Chelliah, M., Ebisuzaki, W., Higgins, W., Janowiak, J., Mo, K.C., Ropelewski, C.,

Wang, J., Leetmaa, A., Reynolds, R., Jenne, R. and Joseph, D. (1996). The NCEP/NCAR 40-year reanalysis project, *Bulletin of the American Meteorological Society*, **77**, 437–471.

Kang, C.-M., Lee, H. S., Kang, B.-W., Lee, S.-K. and Sunwoo, Y. (2004). Chemical characteristics of acidic gas pollutants and $PM_{2.5}$ species during hazy episodes in Seoul, South Korea, *Atmospheric Environment*, **38**, 4749–4760.

Koe, L. C. C., Arellano Jr., A. F. and McGregor, J. L. (2001). Investigating the haze transport from 1997 biomass burning in Southeast Asia: Its impact upon Singapore, *Atmospheric Environment*, **35**(15), 2723–2734.

Kunii, O., Kanagawa, S., Yajima, I., Hisamatsu, Y., Yamamura, S., Amagai, T. and Ismail, I. T. S. (2002). The 1997 haze disaster in Indonesia: Its air quality and health effects, *Archives Environmental Health*, **57**(1), 16–22.

Langmann, B. and Heil, A. (2004). Release and dispersion of vegetation and peat fire emissions in the atmosphere over Indonesia 1997/1998, *Atmospheric Chemistry and Physics Discussions*, **4**, 2117–2159.

Levine, J. S. (1999). The 1997 fires in Kalimantan and Sumatra, Indonesia: Gaseous and particulate emissions, *Geophysical Research Letters*, **26**(7), 815–818.

Mahmud, M. (2009). Simulation of equatorial wind field patterns with TAPM during the 1997 haze episode in Peninsular Malaysia, *Singapore Journal of Tropical Geography*, **30**, 312–326.

Matsueda, H., Taguchi, S., Inoue, H. Y. and Ishii, M. (2002). A large impact of tropical biomass burning on CO and CO_2 in the upper troposphere, *Science in China Series C*, **45**, 116–125.

McPhaden, M. J. (1999). Genesis and evolution of the 1997–1998 El Niño, *Science*, **283**, 950–954.

Miettinen, J. and Liew, S. C. (2010). Status of peatland degradation and development in Sumatra and Kalimantan, *Ambio*, **39**, 394–401.

Mohd Shahwahid, H. O. and Othman, J. Causes and impacts of the fire: Malaysia, in Glover, D. and Jessup, T. (eds.), *Indonesia's Fires and Haze: The Cost of Catastrophe* (Institute of Southeast Asia Studies, Singapore).

Nakajima, T., Higurashi, A., Takeuchi, N. and Herman, J. R. (1999). Satellite and ground-based study of optical properties of 1997 Indonesian forest fire aerosols, *Geophysical Research Letters*, **26**, 2421–2424.

Narayanan, S. (2002). Assessing the economic damage from Indonesian fires and the haze: a conceptual note, *The Singapore Economic Review,* **47**(2), 229–241.

Narukawa, M., Kawamura, K., Takeuchi, N. and Nakajima, T. (1999). Distribution of dicarboxylic acids and carbon isotopic compositions in aerosols from 1997 Indonesian forest fires, *Geophysical Research Letters,* **26**, 3101–3104.

Nasir, M. H., Choo, W. Y., Rafia, A., Md, M. R., Theng, L. C. and Noor, M. M. H. (2000). Estimation of health damage cost for 1997-haze episode in Malaysia using the Ostro Model, *Proceedings Malaysian Science and Technology Congress Kuala Lumpur.*

Nichol, J. (1997). Bioclimatic impacts of the 1994 smoke haze event in Southeast Asia, *Atmospheric Environment,* **31**(8), 1209–1219.

Nichol, J. (1998). New directions smoke haze in southeast Asia: A predictable recurrence, *Atmospheric Environment,* **32**(14/15), 2715–2716.

Odihi, J. O. (2001). Haze and health in Brunei Darussalam: The case of the 1997–1998 episodes, *Singapore Journal of Tropical Geography,* **22**(1), 38–51.

Odihi, J. O. (2003). Haze in Southeast Asia: needed local actions for a regional problem, *Pure and Applied Geophysics,* **160**, 205–220.

Odman, M. T., Hu, Y., Russell, A. G., Hanedar, A., Boylan, J. W. and Brewer, P. F. (2009). Quantifying the sources of ozone, fine particulate matter, and regional haze in the Southeastern United States, *Journal of Environmental Management,* **90**, 3155–3168.

Okada, K., Ikegami, M., Zaizen, Y., Makino, Y., Jensen, J. B. and Gras, J. L. (2001). The mixture state of individual aerosol particles in the 1997 Indonesian haze episode, *Aerosol Science,* **32**, 1269–1279.

Othman, J., Sahani, M., Mahmud, M. and Ahmad, M. K. S. (2014). Transboundary smoke haze pollution in Malaysia: inpatient health impacts and economic valuation, *Environmental Pollution,* **189**, 194–201.

Page, S. E., Siegert, F., Rieley, J. O., Boehm, H.-D. V., Jaya, A. and Limin, S. (2002). The amount of carbon released from peat and forest fires in Indonesia during 1997, *Nature,* **420**(6911), 61–65.

Pavagadhi, S., Betha, R., Venkatesan, S., Balasubramanian, R. and Hande, M. P. (2013). Physicochemical and toxicological characteristics of urban aerosols during a recent Indonesian biomass burning episode, *Environmental Science Pollution Research,* **20**(4), 2569–2578.

Pinto, J. P., Grant, L. D. and Hartlage, T. A. (1998). Report on U.S. EPA Air Monitoring of Haze from S.E. Asia Biomass Fires, (U.S. Environmental Protection Agency, U.S. Environmental Protection Agency).

Pulliam, J. R. C., Epstein, J. H., Dushoff, J., Rahman, S. A., Bunning, M., Jamaluddin, A. A., Hyatt, A. D., Field, H. E., Dobson, A. P., Daszak, P. and Group, H. E. R. (2012). Agricultural intensification, priming for persistence and the emergence of Nipah virus: A lethal bat-borne zoonosis, *Journal of The Royal Society Interface*, **9**(66), 89–101.

Radojevic, M. (2003). Haze research in Brunei Darussalam during the 1998 episode, *Pure and Applied Geophysics*, **160**, 251–264.

Radojevic, M. and Hassan, H. (1999). Air quality in Brunei Darussalam during the 1998 haze episode, *Atmospheric Environment*, **33**, 3651–3658.

Radojevic, M. and Tan, K. S. (2000). Impacts of biomass burning and regional haze on the pH of rainwater in Brunei Darussalam, *Atmospheric Environment*, **34**, 2739–2744.

Rein, G., Cleaver, N., Ashton, C., Pironi, P. and Torero, J. L. (2008). The severity of smouldering peat fires and damage to the forest soil, *Catena*, **74**(3), 304–309.

Risk, M. J., Sherwood, O. A., Heikoop, J. M. and Llewellyn, G. (2003). Smoke signals from corals: Isotopic signature of the 1997 Indonesian 'haze' event, *Marine Geology*, **202**, 71–78.

Ruitenbeek, J. (1999). *Indonesia*, in Glover, D. and Jessup, T. (ed.), *Indonesia's Fire and Haze: The Cost of Catasrophe* (Institute of Southeast Asian Studies: Singapore), pp. 86–129.

Sahani, M., Zainon, N. A., Wan Mahiyuddin, W. R., Latif, M. T., Hod, R., Khan, M. F., Tahir, N. M. and Chan, C.-C. (2014). A case-crossover analysis of forest fire haze events and mortality in Malaysia, *Atmospheric Environment*, **96**, 257–265.

Sastry, N. (2002). Forest fire, air pollution, and mortality in Southeast Asia, *Demography*, **39**, 1–23.

Sawa, Y., Matsueda, H., Tsutsumi, Y., Jensen, J. B., Inoeu, H. Y. and Makino, Y. (1999). Tropospheric carbon monoxide and hydrogen measurements over Kalimantan in Indonesia and northern Australia during October, 1997, *Geophysical Research Letters*, **26**(10), 1389–1392.

See, S. W., Balasubramanian, R. and Wang, W. (2006). A study of the physical, chemical, and optical properties of ambient aerosol particles in

Southeast Asia during hazy and nonhazy days, *Journal of Geophysical Research: Atmospheres*, **111**, D10S08, doi:10.1029/2005JD006180.

Siegert, F., Ruecker, G., Hinrichs, A. and Hoffmann, A. (2001). Increased damage from fires in logged forests during droughts caused by El Niño, *Nature*, **414**(6862), 437–440.

Suyanto, S., Applegate, G., Permana, R.P., Khususiyah, N. and Kurniawan, I. (2004). The role of fire in changing land use and livelihoods in Riau-Sumatra. Ecology and Society, (1): 15. Available at: http://www.ecologyandsociety.org/vol9/iss1/art15/.

Tacconi, L. (2003). *Fires in Indonesia: Causes, Costs and Policy Implications* (CIFOR, Bogor, Indonesia).

Tacconi, L., Jotzo, F. and Grafton, R. Q. (2007). Local causes, regional co-operation and global financing for environmental problems: the case of Southeast Asian Haze pollution, *International Environmental Agreements*, **8**, 1–16.

Tan, J., Duan, J., He, K., Ma, Y., Duan, F., Chen, Y. and Fu, J. (2009). Chemical characteristics of $PM_{2.5}$ during a typical haze episode in Guangzhou, *Journal of Environmental Sciences*, **21**, 774–781.

Tan, W. C., Qiu, D., Liam, B. L., Ng, T. P., Lee, S. H., Eeden, S. F., D'yachkova, Y. and Hogg, J. C. (2000). The human bone marrow response to acute air pollution caused by forest fires, *American Journal Respiratory Critical Care Medicine*, **161**, 1213–1217.

Usup, A., Hashimoto, Y., Takahashi, H. and Hayasaka, H. (2004). Combustion and thermal characteristics of peat fire in tropical peat-land in Central Kalimantan, Indonesia. *Tropics*, **14**(1), 1–19.

van der Werf, G. R., Randerson, J. T., Giglio, L., Collatz, G., Mu, M., Kasibhatla, P. S., Morton, D. C., DeFries, R., Jin, Y. V. and van Leeuwen, T. T. (2010). Global fire emissions and the contribution of deforestation, savanna, forest, agricultural, and peat fires (1997–2009), *Atmospheric Chemistry and Physics*, **10**(23), 11707–11735.

Wang, Y., Field, R. D. and Roswintiari, O. (2004). Trends in atmospheric haze induced by peat fires in Sumatra Island, Indonesia and El Niño phenomenon from 1973 to 2003, *Geophysical Research Letters*, **31**(L04103), 1–4.

Wooster, M., Perry, G. and Zoumas, A. (2012). Fire, drought and El Niño relationships on Borneo (Southeast Asia) in the pre-MODIS era (1980–2000), *Biogeosciences*, **9**(1), 317–340.

Xiao, Z., Zhang, Y., Hong, S., Bi, X., Jiao, L., Feng, Y. and Wang, Y. (2011). Estimation of the main factors influencing haze, based on a long-term monitoring campaign in Hangzhou, China, *Aerosol and Air Quality Research*, **11**, 873–882.

Xie, P. and Arkin, P. (1997). Global Precipitation: A 17-year monthly analysis based on gauge observations, satellite estimates, and numerical model outputs, *Bulletin of the American Meteorological Society*, **78**(11), 2539–2558.

Yanhong, T., Naoki, K., Akio, F. and Awang, M. B. (1996). Light reduction by regional haze and its effect on simulated leaf photosynthesis in a tropical forest of Malaysia, *Forest Ecology and Management*, **89**, 205–211.

Yokelson, R. J., Susott, R., Ward, D. E., Reardon, J. and Griffith, D. W. T. (1997). Emissions from smoldering combustion of biomass measured by open-path Fourier transform infrared spectroscopy, *Journal of Geophysical Research Atmospheres*, **102**, 18865–18877.

Yoneda, T., Nishimura, S. and Chairul (2000). Impacts of dry and hazy weather in 1997 on a tropical rainforest ecosystem in West Sumatra, Indonesia, *Ecological Research*, **15**, 63–71.

Chapter 11

World Trade Center (2001): Health Effects of the Dust

Morton Lippmann

*Nelson Institute of Environmental Medicine,
New York University, USA*

11.1 Introduction

This chapter provides a holistic reappraisal on the health impacts of inhalation exposures to dust created by the collapses of the World Trade Center (WTC) Twin Towers on 11 September 2001, and reviews the work of many investigators on the characterisation, toxicology and health effects of inhaled WTC Dusts.

I did not get personally involved in the research on exposures to, and the health effects of WTC Dust until 2014, when I undertook an investigation in response to the overwhelming evidence in the literature of substantial health impacts among populations exposed to WTC Dusts in Lower Manhattan. My initial doubts about major public health impacts had derived from my expectation that the inhaled doses of long chrysotile asbestos fibres and of other trace-level components in the fine particulate matter, i.e. particles less than 2.5 μm in aerodynamic diameter ($PM_{2.5}$) would be too small to produce measurable effects. I had not adequately considered the unique aspects of the initial airborne dust cloud that became known in the

227

first month after the disaster, nor had I appreciated the ease of subsequent resuspension of WTC Dust from the settled dust layers. The WTC Dust was unique in terms of: (1) overall airborne mass concentrations; and (2) the dominance by coarse-sized particles in the airborne dust in terms of: (a) alkalinity; and (b) high concentrations of crushed synthetic vitreous fibres (SVFs) as well as cement, and gypsum. Finally, the initial paucity of toxicological literature on dusts having such characteristics made it difficult to envision, in advance, the toxic effects. For more on these factors, and of my interpretations of the evidence, see Lippmann *et al.* (2015). This chapter summarises the evidence presented in that comprehensive review of effects reported in the respiratory tract and oesophagus, and their public health significance.

11.2 WTC Dust

Any adult alive today has a vivid memory of the morning of 11 September, 2001, when there were separate high-speed impacts of commercial airliners on each of the Twin Towers of the WTC in New York City (NYC). Both towers collapsed within the next few hours, and dense, light-obscuring, clouds of dust spread throughout the southern end of Manhattan Island. The bulk (about 80–90%) of the WTC Dust that was generated and dispersed by the successive collisions of the concrete floor slabs of the 110-story towers as they cascaded down on each lower floor, crushing the concrete slabs, the gypsum (calcium sulphate) wallboard, and the glass- and slag-wool (SVF thermal insulation) into a finely divided mixed powder with aerodynamic diameters ranging upward from about 10 μm. The dust layers on the streets, sidewalks, and building setbacks of Lower Manhattan (Fig. 11.1(a)), and within buildings with shattered or open windows (Fig. 11.1(b)) could be inches thick in the nearby buildings. The settled dusts were readily resuspended by air currents or mechanical disturbance during the subsequent cleanup activities by: (1) rescue and recovery workers and volunteers on the debris pile at what became known as "Ground Zero"; (2) cleanup workers for the streets and building exteriors; (3) indoor cleanup workers; (4) residents (of all ages) and (5) commercial office workers.

(a)

(b)

Fig. 11.1. (a) View along plaza heading toward Chase Building on 9/13/11. (b) A scene from an indoor residence near Ground Zero. (photos property of C. Prophete, M. Cohen, and L. Chen).

The collapse of the WTC was a unique primary source of ambient airborne and settled dust. Gregory Meeker of the United States Geological Survey (USGS), a participant in the early dust collection and analyses, was quoted in Chemical and Engineering News (10/20/2003) as saying "Six million ft^2 of masonry, 5 million ft^2 of painted surfaces, 7 million ft^2 of flooring, 600,000 ft^2 of window glass, 200 elevators, and everything inside came down as dust when the towers collapsed." The dust generated by the collapse of the

WTC towers, hereafter designated herein as "WTC Dust", was unique in terms of its particle size distribution, especially in terms of its lower and upper ends. Of the total mass in the settled WTC Dust, ~1% was in the fine size range (<2.5 μm in aerodynamic diameter), and ~90% was between 10 m and 100 m (Lioy *et al.*, 2002, 2006; McGee *et al.*, 2003; Yiin *et al.*, 2006). Thus, virtually all of the WTC settled dust was in the coarse particle size range, and differed in sources and composition from those of fine particles. In the humid Eastern US, most of the ambient PM is usually smaller than 2.5 μm in aerodynamic diameter ($PM_{2.5}$), originates from both organic and inorganic vapour-phase pollutants, and is carbonaceous and/or acidic and hydroscopic. Most of the PM2.5 penetrates through the lung conductive airways during inhalation to reach the gas exchange airways in the deep lung. In the more arid parts of the Western US, there are often more equal coarse PM mass concentrations of particles between 2.5 μm and 10 μm as of $PM_{2.5}$. Most particles in the thoracic particle size range (2.5–10 μm) penetrate into the conductive airways within the human and deposit in the larger lung airways. Supercoarse particles, that is, those larger than 10 μm, have less than 50% penetration through the upper respiratory tract (URT), but can still have appreciable penetration through the URT, and deposition within the lungs' conductive airways when the airborne concentrations are high, as they were for the thoracic coarse and supercoarse WTC Dust in Lower Manhattan on 11 September 2001.

The WTC Dust differed substantially from typical dusts found within building construction and demolition sites, where the debris is generally composed of much larger pieces that cannot be readily resuspended into the air, or as rapidly dissolved in the aqueous surface fluids of the respiratory and/or gastrointestinal (GI) tracts as the more finely ground WTC Dust. The WTC Dust was also unique in terms of its more uniform chemical composition, with 80–90% of it being a well-blended mixture of nearly equal parts of concrete, gypsum, and SVFs. Further, the WTC Dust was also unique in terms of its geographic extent of spatial dispersion, both outdoors and indoors. Pressure waves created by the collapses of the WTC Towers created hurricane-force winds that radiated out from the WTC Towers through the urban street canyons of Lower Manhattan, with relatively little WTC Dust reaching beyond Canal Street to the north in Manhattan, or beyond the nearest parts of Brooklyn to the southeast.

11.3 WTC Primary Dust was a Secondary Dust Source

WTC Dust was formed on the morning of 11 September 2001, a day with an unusually low humidity. However, the potential for airborne resuspension of settled WTC primary dust deposits was quite different for the outdoor and indoor deposits, especially after the heavy rains of 14 and 24 September that washed away much of the settled outdoor dust into the storm sewers leading to the adjacent waterways. The WTC destruction on 9/11 exposed commercial buildings within ~1,000 feet of the Twin Towers.

The settled WTC settled Dust was, as suggested by Chatfield and Kominsky (2001), unique in its propensity to become airborne, even under "passive conditions." In 2003, the RJ Lee Group (Lee, 2014) undertook a laboratory study to investigate: (1) the relationship between indoor surface dust and airborne dust concentrations and to evaluate the potential for WTC Dust that was resuspended as a result of building maintenance activities in a heavily damaged building; and (2) the potential for residual WTC Dust to become easily resuspended. They used low-velocity air pulses to resuspend dust from a variety of WTC Dust contaminated indoor surfaces into a portable vertical elutriator. Tests were conducted on concrete floors, ceiling tiles, fireproofing materials, floors and carpets, induction unit covers, and office equipment. The elutriator was sealed to the test surface with weather stripping, and used to resuspend, separate, and collect the dust particles that were resuspended. A standardised pulse of air 8 miles per hour (mph) was used, a velocity that was ~1/20 of that used in the "aggressive" air leaf-blower test employed during asbestos cleanups, and within the range encountered during normal office building activities. The key findings were:

- Resuspended dust was identified as WTC Dust in 75% of the sample locations;

- ~98% of resuspended PM collected by the elutriator was >10 μm in aerodynamic diameter;
- The propensity of WTC Dust to aerosolise was 10–10,000 times or more than that of ordinary office building surface dust.

11.4 Inhalation Exposures

11.4.1 *WTC Dust as a source of alkaline dust exposure*

Chemical leach tests were performed by USGS (2002) and by NYU (Maciejczyk *et al.*, 2004). The WTC Dusts were chemically reactive when dispersed in water (Fig. 11.2 for USGS tests). The moisture that came into contact with the dusts initially became highly alkaline, i.e. caustic, due to dissolution of calcium hydroxide in the crushed concrete in the particles and calcium sulphate from the gypsum in the crushed wallboard. Thus, from a health perspective, there should have been concern that inhalation of the coarse particle dusts by individuals who were at/near Ground Zero could create potentially hazardous alkaline conditions in their respiratory and gastrointestinal (GI) tracts. While the rains that occurred on 14 and 24 September, as well as continued reactions with atmospheric CO_2,

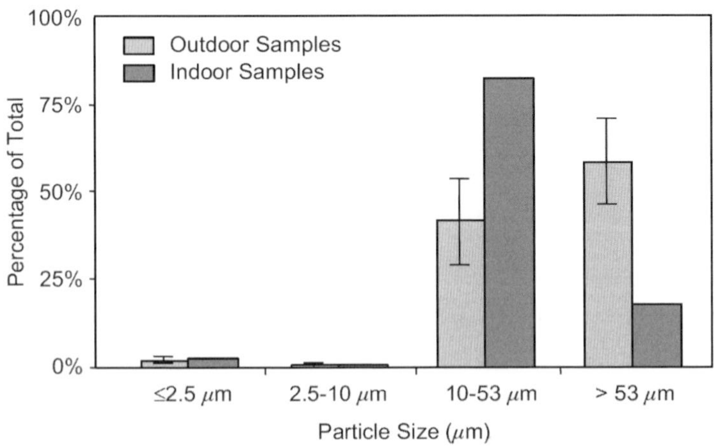

Fig. 11.2.   Particle size distribution from indoor and outdoor dust samples collected in the period of 9/12-13/11 (modification of original figure from Maciejczyk *et al.*, 2006).

helped to neutralise the alkalinity of the dusts on outdoor surfaces, the same remediation occurred with indoor dusts that were protected from direct contact with rainwater.

11.4.2 *Post-collapse human exposures to WTC Dust in Lower Manhattan, in terms of potential for inhalation exposures — Dust as a source of alkaline dust exposure*

Most residents and workers from residential and commercial buildings were not permitted into their Lower Manhattan buildings until the sites were certified as cleaned. For most residents, the reoccupation of their apartments took place over many weeks, while the many office workers were back at work within a few weeks (Reibman *et al.*, 2009). However, the adequacy of the initial cleanings was often disputed, leading to many second-stage cleanings and concerns about exposures occurring between the cleanings.

11.4.3 *Actions taken/not taken to analyse and/or minimise indoor exposures to WTC Dusts*

The building owners and their managers, lacking experience with such an unprecedented disaster, generally followed government agency mandates and guidelines for minimising inhalation exposures to WTC Dust (GAO, 2004). Unfortunately, as these agencies also lacked experience with such a disaster, the information they provided has, in retrospect, proven to be inadequate with respect to health protection.

11.5 Gaining Some Insights, Retrospectively, on the Roles of WTC Dust Components in Disease Causation from the Post-11 September 2001 Air Quality Measurements

11.5.1 *Characterisation of inhalation exposures*

In the days after 9/11, many Federal agencies brought their technical and scientific expertise to bear on addressing the emergency in NYC.

The US Environmental Protection Agency (EPA), other federal agencies, and NYC and New York State (NYS) public health and environmental authorities initiated numerous air monitoring activities to understand the nature and extent of ongoing toxicant exposures.

An overall preliminary evaluation, summarised in EPA (2002), relied primarily on the analyses of ambient air samples of $PM_{2.5}$ collected by newly-installed fixed-site monitors that were located at the perimeter of the WTC Ground Zero area and at various other sites in Lower Manhattan and the surrounding areas to assess the $PM_{2.5}$ inhalation exposure and potential human health risk incurred by the general population residing and those working in the vicinity of the WTC. Numerous other efforts were conducted that addressed other aspects of exposure and potential risk associated with the collapse of the WTC towers, including: (1) Ground Zero worker exposures on the debris pile and those participating in its disposal; (2) inhalable WTC Dust exposures faced by rescue workers; (3) indoor exposures; and (4) epidemiology studies. EPA (2002) concluded that persons exposed to the extremely high levels of ambient $PM_{2.5}$ and its components during the WTC Towers' collapse, and for several hours afterwards, were at risk for immediate acute (and possibly chronic) respiratory and other types of symptoms (e.g. cardiovascular). However, because the first measurements were not made until 14 September, and those of other contaminants were not measured until 23 September, exposures and potential health impacts could not be evaluated.

11.5.2 *Alkalinity of WTC Dust*

The issue of the alkalinity of WTC Dust, and its potential as a possible health concern for exposed individuals, was raised by observations by the USGS and academic researchers of high pH (~10.0) values for aqueous solutions of the settled WTC Dusts that had not been leached by rainfall. After late September, indoor exposures to such dust probably warranted more concern than outdoor exposures for possible acute irritating effects or more chronic health effects.

11.5.3 *Characterisation of WTC Dust*

There were, and remain, key problems with the interpretation of the WTC-related air quality monitoring. The main problem associated with these assessments was the failure to first adequately consider: (1) the unique particle size distribution and chemical composition of the WTC Dust; (2) the likelihood of excessive inhalation exposures; and (3) the optimal means for preventing and/or controlling excessive exposures. Most importantly, there was "tunnel vision" by all responsible parties that focused excessively on known toxicants that were present as very small mass fractions. These known toxicants, collectively, made up only a small percentage of the WTC Dust and of the $PM_{2.5}$ in the air samples that were analysed; with nearly all of the WTC Dust components, consisting of particles >10 μm in diameter, were not monitored, including the crushed concrete, gypsum, and SVFs, each of which is a known irritant. These three major mass components were assumed, incorrectly, to be nuisance dusts. These problems were compounded by the intense initial focus by EPA and the public media on asbestos fibres as the index toxicant for WTC Dust, and in technical guidance for exposure monitoring and control. There was a secondary focus on other trace components that were considered to pose health risks, that is, metals, molds, polyaromatichydrocarbons (PAHs), polychlorinated biphenyls (PCBs), and dioxins.

11.6 Technical Topics Influencing Adverse WTC Dust Inhalation Exposures

11.6.1 *Ease of airborne resuspension of WTC Dusts*

Figure 11.1(a) illustrates that thick deposits of WTC Dust covered outdoor surfaces in Lower Manhattan during the first day after the attack on 9/11. WTC Dust covered indoor surfaces for much longer periods of time (Fig. 11.1(b)). To characterise the chemical composition, particle size distribution, and potential for resuspension of the settled dust particles into human breathing zones, Chen *et al.* at

NYU, size-fractionated seven WTC settled dust samples that were collected by NYU personnel in Lower Manhattan on 12 and 13 September, 2001 (Yiin *et al.*, 2006). Aliquots of these settled dust samples were sieved through a 53 μm mesh screen. The PM_{53} fraction passing through the screen was then aerosolised into a vertical elutriation chamber and passed through a size-selective air sampler inlet with a 10 μm aerodynamic diameter cut to remove particles in the 10–53 μm range. The PM_{10} fraction was then passed through a cyclone with a cut at 2.5 μm to remove the coarse fraction ($PM_{10-2.5}$). The $PM_{2.5}$ fraction was collected on Teflon filters. Each individual size-fractionated sample was then subjected to physical and chemical analyses. Most key findings have previously been described (Lioy *et al.*, 2002, McGee *et al.*, 2003, Yiin *et al.*, 2006). Similar work on elutriation and chemical and physical analyses was performed on settled indoor WTC Dust samples that were collected in June of 2002 for a private client (Lee, 2014).

11.6.2 *Dosimetry of inhaled Dusts in humans*

The WTC Dust, as summarised in Sec. 11.1, differed greatly from conventional airborne dusts encountered in occupational and community settings in terms of particle size distribution and chemical composition. Therefore, it is important to discuss the special circumstances of WTC Dusts in terms of the dosimetry of inhaled PM.

11.6.3 *Conventional review of PM deposition in, and clearance from, the lung*

The penetration of inhaled particles into the thorax is limited by their deposition in the URT during inspiration, which varies with particle size distribution; flow rate and tidal volume; the fraction passing through the oral pathway; and *in vivo* airway dimensions. All of these factors are quite variable from person-to-person, depending on age, transient illness, cigarette smoke exposure and other short-term toxicant exposures that cause transient airway constriction, as well as, elements of occupational histories associated with loss of lung function or cumulative injury.

For inhaled particles that penetrate into the lower respiratory tract (LRT), that is, the airways within the thorax, the deposition patterns and efficiencies within the thoracic conductive airways (trachea, bronchi, and bronchioles) by impaction and sedimentation, and those of particles that deposit in the smaller gas-exchange airways by sedimentation and diffusion, are also highly variable and dependent on particle size, flow rate, and tidal volume, and on *in vivo* airway dimensions (Lippmann and Albert, 1969). Some inhaled particles remain airborne during the inhalation and exhalation phases of a tidal breath, and are exhaled without having been deposited. There is a deposition minimum of $\approx 15\%$ for particles between 0.1 and 1.0 μm that have little intrinsic mobility, as they are too large to deposit efficiently by sedimentation or diffusion, and too small to deposit efficiently by impaction on the surface coatings of larger airway branchings. In fact, the tidal flow in the airways is not completely reversible, with $\approx 15\%$ of the inhaled air remaining in the deep lung over multiple breaths, with a corresponding volume of residual lung air being exhaled. The laminar flow in the smaller airways leads to deeper penetration of particles near the airway axis than nearer the airway walls, and a flatter flow profile during exhalation, with more of the flow nearer the airway walls being exhaled residual lung air that is depleted of particles, and therefore with much lower particle deposition during exhalation than during inhalation (Briant and Lippmann, 1992, 1993). Most of the particles (or aggregates of particles) larger than ≈ 2 μm in aerodynamic diameter deposit in the conductive airways.

Those components that are insoluble or poorly soluble on the mucous layer of the airway lining fluid are carried proximally toward the larynx within a day, swallowed, pass through the GI tract, and are excreted in the faeces. Particles that deposit within oral passages and ciliated nasal passages of the URT are also swallowed and pass through the GI tract where they could contribute to gastroesophageal reflux disease symptoms (GERS) and gastroesophageal disease

(GERD). Soluble particles, components of particle aggregates, and insoluble particles with diameters <0.1 μm can all gain access to the bloodstream and be transported to more distal sites, where their chemical composition determines any long-term retention and toxic potential (Lippmann, 2014b). Most of the insoluble particles and PM components that reach the non-ciliated deep lung airways are phagocytosed by alveolar macrophages (AMs) within the few weeks that the macrophages are resident on airway surfaces. These surface macrophages, and the particles within them, are drawn into the terminal bronchioles due to the high surface tension of the alveolar lining fluid, and then cleared from the lung conductive airways by mucociliary clearance (Lippmann and Chen, 2007).

Aspiration of airborne particles into the human respiratory tract is highly particle size dependent. The penetration of inhaled particles into the tracheobronchial tree is different for inhalation via the nose versus the mouth, with the extent of the greater particle penetration of the oral pathway varying with the cross section of the pathway. Most particles with aerodynamic diameters <2.5 μm that originate from fossil fuel combustion sources can reach the respiratory bronchioles and alveoli. By contrast, few WTC Dust particles, having much larger sizes, penetrate that far. The particle-size-dependent differences in deposition sites within the lung airways strongly affect the residence times of the particles at the deposition sites, and their pathways and rates of clearance to other sites within the body.

In summary, for conventional occupational and ambient air PM, most of the particles with size between 2.5 μm and 10 μm deposit in the ciliated larger thoracic airways, where they can elicit bronchoconstriction and bronchospasm that can lead to bronchitis and exacerbations of asthma. By contrast, most particles <2.5 μm that deposit within the thorax are deposited in the more distal non-ciliated gas-exchange region where they are more likely to dissolve or be translocated to distal body sites wherein they can then exert organ-system-related toxicities, including cardiovascular, hepatic, renal, and nervous systems damage.

11.6.4 *WTC Dust deposition in, and clearance of WTC Dusts, and other PM, from the lung*

Studies of effects on deposition of PM can provide a basis for understanding how the inhaled WTC Dusts were deposited in/cleared from the URT and lungs of exposed individuals. WTC Dust contained much more of its PM in the coarse thoracic size range ($PM_{10-2.5}$), and even greater mass fractions of PM_{10-53} and $PM_{>53}$. Furthermore, the alkalinity of the PM in the larger size ranges was much greater than in the $PM_{2.5}$. As a result, deposition of highly alkaline particles in the URT, and in the tracheobronchial airways in the LRT represents a unique challenge to their ability to clear the airways without eliciting adverse effects (such as erosion and death of epithelial cells at airway surfaces). Disruption of the lung capacity for mucociliary clearance of the tracheobronchial airways by chemical irritants could explain the excess incidences of cough and diseases of the respiratory and GI regions to populations exposed to elevated concentrations of the large-diameter WTC Dust particles.

Very high acute exposures and/or lower levels of chronic exposure to chemical irritants can also destroy ciliated and mucus-secreting epithelial cells within the tracheobronchial airways, as has been seen for long-term cigarette smokers, leading to increased retention of all kinds of PM within the lungs, and not just the irritant particles themselves (Schlesinger and Lippmann, 1978). There is recent evidence that human exposure to WTC Dust can cause disruption of mucociliary particle clearance (McMahon *et al.*, 2011), who reported unusual ultrastructural ciliary abnormalities in three WTC response workers. Their ciliary abnormalities may have a genetic basis and a phenotypic expression that is prompted at the cellular level. Also, there is evidence that some very large WTC particles were able to reach deep in the lung, as demonstrated by the electron micrographs of material that was obtained from a firefighter's lung after bronchoalveolar lavage (Rom *et al.*, 2010).

In view of the unique particle size distributions and high alkalinity of the WTC Dusts, the standard dosimetry models and data tabulations have proven to be of little value in describing either the acute or cumulative WTC Dust dosages. The airborne levels of WTC Dust were highly variable both temporally and spatially, with localised hotspots of exposure being related to human activities causing WTC Dust resuspension. The particle size distribution of the WTC Dust clouds changed with distance due to more rapid settling of the largest particles, and due to the different paths of the high-velocity airflow radiating out from Ground Zero through the surrounding buildings. Further dose variations were caused by the sparse application of respiratory protection to capture the inhalable particles. After the first few days, those people having, and using, well-fitting negative-pressure respirators could avoid inhaling most of the dust in their breathing zone for as long as the dust layer on the filter caused minimal air path clogging. However, the filters in respirators worn by cleanup workers often clogged during the course of a 10-h workday, and replacement cartridges were not always available.

11.6.5 *WTC Summary of key factors affecting the human dosimetry of inhaled WTC Dusts*

(1) *Oro-nasal Breathing and Airway Anatomy*: Humans differ from almost all other mammals in terms of being able to inhale via the oral passages as well as through the nose, and in having nearly dichotomous bronchial airway branchings within the lungs rather than the highly asymmetric branching in nearly all other mammals. Both of these differences contribute to enhanced tracheobronchial airway particle deposition in humans (Lippmann, 1977). The greater flow resistance through the nasal pathway leads adult humans to switch to oral breathing when engaged in strenuous physical activity at inspiratory flow rates >40 L min^{-1}. When wearing a respirator with its own breathing resistance, there is likely to be a switchover to at least partial oral breathing (oronasal breathing) at a lower flow rate, especially as the respirator filter begins to clog. Partial or total inhalation via the oral passage increases particle penetration into the lungs. The asymmetric branching of the human bronchial airways results in enhanced particle deposition, especially

by impaction on the airway bifurcations within them than is characteristic in animals with more asymmetric airway branching (Lippmann and Schlesinger, 1984).

(2) *Clearance Pathways for Deposited Compact Particles and Fibres*: The insoluble particles that deposit on the ciliated mucosal surfaces in the nasal and bronchial airways are carried along the surface mucus toward the larynx, where they are swallowed and pass into the GI tract (Albert *et al.*, 1969). This is the pathway for SVFs within the WTC Dust that deposit within the nasal and oral pathways and the tracheobronchial airways during the first few days after their inhalation. For the cement and gypsum in the WTC Dust, which are soluble within the epithelial surface liquids, they added alkalinity to the mucus as it moved toward the larynx. As such, both the WTC Dusts and the alkaline mucus were likely swallowed and passed into the oesophagus.

(3) *Laceration of Epithelia by SVFs*: It has long been established that people handling fibrous glass in production facilities and at installation sites can develop dermatitis on exposed skin (Siebert, 1942, ATSDR, 2002). Stokholm *et al.* (1982) reported that workers exposed to SVF in the form of rockwool had more eye symptoms, and changes in the cellular and mucous conjunctival fluid, breakup time of the precorneal film, the number of microepithelial defects, and the number of dead and degenerated cells on the cornea and bulbar conjunctiva than control workers. One would expect that crushed SVFs within the WTC Dust had a much greater proportion of SVF shards and sharp ends than SVFs handled by production workers and insulation installers. The sharp ends of the SVF fibres and the shards resulting from the crushing of the vitreous fibres could lacerate phagocytic and epithelial cells in and on the lung and GI epithelia, subsequently leading to release of digestive enzymes and further epithelial damage.

(4) *High pH of Surface Fluids on Airway and Gastroesophageal Tract Epithelia*: The combination of chemical irritation by the highly alkaline mucus that can damage the ciliated cells of the conductive airways, and the physical irritation caused by the sharp

ends of the SVFs, as well as their interactions is a highly likely causal factor for the excess respiratory and GI tract disorders documented in WTC Dust-exposed individuals.

11.7 Evidence of Associations Between WTC Dust Inhalation and Human Health Effects

The extensive peer-reviewed literature discussed in this section documents, excesses in both acute and chronic health effects in both: (1) rescue and recovery workers and volunteers who were heavily exposed to WTC Dust out-of-doors during the first few days after 9/11; and (2) in residents and local workers, many of whom were heavily exposed to WTC Dust from the initial dust clouds while evacuating, and subsequently from more chronic exposures from resuspended dusts in the outside streets and from indoor exposures over much longer periods of time. Thus, the people having adverse health effects associated with their exposures to WTC Dust included groups with quite different amounts and temporal patterns of exposure. In spite of their differences in exposure pattern and intensity, they still growing literature on associations of excess clinical disease and dysfunction with known exposure to WTC Dust demonstrates similarities in the effects that they have experienced, and in their causality. The population groups with the highest disease incidence and the most severe responses include those presumed to have been exposed to quite high WTC Dust concentrations and scant use of respirators during the initial rescue and recovery phases, specifically the: (1) occupational groups exposed at Ground Zero; (2) FDNY Employees (Firefighters and Emergency Medical Technicians); and (3) day labourers engaged in building cleanup who did not work at Ground Zero, or live near Ground Zero, but were engaged in cleaning WTC-Dust contaminated office/residential buildings in Lower Manhattan.

While the incidence rates and severity of WTC Dust-related diseases were generally lower among people going about routine activities in Lower Manhattan, they still had statistically significant excesses in some of the same adverse health responses. Less is known about the local residents and office workers; however, doseresponse relationships for adverse health effects have been shown for these populations as well (Lin *et al.*, 2007; Maslow *et al.*, 2012). There were also clinical examinations of additional population groups who were presumed to have had exposures to lesser concentrations of WTC Dust initially, but whose periods of exposure extended into subsequent months. These groups included: (1) residents and office workers in Lower Manhattan studied at NYU; and (2) the NYC Department of Health and Mental Hygiene health registry of over 71,000 people for long-term follow-up studies to document medical problems of NYC responders and volunteers (WTC Health Registry, 2012). In addition, there were a variety of population groups that were not given clinical examinations whose exposures to WTC Dusts were less well defined, and whose adverse effects were elicited by questionnaires.

Despite different temporal patterns and amounts of WTC Dust exposure, and different frequencies of disease expression, both in terms of short-term and longer-term responses, all of these population groups exhibited quite similar clusters of specific disease categories within three specific anatomic regions, that is, the: (1) URT; (2) LRT; and (3) oesophagus. One of the clear early results of working in indoor or outdoor areas where WTC Dust contamination was a persistent cough, called "WTC cough" (Chen and Thurston, 2002). This "cough" was accompanied by bronchial hyperreactivity and respiratory distress (Prezant *et al.*, 2002) in 8% of the highly exposed firefighters examined. Of the 332 firefighters having "WTC cough," 95% had symptoms of shortness of breath; 87% had GERD; and 54% had nasal symptoms. The other population groups also had elevated incidences of chronic diseases that developed in the same three anatomic regions, as discussed below.

The World Trade Center Health Program (WTCHP) was mandated to identify health conditions linked by epidemiology and health research studies to occupational exposures to WTC Dusts (Howard, 2014). It is an authoritative source for judging what health problems suffered by WTC workers and volunteers who are eligible to receive medical care. The WTCHP developed its list of aerodigestive illnesses based on epidemiological research in the peer-reviewed literature that could be linked to exposure to WTC Dust exposure: i.e. asthma, chronic cough syndrome, chronic laryngitis, chronic nasopharyngitis, chronic respiratory disorder, chronic rhinosinusitis, GERD, interstitial lung diseases, reactive airways dysfunction syndrome (RADS), sleep apnea, upper airway hyperreactivity, and WTC-exacerbated chronic obstructive pulmonary disease (COPD).

This section summarises the health effects for people exposed to WTC Dust in Lower Manhattan. It focuses on objective findings of adverse health effects in the peer-reviewed literature. Some of the published reports were narrowly focused on a specific population, and had limited statistical power.

11.7.1 *Occupational groups exposed to WTC Dust at Ground Zero with clinical examinations*

The Mount Sinai School of Medicine (MSSM) established a clinically based research centre in NYC to measure the health risks among workers and volunteers in the WTC Worker and Volunteer Medical Screening Program (WVMS). The WVMS provided free, standardised medical assessments, clinical referrals, and occupational health education for workers and volunteers exposed to hazards during the WTC rescue and recovery effort). During 16 July, 2002–6 August, 2004, the WVMS evaluated 11,768 non-FDNY workers and volunteers (Levin *et al.*, 2002). During 16 July–31 December, 2002, a substantial proportion of participants experienced new-onset or worsened pre-existing lower and upper respiratory symptoms, with frequent persistence of symptoms for months after their WTC work stopped. The median length of time worked on the WTC effort was 966 (range: 24–4,080) h. Of the 610 examinees present in Lower Manhattan on 11 September, 51% reported being directly in the cloud of dust created by the collapse of the WTC buildings, and an additional 31% reported exposure to substantial amounts of dust. WTC-related lower respiratory symptoms (LRS) were reported by 60%, and 74% reported WTC-related upper respiratory symptoms: 40% had WTC-incident LRS that persisted to the month before screening, and 50% reported WTC-incident and persistent upper respiratory symptoms. Among the 851 participants who reported persistent WTC-related symptoms, an average of 32 (range: 7–63) weeks had elapsed since they stopped working at the site or the end of May 2002. Of all the participants, 46% had nasal mucosal inflammation, but other respiratory abnormalities were less common.

Kleinman *et al.* (2011a, 2011b) compared pulmonary function since the WTC disaster, with pre-exposure function data, in a NYC Police Department (NYPD) Emergency Responder cohort without history of repetitive respiratory exposures. For the 206 unit members who reported arrival time, exposure location, duration, smoking history, respirator mask usage, and respiratory symptoms, there was a mean long-term decline in FVC of 190 ml (3.7%) one-year post-exposure in 2002, and 330 ml (6.4%) in 2007, compared with baseline data. Li *et al.* (2011) examined new-onset GERS following 9/11 that persisted up to 5–6 years in relation to 9/11-related exposures. The cumulative incidence was 20% for post-9/11 GERS and 13% for persistent GERS.

Wisnivesky *et al.* (2011) described a cohort of more than 50,000 people who participated in the rescue and recovery work that followed the 9/11, reported on the incidence and prevalence rates of physical and mental health disorders during the 9 years since the attacks, examined their associations with occupational exposures, and quantified the physical and mental health comorbidities. The 9-year cumulative incidence of asthma was 27.6%, sinusitis was 42.3%, and GERD was 39.3%. In police officers, cumulative incidence of depression was 7.0%, PTSD was 9.3%, and panic disorder was 8.4%. In other rescue and recovery workers, cumulative incidence of depression was 27.5%, PTSD was 31.9%, and panic disorder was 21.2%. 9-Year cumulative incidence of spirometric abnormalities was 41.8%; three-quarters of these abnormalities were low FVC. Incidence of most disorders was highest in workers with the greatest WTC exposure. Thus, 9 years after the 9/11 WTC attacks, rescue and recovery workers continued to have a substantial burden of physical and mental health problems.

Kim *et al.* (2012) studied the risk of asthma among 20,834 participants in the MSSM WTC Medical Monitoring and Treatment Program (MMTP) between July 2002 and December 2007. They calculated the lifetime prevalence and standardised morbidity ratios (SMRs) by comparing the asthma risk among WTC responders with those from the National Health Interview Survey (NHIS). The study subjects underwent a baseline clinical examination, with follow-up every 12–18 months from 2002 to 2007. Rates of asthma in the NHIS served as a comparison group for the WTC responders. There was a major increase in lifetime asthma prevalence. Prior to 11 September, it was 2.9%, and rose to 12.8% in 2002, and 19.4% in 2007, a 6-fold increase. WTC The increasing trend from 2002 on was the same for both women and men; it increased for all ages, with the highest being for 40- to 49-year-old group. The standardised mortality ratios (SMRs) were elevated for all occupational groups: with protective services having a SMR of 1.5; and installation, maintenance, and repair workers having a SMR of 2.0. Acute asthma SMRs were elevated for all professions, with a 1.6 risk for transportation and material moving and a 2.3-fold risk for installation, maintenance, and repair jobs.

Kazeros *et al.* (2013) hypothesised that persistent, asthma-like, symptoms in WTC Dust-exposed individuals would be associated with systemic inflammation characterised by peripheral eosinophils. Their WTC patients underwent a standardised evaluation, including questionnaires and complete blood count. Between September 2005 and March 2009, 2,462 individuals were available for analysis. 1,517 individuals with pre-existing respiratory symptoms or lung disease diagnoses prior to September 2001, met the inclusion criteria. Patients had a mean age of 47 years, with 51% being female. Respiratory symptoms developed after WTC Dust/fume exposure, and remained persistent. They included dyspnea on exertion (DOE) (68%), cough (57%), chest tightness (47%), and wheeze (33%). A larger percentage of patients with wheeze had elevated peripheral eosinophils compared with those without wheeze (21% vs. 13%, p = 0.0001). Individuals with elevated peripheral eosinophils were more likely to have airflow obstruction on spirometry (16% vs. 7%, p = 0.0003). These data suggest that eosinophils may participate in lung inflammation in this population with symptoms consistent with WTC-related asthma.

The studies on FDNY workers at NYU, who had a pre-9/11 health baseline, provided an ideal benchmark for judging the impacts of WTC Dust on any employed group. Where WTC Dust contamination was substantial, there was persistent cough ("WTC cough" — Chen and Thurston, 2002). "WTC cough" was accompanied by bronchial hyperreactivity and respiratory distress (Prezant *et al.*, 2002) in 8% of the highly exposed firefighters. Of the 332 firefighters having WTC cough, 95% had symptoms of shortness of breath, 87% had GERD, and 54% had nasal symptoms. Banauch *et al.* (2002) reported injuries and illnesses among FDNY rescue workers. During the 48 h after the buildings collapsed, ≈90% of 10,116 FDNY rescue workers exposed at the WTC site reported an acute cough, often accompanied by nasal congestion, chest tightness, or chest burning, but only three of them required hospitalisation. Compared with the numbers of service-connected, respiratory medical leave incidents (n = 393) during the 11 months preceding the attacks, the number of respiratory medical leave incidents (n = 1,876) increased 5-fold during the 11 months after the attacks. During February 2002, the incidence of new respiratory illness requiring either medical leave or light duty began to decrease and during May 2002, began to approach pre-attack incidence. During the 6 months after the disaster, 332 firefighters and one EMS worker had WTC-related cough severe enough to require four consecutive weeks of medical leave. Despite treatment of upper and lower aerodigestive tract irritation, 52% showed only partial improvement of WTC-related cough and remained either on medical leave or light duty or were pending a disability retirement evaluation. During the collapse, 52% of workers did not wear respirators, and 38% did not wear respirators for the rest of the first day. In addition, most of those reporting the use of a respirator during the first day used only a disposable paper dust mask that was neither NIOSH-certified nor fit-tested.

A prospective study of a representative sample of 179 FDNY fire/ rescue workers examined links between WTC Dusts and pulmonary hyperreactivity and WTC cough at 1, 3, or 6 months after 9/11 (Banauch *et al.*, 2003). In both the highly and moderately exposed groups, there were significant declines in FVC, FEV_1, and $FEV_1/$FVC, when contrasted with control workers. Bronchial hyperreactivity showed a dose-response trend at 1, 3, or 6 months follow-up. For the highest exposure group, the median methacholine (Mch) induced bronchial hyperreactivity at 6 months rose 46%. After adjusting for smoking and airflow obstruction, highly exposed workers at one month were 7.3 times more likely to have hyperreactive airways than controls. For moderately exposed workers at 1 month, the risk was 6.3 times; at 6 months, it was 6.8 times. Among subjects who had hyperreactive airways at 6 months, respiratory symptoms were more frequent ($p = 0.05$); medical leave for respiratory illnesses was significantly longer ($p = 0.02$); medical leave days among hyper-reactive subjects was 45 versus 12 days in non-reactive workers; and FEV_1 was 80% of predicted was 18% versus 3% ($p = 0.03$) in comparison to non-reactive subjects. Compared with non-reactive workers, workers with hyperreactive airways at 1 or 3 months were 7.3 times more likely to develop WTC cough. A presence of hyperreactive airways is indicative of chronic and acute asthma. This research study began the process of seeing asthma as one of the WTC Dust respiratory hallmarks. Aerodigestive inhalation lung injuries resulting from WTC Dust exposure and the persistence of non-specific bronchial hyperreactivity in a representative sample of 179 FDNY rescue workers were stratified by exposure intensity who underwent challenge testing at 1, 3, 6, and 12 months post-collapse (Banauch *et al.*, 2005a, 2005b). Declines in pulmonary function, RADS, asthma, reactive upper airways dysfunction syndrome (RUDS), GERD, and rare cases of inflammatory pulmonary parenchymal diseases, were documented. In FDNY rescue workers, there was persistent hyperreactivity associated with exposure intensity, independent of airflow obstruction. At one-year post-collapse, 23% of highly exposed subjects were hyperreactive as compared with only 11% of moderately exposed workers and 4% of the controls. At one year, 16% met the criteria for RADS.

Weiden *et al.* (2010) did a followup study of the FDNY cohort and examined 1,720 subjects who were sent for pulmonary medicine evaluations. For the subjects who had PFTs before 9/11, there were significant declines in median FEV_1 and the ratio of FEV_1/FVC, and 59% had obstructive airways disease (OAD) based on a variety of factors. When adjusted for age, race, gender, height, weight, and smoking, the decline in FEV_1 was significantly correlated with predicted residual volume and response to bronchodilators. This set of findings is consistent with injury to the airways, including bronchial wall thickening of the large airways, which was seen in CT scans, consistent with the presence of air trapping in the lung, rather than interstitial lung disease, with the cause being exposure to alkaline dusts and other pollutants from the WTC collapse.

Rom *et al.* (2010) reported that FDNY firefighters had significant respiratory symptoms characterised by cough, dyspnea, gastroesophageal reflux, and nasal stuffiness with a significant one-year decline in FVC and FEV_1. Bronchial hyperreactivity correlated with bronchial wall thickening on CT scans. Compared with the NHANES III data for FVC and FEV_1, 32% of 2,000, WTC Dust-exposed residents and cleanup workers were below the lower 5%. The most common abnormality was a low FVC pattern, a finding similar to that also described for individuals in rescue and recovery activities.

Webber *et al.* (2011) examined physician-diagnosed asthma and other respiratory ailments in the prospective NYU cohort of FDNY firefighters. They studied 14,314 firefighters and emergency medical service (EMS) providers who were also employed by FDNY. After exclusions, there were 9,715 male firefighters and 1,228 male EMS workers studied, and their median age was 40 years. There were also 863 retired firefighters (7.9%), who had returned to assist the rescue efforts. Subjects with a history of OAD (such as asthma, COPD, and emphysema) were excluded from employment in FDNY, and furthermore, they had to have a FEV_1 of at least 80% of predicted (70% for EMS workers). Thus, in this group, there were only 85 asthma cases diagnosed prior to 9/11, and no cases of emphysema. The rates were greatest among the retirees, but there was very little difference between EMS workers and firefighters. Those arriving early at Ground Zero reported an MD-diagnosed asthma of 14%, while those arriving later had an asthma rate of 5.9%. This yielded an OR of 3.3, with the same for all respiratory diagnoses. After adjusting for smoking, age, arrival group and duration of exposure to dust, retirees were 10.2 times more likely to have asthma than active men; and they were 7.4 times more likely to have COPD/emphysema. They also reported a link between the quintile of $FEV_1\%$ and self-reported OAD. Thus, the lowest $FEV_1\%$ predicted was 41.1% for men with asthma and 50.6% for COPD/emphysema, while those firefighters in the highest $FEV_1\%$ had only 10.5% with asthma and 6.8% with COPD/emphysema. By contrast, of those firefighters not reporting any respiratory diagnosis (i.e. those without any respiratory conditions), 23.5% were in the highest quintile and 14.6% were in the lowest $FEV_1\%$. EMS workers showed the same pattern. Sinusitis (9.7%) and asthma (8.8%) were the most common physician diagnoses, and those with the longest exposures had the most reported cases of sinusitis (11.4%) and bronchitis (10.3%). After adjusting for smoking, age on 9/11, and arrival group, the retirees had an OR of 10.0 compared with active duty firefighters. Cough, shortness of breath, and wheeze were clearly correlated with lowest quintiles of $FEV_1\%$, whereas those without respiratory symptoms showed the opposite pattern.

Malievskaya *et al.* (2002) described a free medical screening program provided by Queens College for building cleanup workers who did not work at Ground Zero, but were responsible to clean contaminated office and residential buildings in the vicinity. Many of the building workers were day labourers, hired from street corners on a daily or weekly basis, were of Hispanic ethnicity, did not speak English, lacked health insurance, and did not have training working with hazardous materials. From 2002-01-15 to 2002-02-28, 418 building cleanup workers involved in indoor cleaning operations for at least 1 week were examined in a mobile clinic. Workers were also fitted and provided with respirators and provided instructions about potential work hazards and safety practices. Most of the participants worked in the cleanup for 6–12 weeks, and had stopped such work

4–8 weeks prior to examination. Nearly, all of them examined reported current health symptoms that first appeared or worsened after 9/11, including irritation of the airways (cough, sore throat, nasal congestion, and chest tightness) and systemic symptoms (headaches, fatigue, dizziness, and sleep disturbances). Most of those who reported symptoms experienced little or no improvement despite cessation of work 4–8 weeks prior to examination.

11.7.2 *Residential and working community members with WTC Dust exposures*

Several studies documented adverse health effects in the local community. New-onset symptoms in the residential community were described in a field study of community members. A dose-response relationship was described that was related to home conditions (Lin *et al.*, 2005; Reibman *et al.*, 2005).

Analysis of the WTC registry population demonstrated an increase in upper and LRSs, and an analysis of a subgroup showed a dose-response relationship between the persistence of LRSs, lung function, and exposure (Maslow *et al.*, 2012). Most local workers returned to the surrounding offices one week after the event, when Lower Manhattan was officially reopened for business. Some residents closest to the site were evacuated and returned over the ensuing three months. Many residents remained in their apartments and were never evacuated. Although most individuals cleaned their own residence or worksite, formal cleanup of indoor and outdoor commercial sites, and of some residential sites, was performed by workers hired specifically for the activity.

Reibman *et al.* (2009) described physical symptoms in local residents, local workers, and cleanup workers, had been examined at Bellevue Hospital by NYU physicians, and had reported symptoms and exposure to the dust, gas and fumes released with the destruction of the WTC on 11 September 2001. Of the 1,898 individuals who participated between September 2005 and May 2008, upper and LRSs that began after 11 September 2001 and had persisted at the time of examination. The most common abnormality in each exposure category was low FVC, a finding similar to that for individuals involved in rescue and recovery activities. Between October 2007 and March 2008, individuals with normal spirometry and any LRS, defined as cough, shortness of breath, or wheeze, were referred for Mch challenge studies. Of 68 individuals who completed the examination; 51% of them had airway hyperreactivity. Among patients with any of the abnormal spirometry patterns, lung function improved after bronchodilator administration. There was a significant improvement in FEV_1 in patients with the "obstructed" pattern ($p < 0.0001$), and a significant, but small improvement in FEV_1 in patients with a "low FVC" pattern ($p = 0.0003$). Both the FEV_1 and FVC improved in response to bronchodilator in the "obstructed and low FVC" group ($p = 0.0001$). Dyspnea on exertion (DOE) was the most common symptom identified. A DOE score of "3" or more was associated with a "low FVC," or an "obstructed and low FVC" pattern in the total population ($p < 0.0001$ and $p < 0.0004$, respectively) and in the population with a five pack per year tobacco smoking history ($p < 0.005$ and $p < 0.05$, respectively). In summary, spirometry measurements below the lower limit of normal were found in 31% of those studied. Thus, residents and local workers, who were believed to have had less exposure to WTC Dust compared with those with work associated exposure to WTC Dust, also had new and persistent respiratory symptoms with lung function abnormalities five or more years after the WTC Towers' collapses.

11.7.3 *Populations with work-related or residential exposures to WTC Dust based on questionnaire responses*

While rescue, recovery and cleanup workers were usually regarded as the ones most heavily exposed to WTC Dust, other groups could be heavily exposed simply by being submerged in the heavy dust cloud in Lower Manhattan on 11 September 2001 or soon thereafter. While exposures of other Lower Manhattan groups were never measured, they were generally considered to be less intense and/or shorter in duration than those of the rescue, recovery, and cleanup workers. In any case, the short-term and cumulative exposures of these other populations were to the same WTC Dust constituents and NIEHS supported population-based studies to determine if they would be at risk for the same kinds of adverse health responses as the on-site Ground Zero workers.

Trout *et al.* (2002) examined data on the health status of Federal employees working in an office building close to the WTC on 11th September. A cross-sectional assessment of their health symptoms was conducted 3 months post-11 September on a total of 191 NYC employees and 155 employees in an office building in Dallas that served as a comparison population. The NYC and Dallas cohorts were similar with respect to age, gender, race, history of asthma, history of bronchitis, current smoking status, and degree of social support, and only differed significantly with respect to education ($p <$ 0.01, higher educational attainment in Dallas) and history of allergy, hay fever, or eczema ($p < 0.01$, higher rates in Dallas). Employees in NYC reported multiple constitutional symptoms at a higher frequency than Dallas workers, including a 6-fold higher prevalence of shortness of breath (prevalence ratio [PR] = 6.1, and chest tightness (PR = 6.0, a 3.5-fold higher prevalence of wheezing (PR = 3.5, and a 2-fold higher prevalence of cough (PR = 2.1), among others. For most symptoms, the majority of NYC workers reported worsening of the symptom with exposure to the worksite after 11th September.

Bernard *et al.* (2002) reported the results from a CDC survey of physical and mental health symptoms among workers at four NYC schools. Two schools near the WTC site (High School A, reopened to staff on 20 October 2001; college A, reopened to staff on 26 September 2001) were compared with a high school (High School B) and college (College B) that were located five or more miles from the WTC site. The prevalence of self-reported symptoms was significantly higher for the exposed schools compared with the comparison schools ($p < 0.05$), and strengthened further when considering persistent symptoms 4–6 months after 9/11.

Fagan *et al.* (2002) summarised the results of a telephone survey conducted 5–9 weeks after 9/11 among 988 Manhattan residents, aged >18 years, living south of 110th Street, identified by random digit dialling. Of 134 respondents who self-reported having been told by a doctor that they had asthma prior to 11th September, 34 (27%) reported worsening of asthma symptoms after 11th September. Respondents with asthma who lived or were present south of Canal Street on 11th September were more likely than others to report increased asthma symptoms (OR = 1.7). Exposure to smoke/debris causing breathing difficulty was significantly associated with worsening asthma (OR = 4.6), which strengthened after multivariate adjustment for significant pre- and post-11 September psychological predictors and potential confounders (OR = 7.0).

Green *et al.* (2006) examined changes in health care use among residents of NYC using health care claims data from a large insurance provider. Compared with average usage rates in the New York Consolidated Metropolitan Statistical Area (MSA) between January and August 2001 (i.e. prior to 11th September), average rates of healthcare visits among individuals residing within 10 miles of the WTC site increased significantly in the post-11 September period from October to December 2001 for respiratory diseases and conditions including cough (14.6% increase, $p < 0.01$), asthma (4.1% increase, $p < 0.01$), acute upper respiratory infection (6.2% increase, $p < 0.01$), and bronchitis (1.5% increase, $p < 0.01$). In contrast, during the same time period, for individuals residing 10–50 miles from the WTC site, decreases in usage were seen for the same respiratory diseases and conditions, indicating that the observed increases of these conditions among residents within 10 miles of the WTC were not an artifact of generalised trends, such as seasonal differences.

A series of results from a study showed an increase in the incidence of new-onset and persistent upper and LRSs in residents living near Ground Zero compared with residents of a control area (Reibman *et al.*, 2005; Lin *et al.*, 2005, 2007, 2010b). The affected area was located within 1.5 km of Ground Zero, and included 49 buildings with ≈ 9,200 households. The control area had ≈1,000 households from five Upper Manhattan apartment buildings more than 9 km from the WTC. In households with more than four persons, two adult residents and the two oldest residents under age 18 years were asked to complete the individual questionnaire. At ~1 year after 11th September, the cumulative incidence of upper respiratory symptoms was significantly higher in the WTC-affected area compared with the control area, both for new-onset symptoms and new-onset persistent symptoms. Significantly higher rates were also seen in the WTC-affected area for unplanned medical visits for respiratory problems in the prior month, initiation of respiratory medication post-11 September, affected area for unplanned medical visits for respiratory problems in the prior month, initiation of respiratory medication post-11 September, and use of respiratory medications and/or asthma medications in the past four weeks. Self-reported prevalence of shortness of breath with varying levels of exertion was increased in both the affected and control areas when comparing pre-11 September to post-11 September periods. In the post-11 September period, shortness of breath was significantly worse in the affected area, even after adjustment for potential confounders (Lin *et al.*, 2005).

A total of 30.7% of affected area residents reported some physical damage to their home after 11 September, resulting in significantly higher rates of indoor dust and odours in the affected area compared with those in the control area. Residential damage, the presence of dust, and cleaning activities were associated with elevated risks of any new-onset upper and LRSs. The presence of dust in the residence showed the strongest association with any upper respiratory symptom (CIR = 1.35). Risks were further elevated for persistent upper and LRSs. A significant dose–response relationship was observed with duration of dust or odours in the home for new-onset upper respiratory ($p < 0.05$ for trend), new-onset lower respiratory ($p < 0.05$ for trend), new persistent upper respiratory ($p < 0.05$ for trend), and new persistent LRSs ($p < 0.05$ for trend). In a model that combined residential exposure, work exposure and the presence below Canal Street on 11 September, residential exposure was a significant predictor of upper and LRSs, which strengthened further with the addition of workplace and/or 11 September exposures (Lin, 2007). Lin *et al.* (2010b) reported on LRSs after an additional ~3 years of follow-up. The proportion of WTC-affected residents reporting any LRS had declined from 55.5% to 47.8% at 2- and 4-year post-11 September, but remained considerably higher than the post-11 September rates reported among controls (20.0%). Individual symptoms followed similar trends, and objective indicators (e.g. use of medicines for asthma and new diagnoses of lower respiratory disease) showed improvement over time. In WTC-affected residents who did not have a diagnosis of asthma, COPD, chronic bronchitis, or other lung disease prior to 11th September ("previously healthy"), exposure to WTC Dust in the home was associated with elevated and often statistically significant risks of LRSs, at both the 2- and 4-year time points. Similar to the previous study (Lin, 2007), both living and working in the affected area were associated with a higher prevalence of reported LRSs at 2 years compared with only living in the affected area. Comparable analyses examining upper respiratory symptoms at 2 and 4 years did not observe any major differences between the control and affected areas, nor did they find an association with home conditions. However, this was not unexpected, as most upper respiratory symptoms evaluated represent more short-term acute effects rather than long-term chronic effects of environmental exposure.

Lin *et al.* (2010a) used hospital admission records from 1999 to 2001 with a diagnosis of respiratory, cardiovascular, or cerebrovascular illness and a residential address in Lower Manhattan (the "hot zone" and "near zone") or at least 5 miles from the affected area in Queens to compare changes in admission rates before and after 11th September. The control area in Queens was similar to the affected area with respect to population density, race, ethnicity, median household income and poverty status. Hospital records were identified from the NYS Department of Health or NYSDOH Statewide Planning and Research Cooperative System (SPARCS). Respiratory diseases included: chronic bronchitis, emphysema, asthma, COPD, acute bronchiolitis (in children aged 0–4 years only), and acute bronchitis (in children aged 0–4 years only). For asthma, hospital admissions in the affected area were higher than the control area prior to 11th September (PR = 1.31). Rates peaked during the week of 11th–17th September. A drop-off was observed in the subsequent weeks, but admissions returned to higher levels in the affected area in October. Using the period pre-11 September as reference, hospital admissions in the affected area were higher for both respiratory diseases and asthma from 11th September through early-to-mid October. Compared with the average rates in the affected area from 1991 to 2000, hospitalisation rates in 2001 for respiratory diseases and asthma were observed to peak significantly immediately after 11th September, drop off in the subsequent weeks, and then reemerge later at a lower frequency.

Mauer *et al.* (2010a, 2010b, 2010c) described long-term respiratory symptoms in NYS employees who were WTC responders on or after 11 September 2001. They were initially mailed self-administered questionnaires (initial, Year 1 and Year 2) and then they completed a telephone interview in Year 3. WTC exposure was associated with LRS, including cough symptoms suggestive of chronic bronchitis, 5 years post-11 September. When exposure was characterised using an exposure assessment method, the magnitude of effect was greater in those with exposure scores above the mean. WTC exposure was associated with persistence of LRS over the 3-year study period. Participants with the highest exposures were more likely to experience increased severity of their asthma condition and/or LRS. Even in a moderately exposed responder population, lower respiratory effects were a persistent problem 5 years post-11 September.

Li *et al.* (2011) examined new-onset GERS after 11 September 2001 and persisting up to 5–6 years in relation to WTC-related exposures among 37,118 adult (i.e. RRWs, local residents, area workers and passersby in Lower Manhattan) WTC Health Registry enrollees, and potential associations with comorbid asthma and PTSD. Cumulative incidence was 20% for post-11 September GERS and 13% for persistent GERS. Among enrollees with neither asthma nor PTSD, the aRR for persistent GERS was elevated among workers arriving at the WTC pile on 11th September (aRR = 1.6) or working at the WTC site >90d (aRR = 1.6), residents exposed to the intense dust cloud on 11th September (aRR = 1.5) or who did not evacuate their homes (aRR = 1.7), and area workers exposed to the intense dust cloud (aRR = 1.5).

11.7.4 *Summary and conclusions for population-based studies that compared health status of groups with relatively high WTC Dust exposure groups with that of groups with little/no WTC Dust exposure*

Table 11.1 provides a summary of health-related responses reported in the peer-reviewed literature for cohorts studied in clinical settings

Table 11.1.   Summary of reports of excess respiratory and gastroesophageal responses to WTC Dust exposures.

1st author	Year	Population	Clinic location	Respiratory symptoms	Pulmonary function	Asthma	Wheeze	SOB	Sarcoid	GERD	Other responses
A. Occupational exposures											
Landrigan	2004	Resc.-Recov.	MSSM	*			*				
Skloot	2004	Ironworkers	MSSM	*							
Herbstman	2005	Clean-Recov.	MSSM	*			*				
Herbert	2006	Construction	MSSM	*	*						
Buyantseva	2007	Police	MSSM	*							
Wheeler	2007	Resc.-Recov.	MSSM			*					
de la Hoz	2008	Med. Monitor	MSSM	*				*			
Skloot	2009	Med. Monitor	MSSM		*						
Bowers	2010	Rescue work	MSSM						*		
Brackbill	2009	Med. Monitor	MSSM			*					
Crowley	2011a, b	Med. Monitor	MSSM						*		
Jordan	2011	Med. Monitor	MSSM						*		
Wisnivesky	2011	Resc.-Recov.	MSSM	*	*	*				*	
Li	2011	Work	MSSM							*	
Kim	2012	Med. Monitor	MSSM	*		*					
Prezant	2002	Firefighters	NYU	*				*		*	

Banauch	2002	FDNY	NYU	*	*	*	*	*		*	
Banauch	2005a, b	FDNY	NYU	*	*	*			*		RADS, RUDS
Banauch	2006	FDNY	NYU	*							
Izbicki	2007	Firefighters	NYU		*						
Webber	2009	FDNY	NYU	*							
Weiden	2010	FDNY	NYU		*						
Aldrich	2010	FDNY	NYU		*						
Banauch	2010	Firefighters	NYU		*				*		
Weakley	2011	Firefighters	NYU	*							
Aldrich	2013	Firefighters	NYU		*						
Malievskaya	2002	Cleanup work	Queens Coll.	*							
Kleinman	2011a,b	Police	NYU	*	*						
B. Incidental exposures				*							
Lin	2007	Residents	NYU	*							
Reibman	2009	Residents	NYU	*	*						
Rom	2010	FDNY&Resid.	NYU	*	*			*		*	
Friedman	2011	Residents	NYU		*						
Maslow	2012	Residents	NYU	*							
Kazeros	2013	Residents	NYU	*	*	*	*				

(*Continued*)

Table 11.1. (*Continued*)

1st author	Year	Population	Clinic location	Respiratory symptoms	Pulmonary function	Asthma	Wheeze	SOB	Sarcoid	GERD	Other responses
Bernard	2002	School Employ.	None	*							
Trout	2002	Fed. Employ	None	*							
Fagan	2002	Residents	None	*							
Wagner	2005	Residents	None			*				*	
Green	2006	Residents	None	*		*				*	
Lin and Reibman	2005:2007	Residents	None	*		*		*		*	
Laumbach	2009	Residents	None	*			*				
Lin	2010a	Residents	None			*					

at MSSM, NYU, and Queens College, as well as responses gathered through analyses of data on questionnaire responses among groups with incidental exposures in residences and office work locations. Table 11.1A shows statistically significant responses for groups engaged in rescue and recovery activities out-of-doors, while Table 11.1B provides a summary of health-related responses reported in the peer-reviewed literature for other population groups based on responses to exposures that took place mostly in indoor environments. The plausibility of causal relationships between exposure to WTC Dust and an increased risk of the specific respiratory effects for the data collected in clinical settings is strengthened by the evidence of overall respiratory toxicity associated with exposure to WTC Dust for other population groups whose data were collected via questionnaires. Some of the differences in the response patterns in 1.1 may be primarily due to the differences in the year of publication. The ones published prior to 2007, tended to report more of the shorter-term responses, such as cough and respiratory symptoms, while those published in later years more frequently cited declines in pulmonary function and diagnoses for sarcoid that became more clearly evident with increasing time since the substantial initial exposures. Lastly, it is important to indicate to the readers that for each study shown in Table 11.1, what might appear to be selective positive findings is really reflective of outcomes in some categories not being positive simply because these were not studied at the time or reported in that particular paper.

The similarities and differences in the responses in Table 11.1 between those with occupational exposures and those with incidental exposures in residences and offices are notable. Furthermore, the various studies had somewhat different objectives and different strengths in terms of population size and the sensitivities of its assays, as well as the homogeneity of the WTC Dust exposures within the groupings. What is remarkable, in my view, is that the responses to incidental exposures appear to be so similar to those of the groups engaged in rescue and recovery activities that were believed, initially, to have been more heavily exposed. This suggests that duration of exposure to undisturbed dust residues may be as or more important than the shorter-term exposure to higher concentrations.

Temporal sequence is a critical criterion for attributing causation to WTC Dust exposure, since disease cannot precede exposure or be coincidental with it. In all cases discussed above, asthma and other respiratory disease excesses occurred after 11 September 2001, and this was especially true among workers studied without respiratory ailments prior to 11 September as seen in Herbstman *et al.* (2005). Almost all study authors were aware that smoking and age are potential confounding factors and investigators adjusted for them in their analysis. Every investigative team was aware that the WTC Dust was the specific risk factor they were testing for pulmonary effects, and there were multiple health effects found for this exposure including small airways impacts, impaired PFT, the presence of GERD, and the risk for asthma, COPD, and other chronic lung diseases. Overall, the cumulative evidence fits together and has been confirmed authoritatively by the WTC Health Program (Howard, 2014).

The epidemiological studies are consistent and supportive of a causal association between WTC Dust exposure and adverse respiratory, upper digestive, and cardiac effects. Even short-term exposures were associated with increased risks of adverse conditions, and the risks often increased in a dose-related fashion, with more prolonged or more intensive exposures being associated with greater health risks. The evidence of a temporal association between exposures sustained on and following 11th September, and adverse health effects has been strengthened by the availability of pre-11 September information and longitudinal data within the epidemiological database.

The plausibility of a causal link between WTC Dust and the health outcomes has been strengthened by the consistency of the findings outlined in this critical review. Overall, there is sufficient evidence that, to a reasonable degree of scientific certainty, exposure to WTC Dust is causally associated with increased risk of: (a) COPD; (b) RADS; (c) interstitial lung disease; (d) restrictive lung disease; (e) chronic laryngitis/pharyngitis; (f) chronic rhinosinusitis; and/or (g) specific upper digestive disorders, including GERD.

11.8 Insights Gained from *in vivo* and *in vitro* Studies of Biologic Responses to WTC Dusts

11.8.1 *In vivo studies*

While the evidence that the inhalation of WTC Dust produced adverse health effects in people is substantial, as discussed earlier in this review, there has, to date, been little supporting evidence from inhalation exposure studies in laboratory animals. Overall, conventional rodent studies have proven impractical, since particles with diameters >2.5 μm would not penetrate the nasal passages of these obligatory nose-breathing species. To overcome this, Vaughan *et al.* (2014) developed a novel system to deliver the WTC Dusts, which are all in the coarse particle size range, into the lungs of rats. In a study using that system, WTC Dusts were delivered by intratracheal (IT) instillation by Cohen *et al.* (2015). Specifically, they sought to ascertain if the inhaled WTC Dust caused damage *in situ* that modulated the retention, and thus impact of the WTC Dust. In examining rats exposed to WTC Dust, they isolated the rat lungs over a one-year period post-exposure. The WTC Dust induced a significant decrease in the presence of ciliated cells in the airways and an increase in the levels of hyperplastic goblet cells in the lungs. They also showed that these changes were associated with significant prolonged retention of the dusts ($\approx$ 90–95%) over the one-year period of observation. These findings were consistent with observations by McMahon *et al.* (2011), who noted ultrastructural ciliary abnormalities in the lungs of some Ground Zero workers that corresponded to respiratory and ciliary functional abnormalities, including impaired ciliary motility. Among the ultrastructural abnormalities was a disarray of axonemal microtubules and/or axonemes that were replaced by homogeneously dense cores.

11.8.2 *In vitro studies of potential biological mechanisms*

As with the *in vivo* studies, there has been little information to date in the literature on potential mechanisms of effects of the WTC Dusts based on *in vitro* studies. In recent studies, Naveed *et al.*

(2011a, 2011b) and Weiden *et al.* (2012) compared the effects of WTC Dust size on macrophage inflammatory and/or cytokine/chemokine release *in vitro*. In the *in vitro* studies, cytokines formed two clusters, with granulocyte/macrophage colony-stimulating factor (GMCSF) and macrophage-derived cytokine (MDC, CCL22) as a result of WTC_{10-53} and $WTC_{2.5}$. GM-CSF clustered with IL-6 and IL-12 (p70) at baseline, after exposure to WTC_{10-53} and in sera of WTC Dust-exposed subjects ($n = 70$) with WTC lung injury. Similarly, MDC clustered with the chemokines growth-regulated oncogene (GRO; CXCL1) and monocyte chemotactic protein-1 (MCP-1). WTC_{10-53} consistently induced more cytokine release than $WTC_{2.5}$. WTC_{10-53} induced a stronger inflammatory response by the human AM than did $WTC_{2.5}$. This large particle exposure may have contributed to the high incidence of lung injury in those exposed to PM at the WTC site. The authors concluded that as GM-CSF and MDC consistently clustered separately; these chemokines likely had a key role in the ultimate differential cytokine release seen during WTC Dust-induced lung injuries.

11.8.3 *Key insights from toxicology studies*

From these recent *in vivo* and *in vitro* studies, it is clear that the WTC Dusts can cause toxic responses in various cell types found in the lungs. Whether these changes themselves ultimately give rise to many of the observed pulmonary/cardiovascular health effects that have been noted with the still-increasing frequency among FRs and others exposed to WTC Dusts remains to be determined. Ongoing studies in rodent hosts may provide critical proof that the WTC Dusts were actual causative agents for the noted pathologies. It is also possible, as noted by Cohen *et al.* (2015) in the studies of dust effects on ciliated lung cells, that the WTC Dusts were ultimately contributive to rather than directly causative of many of the reported health effects. This is too the subject of a series of ongoing studies using co-exposure of rodent models to WTC Dusts and major co-pollutants that were also heavily present in the air at Ground Zero, i.e. combustion-related $PM_{2.5}$.

While definitive mechanisms of effect for the WTC Dusts are still not yet fully defined, recent studies have nevertheless built on the above-noted findings and have begun to identify biomarkers of potential health alterations due to exposure to the WTC Dusts. Using serum collected from, and non-invasive measures to examine FRs and others who were exposed at Ground Zero, novel biomarkers of potential lung injury, altered cardiovascular status, and dysregulated inflammatory status have been identified. These have included significant changes in expression of matrix metalloproteinases (MMPs) (i.e. MMP-2, -3 and -12) (Kwon *et al.*, 2013; Nolan *et al.*, 2014), serum immunoglobulin (IgA and IgG) (Ferrier *et al.*, 2011), select cytokines and/or chemokines (IL-6, IL-8, GRO, GM-CSF, granulocyte CSF [G-CSF], and interferon-inducible protein-10 or IP-10) (Nolan *et al.*, 2012; Cho *et al.*, 2014), cardiovascular disease (CVD) markers (apolipoprotein-AII, C-reactive protein, and macrophage inflammatory protein-4) (Weiden *et al.*, 2013; Schenck *et al.*, 2014) that could be correlated with increased risks for CVD or lung injury.

In conclusion, while there has been some significant progress in the identification of potential mechanisms of toxicity for the WTC Dusts as well as in biomarkers of pulmonary/CVDs associated with exposure, it is clear that much more toxicological research with regard to the effects of *in vivo* and *in vitro* exposures to the WTC Dusts is needed. While many immune-system-related endpoints have been examined, it remains unclear if the WTC Dusts were active immunomodulants that allowed nascent diseases (i.e. cancers, auto-immune diseases and asthma) in hosts to flourish when they might not have otherwise. Lastly, there is an unmet need for development, identification and validation of non-invasive methods to quantitate the remaining dust burdens in the lungs of WTC Dust-exposed workers and residents.

11.8.4 *Possible roles of minor mass components as potential causal factors for observed health effects from WTC Dusts*

The initial concerns about possible adverse health effects that might occur, especially in relation to chronic health effects and in relation to cancer in particular were focused on known toxicants and carcinogens that were present at relatively low concentration levels, i.e. asbestos fibres, transition metals (e.g. Cd, Cr and Fe), heavy metals (e.g. Pb, Hg and As), combustion products, and ambient air $PM_{2.5}$. Due to these initial concerns, most of the subsequent air quality analyses both indoors and outdoors involved monitoring of these components based on $PM_{2.5}$ samples. However, there were, at least in retrospect, at least two good reasons why these initial concerns were misplaced. First, the concentrations of these components within the $PM_{2.5}$ fraction were initially quite low, and then had declined markedly over time. Second, most of the excess disease incidences found in the epidemiological studies weren't those most closely associated with these components of the $PM_{2.5}$ samples that were collected. By contrast, the three major mass components of the settled WTC Dust that were present in particles >10 μm are all known irritants, with the highly alkaline cement and gypsum components being chemical irritants, and the SVFs being physical irritants. In any case, their concentrations in PM_{10} and larger particle size fractions were seldom measured. Furthermore, there is a good reason to expect a biological response to these components that were inhaled, based on their interaction following airway deposition. The physical irritation induced by the crushed SVFs is viewed as providing greater access of the alkaline components to the epithelial cells. The chronic diseases in the exposed populations are quite plausibly caused by either the initially high-level exposures of the rescue and recovery workers during the first week, or by the prolonged periods of indoor exposures to resuspended residual dusts of high alkalinity for cleanup workers, residents and commercial workers who were exposed to WTC Dusts that had not been neutralised by rainwater.

The misplaced initial focus on trace-level components within the $PM_{2.5}$ particle size fraction of the WTC Dusts was never reconsidered by the governmental agencies or by those responsible for controlling the indoor exposures to WTC Dust in Lower Manhattan. Many thousands of $PM_{2.5}$ air measurements were made at Ground Zero, and inside buildings contaminated with WTC Dust during the periods that extended for many months before they were declared clean enough for reoccupancy. None of the routine or special purpose $PM_{2.5}$ sampling filters were ever assayed for their alkalinity or SVF content. Some settled dust samples collected in Lower Manhattan were analysed for their percentage of SVFs, and generally showed that 20–25% of the settled dust was SVF. SVF and asbestos fibres are very different in terms of the toxic effects that they are likely to produce when inhaled at concentration normally encountered (Lippmann, 2014a). First, they differ in terms of fibre diameter, which determines aerodynamic diameter. Asbestos are almost always present in the air in fibres having respirable aerodynamic diameters, while conventional SVF have aerodynamic diameters >10 μm, and such fibres are both less likely to be dispersed into the air, and less likely to penetrate into thoracic airways following inhalation. Second, SVF dissolve more rapidly in the fluid linings of the epithelia and to break up into shorter length segments than asbestos fibres. Third, asbestos fibres, and especially chrysotile fibres, are less rigid than SVF, and therefore less likely to lacerate the surface epithelia.

We know now, from an examination of the recent literature on the associations between WTC Dust exposure and adverse respiratory and gastroesophageal effects, as discussed in this review, that inhalation exposures to resuspended SVF, in the presence of highly alkaline co-contaminants, has caused adverse pulmonary and gastroesophageal effects in humans. It is also now recognised that there is some supporting evidence, as discussed in this critical review, in which some of the same kinds of effects can be caused by inhalation exposures of laboratory animals. Thus, it appears that SVF may be a causal factor, as well as, a signature component of the mixture of concrete, gypsum, and SVF that dominated WTC Dusts.

11.9 Limitations of Air Quality and Settled Dust Measurements Made in Lower Manhattan Following WTC Tower Collapses on 11 September 2001

Most of the airborne dust samples, being limited to $PM_{2.5}$, provided no useful information on the health risk potential of inhaled WTC Dust, since almost all of the WTC Dust was in particles >10 μm in aerodynamic diameter. There were too few PM_{10} samples to permit even a crude estimate of WTC Dust exposure to the lower end of its particle size distribution range, and there were no total suspended PM (TSP) samples that could have, at least crudely, suggested the extent of the WTC Dust inhalation for all of the readily resuspendable WTC Dust (Lee, 2014, Personal Communication).

The ease with which WTC Dust could be resuspended varied with venue and time, that is, being easy for (1) outdoor settled WTC Dust, before the first rain; and (2) indoors within commercial buildings for much longer times, extending until the first successful thorough dust removal over succeeding months.

11.9.1 *Early and prolonged failures in exposure assessment, risk assessment, and risk management for WTC Dusts*

It is relatively easy, in hindsight, to fault the governmental agencies that launched investigations into WTC Dust exposures and their potential health effects in terms of the early lack of their: (1) recognition of likely extent and magnitude of inhalation exposures; (2) evaluation of exposures to major mass components of settled dust that were readily resuspended; (3) recognition of very high frequency of cough and other acute responses in relation to likelihood of chronic health effects; (4) application of more appropriate work practice guidelines for prevention of dust resuspension (a major source of inhalation exposures to WTC Dust); and (5) reliance on inappropriate respirators to protect against prolonged exposures to unusually high dust levels. The governmental agencies were faced with an unprecedented disaster with unknown hazards to FRs, recovery workers, and volunteers, as well as a huge number of Lower Manhattan residents and commercial workers with obvious exposures to WTC Dusts and combustion effluents and their reaction products. The agencies, including their staff scientists, had no prior experience or literature to guide them in their well-intentioned efforts to both minimise adverse exposures and their health effects, and at the same time encourage members of the public to not panic and create chaos as efforts were made to get started in a resumption of normal work and lifestyle patterns.

While the deficiencies of the initial agency responses were readily understandable, the obvious lack of recognition that changes in guidance to the public and building owners and managers in Lower Manhattan would be needed, based on early experience, is less so. The best example of a need for more targeted guidance was the widespread and prolonged prevalence of WTC cough among members of the public and the more heavily exposed Ground Zero rescue and recovery workers. The need to determine what components of the WTC Dust were responsible for the cough, and what the prevalence of the cough implied for more chronic health effects, should have been obvious. We should not have had to await the publication of peer-reviewed epidemiology describing excessive pulmonary responses and GERD among WTC Dust-exposed populations to document the need for research or to revise exposure prevention guidelines. In addition, there was no action taken to deal with evidence that the filter canisters of the recommended negative-pressure respirators were, in many cases, overloaded during a workshift, preventing them from providing the protection against dust inhalation that they were intended to provide. To our knowledge, no attempt was made to switch to positive-pressure respirators.

11.9.2 *Discussion of lessons learned in relation to risk assessment, risk management, and risk communication*

Whenever an unprecedented disaster with public health implications occurs in the US, such as Hurricane Katrina in 2005, the Exxon Valdez and BP oil spills, and massive forest fires in western US states, or in foreign crises such as the 2014 outbreak of Ebola in West Africa, we usually find the abilities of Public Agencies to respond to the challenges in effective and timely manners fall at least somewhat short of optimal. Aside from a lack of experience in how best to define the nature and scope of new and different challenge, and how to mobilise existing resources and personnel to limit exposures to people and natural environments on appropriate scales, there is a need to document the effectiveness, or lack thereof, of the actions taken. One consistent lesson we have learned from the past disasters is that there is usually the generic problem of too limited histories of effective coordination of Public Agency responses.

In terms of an attempt to deal with the unprecedented nature and magnitude of WTC Dust exposures of very large numbers of people in Lower Manhattan, there was prompt mobilisation of teams of technical "experts" from many different public agencies, including Federal, State, and Local governmental units. Their on-site activities included the determination of the geographic extent of the WTC Dust dispersion, some of the physical and chemical characteristics of the WTC Dust, and determinations of the airborne concentrations of some WTC Dust components deemed to be risk factors in relation to disease causation, and providing timely information on the perceived health risks and appropriate precautions for limiting future WTC Dust exposures to members of the public and people cleaning WTC Dust contaminated buildings. This included provision of guidance on air quality measurement strategies; what to measure; work practices to limit exposures from resuspended dusts; and respiratory protective devices for limiting inhalation exposures. Unfortunately, the timely responses and guidance that were provided did not prove to be effective at either providing adequate public health protection or of documenting the exposure factors that were causal of the adverse health effects. There is a need to resurrect and update the original tool kit of the hygiene profession and community, whose watchwords are Recognition, Evaluation, and Control, and to institutionalise it within the federal and local governmental agencies.

11.9.3 *Need for a standing interdisciplinary, interagency, and external scientific advisory panel to deal with unanticipated toxicant exposures*

There is a need to anticipate that we are likely to be faced with natural as well as terrorist-sponsored disruptions of infrastructure resources that lead to exposures of the public to chemical, biological, radioactive contaminants and/or mixtures. We could be better prepared for such contingencies by having sets of standing guidelines on contingency plans along the lines of the EPA Science Advisory Board (SAB) Reports on Future Risk (SAB, 1988a), Reducing Risk (SAB, 1988b), and Beyond the Horizon (SAB, 1995). Such a newly constituted Expert Panel Report could provides timely guidance on dealing with occupational and environmental exposures to complex mixtures not previously encountered, with respect to: (1) component concentrations of toxicants that should be monitored; (2) ambient air and personal exposure guidelines; (3) work practices and local exhaust ventilation to minimise exposures; and (4) personal protective devices to be worn when conventional work practices/local exhaust ventilation for minimising exposure are inadequate.

11.9.4 *Implications to risk assessment, risk management, and risk communication*

There were clearly failures to: (1) anticipate the importance of the atypical particle size distribution and chemical composition of the WTC Dust; and (2) recognise the importance of the early reports of high pH of the coarse dust, and/or of widespread persistent cough and other symptoms. These led to misleading communications emanating from Public Agencies, which minimised the potential health risks from exposures to WTC Dust in the period shortly after the collapse of the WTC Towers. As a consequence, both the public and the building owners and managers were never informed about the limitations of the early risk management decisions. We now know that: (1) There was a premature focus on asbestos fibres, trace metals, and organics, such as PAHs, PCBs, and dioxins, as likely causal factors for excessive inhalation exposures and subsequent excesses of preventable diseases; (2) There were premature assurances that airborne exposures were not excessive; (3) Airborne concentrations of the most irritating components, the alkaline coarse particles and SVFs, were not monitored; (4) Adverse respiratory and gastroesophageal health effects observed in people exposed to WTC Dusts were not likely due to $PM_{2.5}$, which was ≈pH 7 and contained most of the asbestos, trace metals and organics that were present at low concentrations. While these agents have been associated with some other adverse health effects, these other effects have occurred in other organs and have only been associated with much higher levels of exposure; (5) Adverse respiratory and gastroesophageal health effects observed in people exposed to WTC Dusts were most likely due to combined effects of a high alkalinity of the concrete and gypsum components and physical irritation attributable to crushed SVF.

The failure of the public agencies to inform the building managers and the public of the consequences of these early risk management decisions, and their inaction in terms of providing revised guidance, which could have minimised later exposures associated with cleanup of WTC Dust in residential and commercial buildings, contributed to at least some of the adverse health effects that did occur. The Public Agencies' failure to adequately identify the health risks to populations exposed to WTC Dusts originated in their failure to consult with an appropriate panel of scientific experts having a broad range of expertise in industrial hygiene, dosimetry, toxicology, epidemiology, and risk assessment that were needed to address the possible health consequences of human exposures to the unique chemical and physical characteristics of WTC Dusts, and to develop adequate guidance on: (1) Appropriate exposure monitoring; (2) Appropriate personal protection for exposure reduction for those living and/or working in Lower Manhattan; and (3) Research needed to identify likely causal factors within the WTC Dusts, and the biological mechanisms that could account for their adverse effects. To be able to provide better risk assessment, risk management, and guidance to the public in the future, there is a need to establish a standing Scientific Committee or Panel that can advise national, statewide and local governments on the management of public health emergencies, including; (1) Monitoring risk factors that can be anticipated to lead to excessive exposures to toxicants;

(2) Guidance on recommendations for effective procedures for minimising toxicant exposures; (3) Maintaining oversight on emerging extent/trends of exposure data and disease incidence; and (4) Issuing timely recommendations for supplemental exposure protection methods and monitoring, and for toxicological research needed to determine factors affecting disease causation.

11.10 Conclusions

The spectacular collapses of the WTC Twin Towers on the morning of 11 September 2001 posed a unique challenge to the US, and especially in terms of the public health risks of hundreds of thousands of people who worked and/or lived in Lower Manhattan. The conversion of the WTC Towers into an enormous cloud of dust that spread throughout Lower Manhattan, and a massive pile of debris at Ground Zero, led to inhalation exposures of an unprecedented nature and amount that was due initially to the dust levels remaining suspended in the air, as well as to the subsequent resuspension of the dust that had settled onto the streets and within the buildings. Within the first few days, the public agencies involved in occupational and environmental risk assessment and risk management sought to help, but were unprepared to provide adequate public health protection in terms of their too limited abilities to:

(1) assess the nature and magnitudes of the risks;
(2) prescribe suitable methods for monitoring the subsequent exposures;
(3) prescribe effective means of minimising the exposures.

Furthermore, the inadequate guidance that the public agencies did provide shortly after 11th September was never adequately reconsidered and revised in light of emerging evidence that there were substantial unanticipated adverse health effects among workers and residents that were attributable to inhalation of components of the WTC Dusts that were not being monitored.

The key findings of this review are:

(1)   The collapses of the WTC Towers on 11 September 2001 created a dense dust cloud that radiated out at very high velocity, creating settled dust deposits, ranging from clearly visible to inches thick on the streets, building exteriors, interior building surfaces, and in building air ducts throughout Lower Manhattan;

(2)   Dust deposition was much lower in other parts of NYC and the adjacent areas;

(3)   The settled WTC Dusts differed in important ways from conventional settled dusts in regard to particle size distributions, chemical composition;

(4)   The WTC Dust was readily redispersed into the ambient air (i.e. by air movement/physical disturbances caused by peoples' activities);

(5)   In terms of particle size distribution of WTC settled dusts, ~1% had <2.5 μm min aerodynamic diameter, 0.3–0.4% ranged from 2.5 μm to 10 μm, ~40% ranged from 10 μm to 53 μm, and 52–63% that did not pass through a screen with 53 μm pores;

(6)   As compared with outdoor settled WTC Dust, the settled dust deposits within buildings were depleted of particles in the upper end of the particle size range;

(7)   In terms of sources of WTC Dusts, 80–90% was attributable to a mixture of SVFs from glass and slag wool insulation, gypsum from wallboards, and cement from accounted for 9–20% and chrysotile asbestos fibres for 0.8–3.0%, with other components having much smaller amounts;

(8)   In terms of particle chemistry, the aqueous solubility of calcium oxide from the cement and calcium sulphate from the gypsum resulted in a very high pH (9–11) for the outdoor settled dust samples collected within the first few days, and even higher pH levels in indoor dust (pH = 12). By contrast, the pH in particles in the fine dust fraction (<2.5 μm) was nearly neutral (pH = 7–8);

(9)   The particle size range of the alkaline dust created by the collapses did not extend down to 2.5 μm;

(10)   Almost all the air monitoring for PM in Lower Manhattan after 11 September 2001 were based on determinations of total gravimetric mass of $PM_{2.5}$, or of specific components of the $PM_{2.5}$ air samples. Thus, there were very few determinations of levels of the major mass components generated by the buildings' collapse, which were in particle sizes >10 μm (SVFs, cement and gypsum);

(11)   Coarse (>2.5 μm) alkaline particles that were inhaled were deposited in conductive airways in the head and LRT tracheobronchial airways, including most that were between 2.5 μm and 10 μm, as well as smaller percentages of particles of 10–30 μm that had not been deposited in the URT airways;

(12)   The high pH of these coarse particles could have overwhelmed the capacity of the conductive airways to maintain the homeostasis that removes debris from airways to the oesophagus by mucociliary clearance due to alkalinity of the surface fluids, thereby inducing acute responses such as cough, chest pain, and other respiratory symptoms, as well as inducing gastroesophageal reflux. The epidemiological literature now shows significant excesses of these responses in residents and workers exposed to WTC Dusts;

(13)   There is overwhelming epidemiologic evidence that individuals working/living in Lower Manhattan on and after 11 September 2001 have also exhibited more chronic disease, such as incident cases/greater prevalence of respiratory and gastroesophageal illness than comparison populations that lived and worked at further distances from Ground Zero;

(14)   The rise of manifestations of adverse acute and chronic respiratory tract and gastroesophageal effects in people working/residing in Lower Manhattan after 11 September 2001, which was not seen in other populations in the region, provides convincing evidence of a causal association attributable to WTC Dust exposures that began on 11th September.

The most plausible causal candidates for inhalation exposures that were not monitored, and were present at extraordinarily high concentrations post-11 September, were coarse particles composed of cement, gypsum and SVFs. All the three components are known irritants to large airways, and the deposition sites for these coarse particles are consistent with locations of the adverse responses. Further, the WTC Dusts, a mixture of all these components, were shown to be unusually easy to resuspend into the air, and the dust that remained inside buildings was not washed away or neutralised by rainwater, unlike the dusts outside of the buildings;

The fact that the concentrations of fine particle toxicants that were monitored were within recognised exposure limits led some people to conclude that exposure to WTC Dust could not have caused the excess disease among workers and residents in Lower Manhattan since 11 September 2001. This conclusion is demonstrably false, since exposure limits cited were specified for concentrations of insoluble/poorly soluble dust overall mass, or the mass concentrations of some specific components, while the concentrations measured were limited to those within the $PM_{2.5}$. In fact, having a $PM_{2.5}$ inlet prior to the filter that collected the sample to be analysed precluded measurements of the larger particles in the ambient air that were most likely to be causal;

Since the monitored airborne $PM_{2.5}$ concentrations were too low to have caused clogging of the filters in respirator cartridges during one day, the fact the respirators often became overloaded provided evidence that the respirator filters were being clogged by coarse particles at much higher concentrations, that is, by particles that could be deposited in URT and tracheobronchial airways;

It is clear that, even almost 14 years on from the WTC disaster, much remains to be done to not only identify how exposures to WTC Dusts may have contributed to diseases/adverse health effects in those who were at/near Ground Zero, but to also devise methodologies to ascertain who is still healthy but at risk for developing exposure-related lung, cardiovascular, and other pathologies.

Acknowledgements

I hereby acknowledge ongoing support services provided by the New York University, and by the National Institute of Environmental Health Sciences under an NIEHS Center Grant (ES 000260).

References

Albert, R. E., Lippmann, M. and Briscoe, W. (1969). The characteristics of bronchial clearance in humans and the effects of cigarette smoking, *Archives of Environmental Health,* **18**, 738–755.

Aldrich, T. K., Gustave, J., Hall, C. B., Cohen, H. W., Webber, M. P., Zeig-Owens, R., Cosenza, K., Christodoulou, V., Glass, L., Al-Othman, F., Weiden, M. D., Kelly, K. J. and Prezant, D. J. (2010). Lung function in rescue workers at the World Trade Center after 7 years. *New England Journal of Medicine,* **362**, 1263–1272.

Aldrich, T. K., Ye, F., Hall, C. B., Webber, M. P., Cohen, H. W., Dinkels, M., Cosenza, K., Weiden, M. D., Nolan, A., Christodoulou, V., Kelly, K. J. and Prezant, D. J (2013). Longitudinal pulmonary function in newly hired, non-World Trade Center-exposed fire department City of New York firefighters: The first 5 years. *Chest Journal,* **143**, 791–797.

ATSDR (2002). New York Department of Health and Mental Hygiene and Agency for Toxic Substances and Disease Registry. Final Technical Report of the Public Health Investigation to Assess Potential Exposures to Airborne and Settled Surface Dust in Residential Areas of Lower Manhattan. Agency for Toxic Substances and Disease Registry, US Department of Health and Human Services, Atlanta, GA (http://www.atsdr.cdc.gov/asbestos/asbestos WTC.html).

Banauch, G., Hall, C., Weiden, M., Cohen, H., Aldrich, T. and Christodoulou, V. (2006). Pulmonary function after exposure to the World Trade Center collapse in the New York City Fire Department. *American Journal of Respiratory and Critical Care Medicine,* **174**, 312–319.

Banauch, G., McLaughlin, M., Hirschhorn, R., Corrigan, M., Kelly, K. and Prezant, D. (2002). Injuries and illnesses among New York City Fire Department rescue workers after responding to the World Trade Center attacks, *Morbidity and Mortality Weekly Report MMWR,* **51**, 1–5.

Banauch, G. I., Alleyne, D., Sanchez, R., Olender, K., Cohen, H. W., Weiden M. *et al.* (2003). Persistent hyperreactivity and reactive airway dysfunction in firefighters at the World Trade Center, *American Journal of Respiratory and Critical Care Medicine,* **168**, 54–62.

Banauch, G. I., Dhala. A., Alleyne, D., Alva. R., Santhyadka. G., Krasko A, *et al.* (2005a). Bronchial hyperreactivity and other inhalation lung injuries in rescue/recovery workers after the World Trade Center collapse. *Critical Care Medicine,* **33**, S102–S1026.

Banauch, G. I., Dhala, A. and Prezant, D. J. (2005b). Pulmonary disease in rescue workers at the World Trade Center site, *Current Opinion in Pulmonary Medicine,* **11**, 160–168.

Banauch, G. I., Brantly, M., Izbicki, G., Hall, C., Shanske, A., Chavko, R., Santhyadka, G., Christodoulou, V., Weiden, M. D. and Prezant, D. J. (2010). Accelerated spirometric decline in New York City firefighters with a1-antitrypsin deficiency. *CHEST Journal,* **138**(5), 1116–1124.

Bernard, B. P., Baron, S. L., Mueller, C. A., Driscoll, R. J., Tapp, L. C., Wallingford, K. M. and Tepper, A. L. (2002). Impact of September 11 attacks on workers in the vicinity of the World Trade Center–New York City, *Morbidity and Mortality Weekly Report MMWR,* **51**, 8–10.

Brackbill, R. M., Hadler, J. L., DiGrande, L., Ekenga, C. C., Farfel, M. R. and Friedman S. (2009). Asthma and posttraumatic stress symptoms 5 to 6 years following exposure to the World Trade Center terroristattack. *Journal of the American Medical Association,* **302**, 502– 516.

Briant, J. K., Lippmann, M. (1992). Particle transport through a hollow canine airway cast by high-frequency oscillatory ventilation, *Experimental Lung Research,* **18**, 385–407.

Briant, J. K. and Lippmann, M. (1993). Aerosol bolus transport through a hollow airway cast by steady flow in different gases, *Aerosol Science and Technology,* **19**, 27–39.

Bowers, B., Hosni, S. and Gruber, B. L. (2010). Sarcoidosis in World Trade Center rescue workers presenting with rheumatologic manifestations. *Journal of Clinical Rheumatology,* **16**, 26–27.

Buyantseva, L. V., Tulchinsky, M., Kapalka, G. M., Chinchilli, V. M., Qian, Z. and Gillio R. (2007). Evolution of lower respiratory symptoms in New York police officers after 9/11: A prospective longitudinal study. *Journal of Occupational and Environmental Medicine,* **49**, 310–317.

Chatfield, E. J. and Kominsky, J. R. (2001). Summary report: Characterization of particulate found in apartments after destruction of the world trade center. Requested by "GZ" Elected Officials Task Force comprised of: U.S. Congressman Jerrold Nadler, Manhattan Borough President Virginia Fields, and 7 other New York State and New York City Officials. Available at http://www.eqm.com.

Chen, L. and Thurston, G. (2002). World Trade Center cough, *Lancet* **360**(suppl.), 37–38.

Cho, S., Echevarria, G., Kwon, S., Naveed, B., Schenck, E. J., Tsukiji, J. *et al.* (2014). One airway: Biomarkers of protection from upper and lower airway injury after WTC Dust exposure, *Respiratory Medicine,* **108**, 162–170.

Cohen, M. D., Vaughan, J. M., Garrett, B., Prophete, C., Horton, L., Sisco, M. *et al.* (2015). Impact of acute exposure to WTC dust on ciliated and goblet cells in lungs of rats, *Inhalation Toxicology,* **27**(7), 354–361.

Crowley, L. E., Herbert, R., Moline, J. M., Wallenstein, S., Shukla, G., Schechter, C., Skloot, G. S., Udasin , I., Luft, B. J., Harrison, D., Shapiro, M., Wong, K., Sacks, H. S. and Teirstein, A. S. (2011a). Sarcoid like granulomatous pulmonary disease in World Trade Center disaster responders. *American Journal of Industrial Medicine,* **54**(3), 175–184.

Crowley, L. E., Herbert, R., Moline, J. M., Wallenstein, S., Shukla, G., Schechter, C., Skloot, G. S., Udasin , I., Luft, B. J., Harrison, D., Shapiro, M., Wong, K., Sacks, H. S. and Teirstein, A. S. (2011b). Response to Dr. Reich's letter:"'Sarcoid-like'granulomatous pulmonary disease in world trade center disaster responders: Influence of incidence computation methodology in inferring airborne dust causation". *American Journal of Industrial Medicine,* **54**(11), 894–895.

de la Hoz, R. E., Christie, J., Teamer, J. A., Bienenfeld, L. A., Afilaka, A. A. and Crane M. (2008). Reflux symptoms and disorders and pulmonary disease in former World Trade Center rescue and recovery workers and volunteers. *Journal of Occupational and Environmental Medicine,* **50**, 1351–1354.

EPA (2002). Exposure and human health evaluation of airborne pollution from the World Trade Center. EPA/600/P-2/002A, National Center for Environmental Assessment, Office.

Fagan, J., Galea, S., Ahern, J., Bonner, S. and Vlahov D. (2002). Self-reported increase in asthma severity after the September 11 attacks on the World Trade Center, Manhattan, New York, 2001, *Morbidity and Mortality Weekly Report,* **51**, 781–784.

Ferrier, N., Nolan, A., Naveed, B., Rom, W. N., Comfort, A. L., Prezant, D. J., Weiden, M. D. (2011). Low serum IgA and IgG4 levels predict accelerated decline in lung function of WTC Dust-exposed firefighters, *American Journal of Respiratory and Critical Care Medicine,* **183**, A4773.

Friedman, S. M., Maslow, C. B., Reibman, J., Pillai, P. S., Goldring, R. M., Farfel, M. R. and Berger, K. I. (2011). Case–control study of lung function in World Trade Center Health Registry area residents and workers. *American Journal of Respiratory and Critical Care Medicine*, **184**(5), 582–589.

Green, D. C, Buehler, J. W., Silk, B. J., Thompson, N. J., Schild, L. A., Klein, M. and Berkelman, R. L. (2006). Trends in healthcare use in the New York City region following the Terrorist Attacks of 2001, *Biosecurity and Bioterrorism*, **4**, 263–275.

Herbert, R., Moline, J., Skloot, G., Metzger, K., Baron. S. and Luft, B. (2006). The World Trade Center disaster and the health of workers: FIve year assessment of a unique medical screening program. *Environmental Health Perspectives*, **14**, 1853–1858.

Herbstman, J. B., Frank, R., Schwab, M., Williams, D., Samet, J., Breysse, P. *et al.* (2005). Respiratory effects of inhalation exposure among workers during the clean-up effort at the World Trade Center disaster site, *Environmental Research*, **99**, 85–92.

Howard, J. (2014). *WTC Health Program, Health Conditions Medically Associated with World Trade Center-Related Health Conditions* (NIOSH, Washington, DC).

Jordan, H. T., Stellman, S. D., Prezant, D., Teirstein, A., Osahan, S.S. and Cone J. E. (2011). Sarcoidosis diagnosed after September 11, 2001, among adults exposed to the World Trade Center disaster. *Journal of Occupational and Environmental Medicine*, **53**, 966–974.

Kazeros, A., Maa, M. T, Patrawalla, P., Liu, M., Shao, Y., Qian. M. *et al.* (2013). Elevated peripheral eosinophils are associated with new-onset and persistent wheeze and airflow obstruction in world trade center-exposed individuals, *Journal of Asthma*, **50**, 25–32.

Kim, H., Herbert, R., Landrigan, P., Markowitz, S. B., Moline, J. M., Savitz, D. A. *et al.* (2012). Increased rates of asthma among World Trade Center disaster responders, *American Journal of Industrial Medicine*, **55**, 44–53.

Kleinman, E. J., Cucco, R. A., Martinez, C., Romanelli, J., Berkowitz, I., Lanes, N. *et al.* (2011a). Erratum: Pulmonary function in a cohort of New York City Police Department emergency responders since the 2001 World Trade Center disaster, *Journal of Occupation and Environmental Medicine*, **53**, 1086.

Kleinman, E. J., Cucco, R. A., Martinez, C., Romanelli, J., Berkowitz, I. and Lanes, N. *et al.* (2011b). Pulmonary function in a cohort of New York City Police Department emergency responders since the 2001 World Trade Center disaster, *Journal of Occupation Environmental Medicine*, **53**, 618–626.

Kwon, S., Weiden, M. D., Echevarria, G. C., Comfort, A. L., Naveed, B., Prezant, D. J. *et al.* (2013). Early elevation of serum MMP-3 and MMP-12 predicts protection from World Trade Center lung injury in New York City firefighters: A nested case-control study, *Plos One*, **8**, e76099.

Landrigan, P. J., Lioy, P. J., Thurston, G., Berkowitz, G., Chen, L. C. and Chillrud, S. N. (2004). The NIEHS World Trade Center Working Group. Health and environmental consequences of the World Trade Center disaster. *Environmental Health Perspectives*, **112**, 731–739.

Laumbach R. J., Harris, G., Kipen, H. M., Georgopoulos, P., Shade, P., Isukapalli, S. S., Efstathiou, C., Galea, S., Vlahov, D. and Wartenberg, D. (2009). Lack of association between estimated World Trade Center plume intensity and respiratory symptoms among New York City residents outside of Lower Manhattan. *American Journal of Epidemiology*, **170**, 640–649.

Lee, R. J. (2014). Personal communication of technical analyses performed by RJ Lee Group, Inc. Monroeville, PA, 2002–2004 for a commercial building impacted by WTC Dust.

Levin, S., Herbert, R., Skloot, G., Szeinuk, T., Fishler, D., Milek, D. *et al.* (2002). Health effects of World Trade Center site workers, *American Journal of Industrial Medicine*, **42**, 545–547.

Li, J., Brackbill, R. M., Stellman, S. D., Farfel, M. R., Miller-Archie, S. A., Friedman, S. *et al.* (2011). Gastroesophageal reflux symptoms and comorbid asthma and posttraumatic stress disorder following the 9/11 terrorist attacks on World Trade Center in New York City, *American Journal of Gastroenterology*, **106**, 1933–1941.

Lin, S., Reibman, J., Bowers, J. A., Hwang, S., Hoerning, A., Gomez, M. L. and Fitzgerald, E. F. (2005). Upper respiratory symptoms and other health effects among residents living near the World Trade Center site after September 11, 2001, *American Journal of Epidemiology*, **162**, 499–507.

Lin, S, Jones, R., Reibman, J., Bowers, J., Fitzgerald, E. F. and Hwang, S. A. (2007). Reported respiratory symptoms and adverse home conditions after 9/11 among residents living near the World Trade Center, *Journal of Asthma*, **44**, 325–332.

Lin, S., Gomez, M. I., Gensburg, L., Liu, W. and Hwang, S. A. (2010a). Respiratory and cardiovascular hospitalizations after the World Trade Center disaster, *Archives Environmental Occupational Health*, **65**, 12–20.

Lin, S., Jones, R., Reibman, J., Morse, D. and Hwang, S.A. (2010b). Lower respiratory symptoms among residents living near the World Trade Center, two and four years after 9/11, *International Journal of Occupational Environmental Health*, **16**, 44–52.

Lioy, P. J., Weisel, C. P., Millette, J. R., Eisenreich, S., Vallero, D., Offenberg, J. *et al.* (2002). Characterization of the dust/smoke aerosol that settled east of the World Trade Center (WTC) in lower Manhattan after the collapse of the WTC 11 September 2001, *Environmental Health Perspectives*, **110**, 703–714.

Lioy, P. J, Pellizzari, E. and Prezant, D. (2006). The World Trade Center aftermath and its effects on health: understanding and learning through human-exposure science, *Environmental Science and Technology*, **40**, 6876–6885.

Lippmann, M. and Albert, R. E. (1969). The effect of particle size on the regional deposition of inhaled aerosols in the human respiratory tract, *American Industrial Hygiene Association Journal*, **30**, 257–275.

Lippmann, M. (1977). Regional deposition of particles in the human respiratory tract, in Lee, D. H. K., Falk, H. L. and Murphy, S. D., (eds.), *Handbook of Physiology, Section 9: Reactions to Environmental Agents* (Bethesda, MD: American Physiological Society), pp. 213–232.

Lippmann, M. and Schlesinger, R. B. (1984). Interspecies comparisons of particle deposition and mucociliary clearance in tracheobronchial airways, *Journal of Toxicology and Environmental Health*, **13**, 441–469.

Lippmann, M. and Chen, L. C. (2007). Particle deposition and pulmonary defense mechanisms, in Rom, W. N. and Markowitz, S. B. (eds.), *Environmental and Occupational Medicine*, 4th Ed. (Philadelphia: Lippincott, Williams & Wilkens), pp. 168–186.

Lippmann, M. (2014). Toxicological and epidemiological studies of cardiovascular effects of ambient air fine particulate matter ($PM_{2.5}$) and its chemical components: Coherence and public health implications, *Critical Reviews in Toxicology*, **44**, 299–347.

Lippmann, M., Cohen, M. D. and Chen, L.-C. (2015). Health effects of World Trade Center (WTC) Dust: An unprecedented disaster's inadequate risk management, *Critical Reviews in Toxicology*, **45**(6), 492–530.

Maciejczyk, P. B., Zeisler, R. I., Hwang, J. S., Thurston, G. D. and Chen, L. C. (2004). Characterization of size-fractionated World Trade Center dust and estimation of relative dust contribution to ambient particulate concentrations, in Jeffrey, S. G. and Nancy, A. M. (eds.), *Urban Aerosols and Their Impacts: Lessons learned from the World Trade Center Tragedy*, Chapter 7, Vol. 919, ACS Symposium Series, pp. 114–131.

Malievskaya, E., Rosenberg, N. and Markowitz, S. (2002). Assessing the health of immigrant workers near Ground Zero: Preliminary results of the World Trade Center Day Laborer Medical Monitoring Project, *American Journal of Industrial Medicine*, **42**, 548–549.

Maslow, C. B., Friedman, S. M., Pillai, P. S., Reibman, J., Berger, K. J., Goldring, R. *et al.* (2012). Chronic and acute exposures to the world trade center disaster and lower respiratory symptoms: Area residents and workers, *American Journal of Public Health*, **102**, 1186–1194.

Mauer, M. P. and Cumming, K. R. (2010a). Impulse oscillometry and respiratory symptoms in World Trade Center responders, 6 years post-9/11, *Lung*, **188**, 107–113.

Mauer, M. P., Cummings, K. R. and Carlson, G. A. (2007). Health effects in New York State personnel who responded to the World Trade Center disaster, *Journal of Occupational and Environmental Medicine*, **49**, 1197–1205.

Mauer, M. P., Cummings, K. R. and Hoen, R. (2010b). Long-term respiratory symptoms in World Trade Center responders, *Occupational Medicine (Lond)*, **60**, 145–151.

Mauer, M. P., Herdt-Losavio, M. L. and Carlson, G. A. (2010c). Asthma and lower respiratory symptoms in New York State employees who responded to the World Trade Center disaster, *International Archives of Occupational and Environmental Health*, **83**, 21–27.

McGee, J. K., Chen, L. C., Cohen, M. D., Chee, G. R., Prophete, C., Haykal-Coates, N. *et al.* (2003). Chemical analysis of World Trade Center fine particulate matter for use in toxicologic assessment, *Environmental Health Perspectives*, **111**, 972–980.

McMahon, J. T., Aslam, R. and Schell, S. E. (2011). Unusual ciliary abnormalities in three 9/11 response workers, *Annals of Otology, Rhinology and Laryngology*, **120**, 40–48.

Naveed, B., Weiden, M., Rom, W. N., Prezant, D. J., Comfort, A. L., Chen, Y. *et al.* (2011a). WTC $PM_{2.5}$ stimulates a more intense inflammatory response in human BAL cells than other ambient $PM_{2.5}$ from NYC and surrounding environs, *American Journal of Respiratory and Critical Care Medicine*, **183**, A1158.

Naveed, B., Comfort, A. L., Ferrier, N., Segal., L. N., Kasturiarachchi, K. J., Kwon, S. *et al.* (2011b). WTC Dust induces GM-CSF in serum of FDNY rescue workers with accelerated decline of lung function and in cultured alveolar macrophages, *American Journal of Respiratory and Critical Care Medicine*, **183**, A4770.

Nolan, A., Naveed, B., Comfort, A. L., Ferrier, N., Hall, C. B., Kwon, S. *et al.* (2012). Inflammatory biomarkers predict airflow obstruction after exposure to World Trade Center dust, *Chest*, **142**, 412–418.

Nolan, A., Kwon, S., Cho, S., Naveed, B., Comfort, A. L., Prezant, D. J. *et al.* (2014). MMP-2 and TIMP-1 predict healing of WTC-lung injury in New York City firefighters, *Respiratory Research*, **15**, 5.

Prezant, D. J., Weiden, M., Banauch, G. I., McGuinness, G., Rom, W. N., Aldrich, T. K., Kelly K. J. (2002). Cough and bronchial responsiveness in firefighters at the World Trade Center site, *New England Journal of Medicine*, **347**, 806–815.

Reibman, J., Lin, S., Hwang, S. A., Gulati, M., Bowers, J. A., Rogers, I. *et al.* (2005). The World Trade Center residents' respiratory health study: New-onset respiratory symptoms and pulmonary function, *Environmental Health Perspectives*, **113**, 406–411.

Reibman, J., Liu, M., Cheng, Q., Liautaud, S., Rogers, L., Lau, S. *et al.* (2009). Characteristics of a residential and working community with diverse exposure to World Trade Center dust, gas, and fumes, *Journal of Occupational and Environmental Medicine*, **51**, 534–441.

Rom, W. N., Reibman, J., Rogers, L., Weiden, M. D., Oppenheimer, B., Berger, K. *et al.* (2010). Emerging exposures and respiratory health — World Trade Center dust, *Proceedings of the American Thoracic Society*, **7**, 142–145.

SAB. (1988a). *Future Risk: Research Strategies for the 1990's.* SAB-EC-88–040. (Science Advisory Board, U.S. Environmental Protection Agency, Washington, DC).

SAB. (1988b). *Reducing Risk: Setting Priorities for Environmental Protection.* SAB-EC-90–021 (Science Advisory Board, U.S. Environmental Protection Agency, Washington, DC).

SAB. (1995). *Beyond the Horizon: Using Foresight to Protect the Environmental Future.* EPA-SAB-EC-95–007 (Science Advisory Board, U.S. Environmental Protection Agency, Washington, DC).

Schenck, E. J., Echevarria, G. C., Girvin, F. G., Kwon, S., Comfort, A. L., Rom, W. N. *et al.* (2014). Enlarged pulmonary artery is predicted by vascular injury biomarkers and is associated with WTC-lung injury in exposed fire fighters: A case-control study, *British Medical Journal*, **4**, e005575.

Schlesinger, R. B. and Lippmann, M. (1978). Selective particle deposition and bronchogenic carcinoma, *Environmental Research*, **15**, 424–431 (1978).

Siebert, W. J. (1942). Fiberglas health hazard investigation, *Industrial Medicine*, **11**, 6–9.

Skloot, G., Goldman, M., Fishler, D., Goldman, C., Schector, C., Levin, S. and Teirstein, A. (2004). Respiratory symptoms and physiologic assessment of ironworkers at the World Trade Center disaster site. Chest, **125**, 1248–1255.

Skloot, G. S., Schechter, C. B., Herbert, R., Moline, J. M., Levin, S. M., Crowley, L. E., Luft, B. J., Udasin, I. G. and Enright, P.L. (2009). Longitudinal assessment of spirometry in the world trade center medical monitoring program. *Chest*, **135**, 492–498.

Stokholm, J., Norn, M. and Schneider, T. (1982). Ophthamologic effects of man-made mineral fibers, *Scandinavian Journal of Work, Environment & Health*, **8**, 185–190.

Trout, D., Nimgade, A., Mueller, C., Hall, R. and Earnest, G. S. (2002). Health effects and occupational exposures among office workers near the World Trade Center disaster site, *Journal of Occupational Environmental Medicine*, **44**, 601–605.

USGS. (2002). September 11, 2001: Studying the Dust from the World Trade Center Collapse. USGS Fact Sheet F#-050–02. United States Geological Survey.

Vaughan, J. M., Garrett, B. J., Prophete, C., Horton, L., Sisco, M., Soukup, J. M. *et al.* (2014). A novel system to generate WTC dust particles for inhalation exposures, *Journal of Exposure Science and Environmental Epidemiology*, **24**, 105–112.

Wagner, V. L., Radigan, M. S., Roohan, P. J., Anarella, J. P. and Gesten, F. C. (2005). Asthma in Medicaid managed care enrollees residing in New York City: Results from a post-World Trade Center disaster survey. *Journal of Urban Health*, **82**, 76–89.

Weakley, J., Webber, M. P., Gustave, J., Kelly, K., Cohen, H. W., Hall, C. B. and Prezant, D. J. (2011). Trends n respiratory diagnoses and symptoms of firefighters exposed to the World Trade Center disaster: 2005–2010. *Preventive Medicine,* **53**(6), 364–369.

Webber, M. P., Gustave, J., Lee, R., Niles, J. K., Kelly, K., Cohen, H. W. and Prezant, D. J. (2009). Trends in respiratory symptoms of firefighters exposed to the World Trade Center disaster: 2001–2005, *Environmental Health Perspectives,* **117**, 975–980.

Webber, M. P., Glaser, M. S., Weakley, J., Soo, J., Ye, F. and Zeig-Owens, R. (2011). Physician-diagnosed respiratory conditions and mental health symptoms 7–9 years following the World Trade Center disaster. *American Journal of Industrial Medicine,* **54**, 661–671.

Weiden, M. D., Ferrier, N., Nolan, A., Rom, W. N., Comfort, A., Gustave, J. *et al.* (2010). Obstructive airways disease with air trapping among firefighters exposed to World Trade Center dust, *Chest,* **137**, 566–574.

Weiden, M. D., Naveed, B., Kwon, S., Segal, L. N., Cho, S. J., Tsukiji, J. *et al.* (2012). Comparison of WTC dust size on macrophage inflammatory cytokine release *in vivo* and *in vitro, PLoS One,* **7**, e40016.

Weiden, M., Naveed, B., Kwon, S., Cho, S. J., Comfort, A. L., Prezant, D. J. *et al.* (2013). Cardiovascular biomarkers predict susceptibility to lung injury in World Trade Center dust-exposed firefighters, *European Respiratory Journal,* **41**, 1023–1030.

Wheeler, K., McKelvey, W., Thorpe, L., Perrin, M., Cone, J., Kass, D. Farfel, M., Thomas, P. and Brackbill, R. (2007). Asthma diagnosed after 11 September 2001 among rescue and recovery workers: Findings from the World Trade Center Health Registry. *Environmental Health Perspectives,* **115**, 1584–1590.

Wisnivesky, J. P., Teitelbaum, S. L., Todd, A. C., Boffetta, P., Crane, M., Crowley, L. *et al.* (2011). Persistence of multiple illnesses in World Trade Center rescue and recovery workers: A cohort study, *Lancet,* **378**, 888–897.

Yiin, L.-M., Millette, J. R., Vette, A., Ilacqua, V., Quan, C., Gorczynski, J. *et al.* (2006). Comparisons of the dust/smoke particulate that settled inside the surrounding buildings and outside on the streets of southern New York City after the collapse of the World Trade Center, September 11, 2001, *Journal of the Air & Waste Management Association,* **54**, 515–528.

Chapter 12

North Atlantic Volcanic Ash (2010): Contemporary Social Vulnerability to a Natural Event

Baerbel Langmann

Institute of Geophysics, University of Hamburg, Germany

12.1 Introduction

Volcanic eruptions cover a wide range of explosiveness, from passive degassing to Plinian eruptions. Worldwide, 50–70 volcanoes erupt throughout a year (Simkin and Siebert, 1994; Global Volcanism Program, http://www.volcanic.si.edu). On average, at least one large eruption with a volcanic explosivity index (VEI) greater than four occurs annually. VEI is a relative measure of the explosiveness of a volcanic eruption defined in Newhall and Self (1982).

Volcanic ash, which is generated by fragmentation processes of the magma and the surrounding rock material of a volcanic vent (Sparks *et al.*, 1997; Zimanowski *et al.*, 2003), represents a major product of volcanic eruptions (Rose and Durant, 2009; Sparks *et al.*, 1997). Dependent on magma composition and viscosity, eruption conditions and the availability of volatiles during the eruption, the production of very fine volcanic ash particles with diameters less than 30 μm varies from a few percent to more than 50% of the total

287

erupted mass flux. The total mass flux consists otherwise of volatiles, coarse ash, lapilli, volcanic bombs and lava. Phreatomagmatic volcanic eruptions (interaction of erupting hot magma with water from e.g. volcanic lakes or glaciers) and eruptions followed by the formation of coignimbrite clouds (secondary convective volcanic plumes out of pyroclastic flows) are especially effective in the generation of fine volcanic ash (Langmann *et al.*, 2012). All fragmentation processes can take place simultaneously and eruption conditions may be highly variable in time.

Depending on the strength of a volcanic eruption, volcanic ash is released into the troposphere or stratosphere (Robock, 2000; Sparks *et al.*, 1997), where the very fine ash particles may be carried for hundreds to thousands of kilometres by the prevailing winds before settling onto land or into the ocean by gravitational sedimentation and wet deposition (Brown *et al.*, 2012). During atmospheric dispersion, such fine volcanic ash particles can affect air quality, reduce visibility and endanger aviation because of failures of jet engines. Additionally, land deposits of volcanic ash can be remobilised into the atmosphere under windy conditions (Leadbetter *et al.*, 2012; Thorsteinsson *et al.*, 2012).

12.2 Icelandic Volcanic Eruptions and Other Sources of Volcanic Ash over the North Atlantic

The spreading of the Eurasian and North American plates at the Mid-Atlantic Ridge generates volcanism on Iceland. Thordarson and Larsen (2007) reconstructed volcanic activity over the past 12,000 years on Iceland and identified 205 eruptive events, about 15–25 eruptions per century (Fig. 12.1). About 120 of these evolved significant explosive activity. The four most active volcanic systems are Grimsvötn, Bardarbunga, Hekla and Katla. Between 1970 and 2010 (prior to the Eyjafjallajökull eruption) seven volcanic eruptions with VEI = 3 occurred on Iceland lasting from about an hour to 1.5 days (Gudmundsson *et al.*, 2012).

At Eyjafjallajökull (63.63°N, 19.62°W, 1666 m a.s.l.), a glacier-covered volcano located in South Iceland, four eruptions have been registered until now, namely in 920, 1612, 1821–1823 and 2010. In

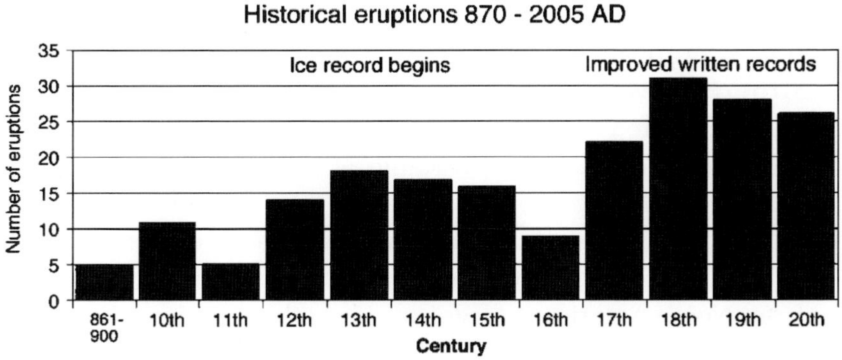

Historical eruptions 870 - 2005 AD

Fig. 12.1.   Volcanic activity over the past 12,000 years on Iceland. Modified form Thordarson and Larsen (2007).

2010, the Eyjafjallajökull eruption on Iceland was one of ten volcanic eruptions worldwide with maximum ash plume heights exceeding 8 km above sea level (http://www.volcano.si.edu). The three phases of the 2010 eruption started with an explosive phreatomagmatic eruption from 14 April until about 18 April accompanied by considerable volcanic ash release and dispersion (Sec. 12.3). The following more effusive eruption phase with lava flows was less ash loaded. During the third phase from about 5 May, the explosive activity and associated volcanic ash release increased again before the continuous eruption stopped on 23 May (Gudmundsson *et al.*, 2012, Langmann *et al.*, 2012). In comparison to previous volcanic eruptions on Iceland, neither the eruption style of the 2010 Eyjafjallajökull eruption is unusual (Davies *et al.*, 2010), nor the duration of the eruption, e.g. during 1821–1823 Eyjafjallajökull erupted for 14 months.

The Grimsvötn eruption on Iceland from 21 May, 2011 to May 28, 2011 was shorter but of greater intensity than the 2010 Eyjafjallajökull eruption, as the plume height reached more than 15 km (Marzano *et al.*, 2013). As the Grimsvötn volcano is covered by the Vatnajökull glacier, the interaction of the ejected hot magma and the melt water form the glacier causes a phreatomagmatic explosion with efficient volcanic ash production. The 2014–2015 eruption of the Bardarbunga volcanic system on Iceland started on 29 August 2014 and lasted until 27 February, 2015. Large amounts of sulphur dioxide were released in near surface air masses during this fissure eruption impacting air quality on Iceland (Gauthier *et al.*, 2016). However, volcanic ash release was negligible so that aviation was unaffected (see Sec. 12.5.3).

In addition to Icelandic volcanic activity, Jan Mayen (~70°N, 8°W), a small volcanic island in the Artic Ocean represents another source of volcanic ash for the North Atlantic. Furthermore, the North Atlantic region may receive volcanic ash and SO_2/sulphate also from North American volcanism, when the eruption plume penetrates at least into the upper troposphere/lower stratosphere. However, volcanic ash from North American volcanism occurs in much lower concentrations over the North Atlantic compared to Icelandic volcanic ash, as most of the ash is already removed from the atmosphere during the long-range transport. One example is the 2008 eruption of Kasatochi during 8–9 August on the Aleutian Islands. Interestingly, a major volcano scientific conference took place in Reykjavik on Iceland shortly after the eruption (IAVCEI 2008 General Assembly), where the American colleagues reported that their flights crossed the SO_2 clouds from Kasatochi — the diluted ash concentrations (Martinsson *et al.*, 2009) were already out of danger for aircraft (see Sec. 12.5.3). Other examples of North American volcanic signals detected during 2004–2011 in Europe are given in Trickl *et al.* (2013), including the volcanic eruptions of e.g. Cleveland, Bezymianny, Shiveluch, Redoubt, Sarychev on the Aleutian Islands. Webley *et al.* (2012) report an ash cloud from the Crater Peak, Mt Spurr eruption in September 1992 in Alaska detected by satellite sensors reaching Greenland. At the time of writing this article, the Aleutian volcano Pavlof erupted (28 March 2016). According to the Montreal Volcanic Ash Advisory Centre (VAAC, see Sec. 12.5.3), the ash cloud was transported over the Canada, reaching the Labrador Sea on 1 April 2016. Further tracing of the ash cloud did not take place because of the low concentration level.

12.3 Volcanic Ash Over the North Atlantic and Europe

Ash layers preserved in historical archives like ice, peat, lake and ocean sediments give indications for the timing of volcanic eruptions, the strength of a volcanic eruption and the dispersion directions. Wastegard and Davies (2009) present an overview of ash layers in historical archives for northern Europe that cover the last 1,000 years. Volcanism on Iceland is characterised as the major source for the volcanic ash, although Jan Mayen is mentioned as well. Ash layers from at least 16 historic volcanic eruptions on Iceland have been traced on Greenland and Spitsbergen, in mainland Norway, Sweden, UK and Ireland, indicating various dispersion directions of volcanic ash released from Iceland, including northern Europe.

Figure 12.2 illustrates the dispersion of volcanic ash as detected from satellite during the eruption of Eyjafjallajökull in spring 2010 according to Gudmundsson *et al.* (2012). During the first phase of the eruption, the ash was transported in a stable jet stream towards

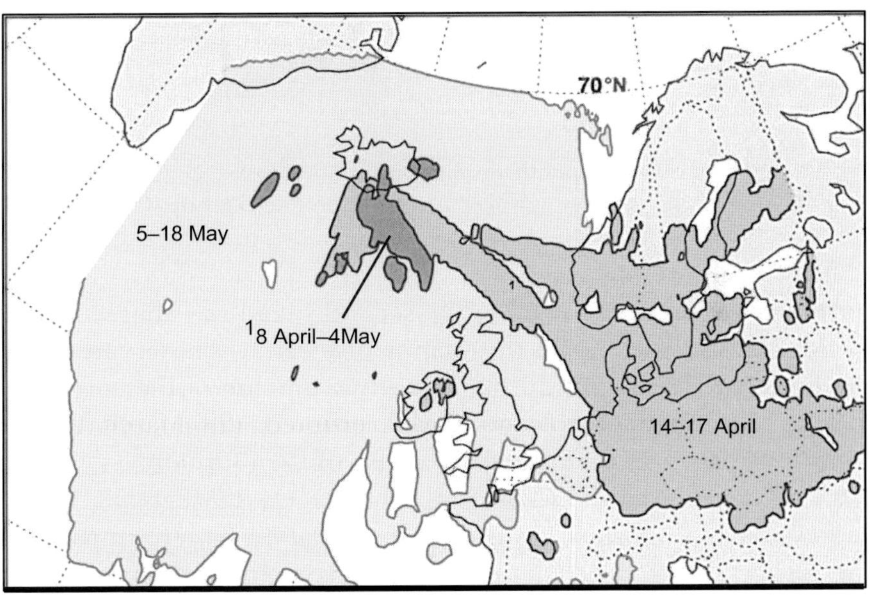

Fig. 12.2. Airborne volcanic ash from SEVIRI retrievals during the eruption of Eyjafjallajökull in spring 2010. Daily ash mass loadings that exceed 0.2 gm^{-2} within a 0.25° × 0.25° grid cell are summed up. Figure modified from Gudmundsson *et al.* (2012). Median grey colour stands for the ash cloud during 14–17 April, dark grey: 18 April–4 May and light grey: 5–18 May.

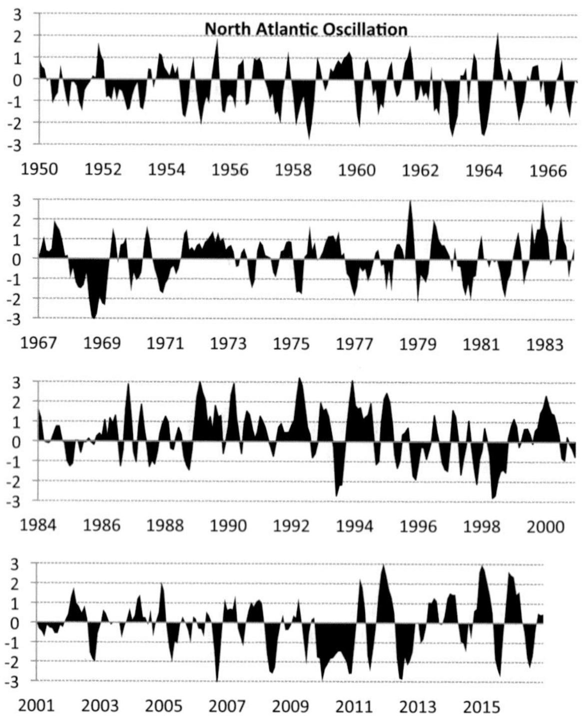

Fig. 12.3.   Standardised 3-month running mean NAO index from 1950 to 2016 (raw data from http://www.cpc.ncep.noaa.gov/products/precip/CWlink/pna/nao_index.html).

northern and central Europe. The second phase was characterised by ash clouds dispersed in the vicinity of Iceland. This is caused by the low eruption height during this effusive phase of the eruption. Large-scale dispersion, all over the Northern Atlantic and Western Europe occurred during the third phase of the eruption.

The North Atlantic Oscillation (NAO) (Marshall *et al.*, 2001) represents an important atmospheric circulation pattern impacting the dispersion of volcanic ash released from Icelandic volcanism. During spring 2010, when Eyjafjallajökull erupted, NAO was in its negative phase (Fig. 12.3), directing transport more strongly to the east and southeast (Christoudias *et al.*, 2012). During the positive NAO phase, when the Icelandic Low and the Azores High are more pronounced, flow over Europe is separated into a northward branch over Northern Europe towards the Arctic and a southward branch in the south. One example for an Icelandic eruption occurring during the positive NAO phase is the 2000 eruption of Hekla. As expected, the ash was transported towards the North (Hunton *et al.*, 2005).

12.4 Impact of Air Pollution

Atmospheric concentrations of volcanic ash in the higher atmosphere in the volcanic plume directly above the crater, but also those in the volcanic cloud exceed thresholds as e.g. defined by EU environmental and health standards for PM10 by far (50 μg m^{-3} as daily average, 2008/50/EG available from http://eur-lex.europa.eu/legal-content/DE/TXT/?uri=celex%3A32008L0050). However, besides differences in toxicity of PM10 from different sources (e.g. urban and biogenic particles) huge uncertainties in atmospheric ash concentrations exist for various reasons: (a) On one hand site, *in-situ* measurements are dangerous for humans, while on the other hand site, it may lead to destruction of the measurement equipment. (b) Remote sensing measurements from satellite or other platforms lack vertical information, so that often only column concentrations are available (e.g. Francis *et al.*, 2012). In addition, the necessary retrieval assumptions introduce huge uncertainties. (c) Model simulation results are associated with their own uncertainties, and cannot be convincingly evaluated against measurements, because of (a) and (b). Nevertheless, based on modelling studies Webster *et al.* (2013) report maximum ash concentrations of at least 4,000 μg m^{-3} during events, when an aircraft encounters problems with volcanic ash clouds, e.g. during the 1982 Galunggung eruption in Indonesia. Nearby maximum concentrations are even simulated in the order of several 100,000 μg m^{-3} indicating that the aircraft luckily missed crossing the main volcanic ash cloud.

Near surface atmospheric concentrations of volcanic ash are highest close to the volcanic eruption. This does not only hold for the period when the eruption is on-going, but may last after the volcanic eruption stops. The reason is that volcanic ash from the ash deposits may be remobilised at higher wind speeds, in particular in arid and semi-arid regions. During such events on Iceland, hourly concentrations may exceed 10,000 μg m^{-3} in near surface air masses (Dagsson-Waldhauserova *et al.*, 2014; Thorsteinsson *et al.*, 2012), with daily averages still up to 1,000 μg m^{-3} PM10 and therefore far beyond healthy conditions.

After long-range transport, the volcanic ash concentrations are strongly reduced. Measurements over continental Europe during the 2010 Eyjafjallajökull eruption allowed tracing the dispersion of the volcanic ash cloud. According to Lidar measurements reported by Ansmann *et al.* (2010), the volcanic ash cloud arrived at Hamburg in the early morning hours of 16 April. It subsided and spread southward on 16–17 April and was entrained into the planetary boundary layer by noon of 17 April. First traces of volcanic ash were observed in Leipzig shortly after midnight on 17 April and at about 18 UTC in Munich (Ansmann *et al.*, 2010). Peak ash concentrations of 1,000 (± 350) μg m^{-3} between 3 km and 4 km height above sea level were found on 16 April shortly after noon at Leipzig and shortly after midnight on 17th April in Munich (Ansmann *et al.*, 2010). The analysis of Gasteiger *et al.* (2011) suggests even higher concentrations at Munich (1,100 μg m^{-3} with an uncertainty range from 650 μg m^{-3} to 1700 μg m^{-3}). At Hohenpeissenberg, southwest from Munich, Flentje *et al.* (2010) estimate peak ash concentrations of 500–750 (±300) μg m^{-3} during 17 April from ceilometer measurements suggesting the possibility of lower ash mass concentration in this region.

Ansmann *et al.* (2010) report a considerable amount of volcanic ash with particle diameters exceeding 20 μm in the ash layers over Leipzig on 16th April whereas other observations suggest smaller particle diameters, e.g. lower than 10 μm at Jungfraujoch on 17/18th April (Baltensperger *et al.*, 2010), between 1 μm–4 μm during the first eruptive phase (Flentje *et al.*, 2010), an effective diameter of 1.5 μm–3.7 μm at Munich on 17 April (Gasteiger *et al.*, 2011) and effective diameters between 3.1 μm–7.5 μm at Leipzig and 2.4 μm and 4.8 μm at Munich on 19th April (Schumann *et al.*, 2011).

Generally, Schumann *et al.* (2011) determined that effective particle diameters decreased from about 5 μm in the fresh ash plume to about 1 μm for plumes aged up to 6 days and no particles beyond 30 μm.

Published volcanic ash particle number concentrations over Europe during the eruption of Eyjafjallajökull are sparse. In Flentje *et al.* (2010) ash particle number concentrations up to about 1 cm^{-3} are reported at Hohenpeissenberg for particles with diameters of 1–2 μm. The airborne measurements on 19th April by Schumann *et al.* (2011) over Leipzig (Munich) resulted in particle number concentrations of 6.3 cm^{-3} for accumulation mode particles (0.25–1 μm in diameter) and 0.3 cm^{-3} for coarse mode particles (2–50 μm). Unfortunately, particle number concentrations for particles in the size range of 1–2 μm in diameter are not available in Schumann *et al.* (2011), although volcanic ash particles in this size range may contribute considerably to the particle number concentration.

12.5 Impact on Society

12.5.1 *Infrastructure, vegetation and soils*

When volcanic ash buries vegetation, photosynthesis and transpiration may be inhibited (Ayres and Delmelle, 2012; Langmann, 2013), thereby altering plant growth and affecting agricultural production. Contrarily, the soil may benefit from the volcanic ash load due to mulching and nutrient supply deriving from volcanic ash.

Important elements of the infrastructure required for the modern society may be affected by volcanic ash, particularly in the vicinity of an erupting volcano. Wilson *et al.* (2012) provide various examples. They report flashovers and subsequent power outage for electricity networks. Roofs may collapse especially when the dry ash is wetted — therefore, it is always recommended to remove the fresh ash from the roofs of the houses. Fresh water supply may be affected leading to drinking water shortages. Reduced visibility due to volcanic ash impacts on ground transportation including airport runways. Even during remobilisation events after the main eruption, airport closures have been necessary because of low visibility and ash cover on airplanes and runways. The impact of volcanic ash on aviation is discussed in detail in Sec. 12.5.3.

12.5.2 *Human health*

Close to a volcanic eruption, elevated atmospheric volcanic ash concentrations may cause acute and/or aggravate pre-existing respiratory and lung diseases, asthma and allergic alveolitis (Horwell and Baxter, 2006; Gudmundsson, 2011). Coarse particles affect the upper airways, whereas fine ash particles are able to penetrate deep into the lungs. However, long-term respiratory effects are not reported in the literature (Gudmundsson, 2011). Usually sharp surface structures of volcanic ash can abrade front of the eye leading to eye irritation. Staying indoors as far as possible, wearing breathing masks and eye protection are easy protection measures in countries of the civilised world, but hardly practicable in tropical/subtropical countries of the developing world. According to Gudmundsson (2011) worldwide over 500 million people live within a potential exposure radius (100 km) around a volcano.

12.5.3 *Aviation*

Volcanic ash poses a risk for aviation for several reasons (Miller and Casadevall, 2000). Air quality in the cabin and the cockpit may be reduced due to volcanic gases and fine ash. Pilot instrumentation, e.g. air speed sensor, can get damaged and fail. Volcanic ash may lead to abrasion and contamination of the aircraft exterior, thereby affecting amongst other objects the cockpit windows, pneumatic techniques and wind leading edges. Most importantly, however, is the threat to the jet engines from volcanic ash, as the melting temperatures of volcanic ash[1] are at or below the operating temperature (~1,400°C) of jet engine turbines (Dunn *et al.*, 1996). When the molten ash is deposited on cooler parts of the engine, it may cause

---

[1]Volcanic ash is made up of minerals and glass in varying compositions, leading to melting temperature from about 650°C to 1200°C (Langmann *et al.*, 2012).

flame-outs through loss of compression. When the aircraft descends after an engine stalls, the increasing atmospheric pressure has allowed in several cases to restart the engines. From the 129 flights affected by volcanic ash during the last 60 years (Guffanti *et al.*, 2010), a temporarily failure of the jet engine turbine due to melting ash is reported for nine flights.

To prevent unsafe flights in the presence of volcanic ash, worldwide nine Volcanic Ash Advisory Centers (VAACs) have been set-up by the International Civil Aviation Organization (ICAO) in cooperation with the World Meteorological Organization (WMO). VACCs have the mandate to issue regular volcanic ash advisories on the location and dispersion of volcanic ash in the atmosphere. They rely on information from local volcanological agencies, pilot reports, satellite observations and dispersion models. The London VACC responsible for Northern Europe, followed the official ICAO zero tolerance guidance at the beginning of the 2010 eruption of Eyjafjallakökull: if for good reasons the presence of volcanic ash in the atmosphere is expected, no flights should take place in that area. In April 2010, a new procedure for aviation security was introduced with three threshold levels for volcanic ash concentrations (EU Commission, 2010; Zehner, 2010). Flights were no longer permitted only in regions where volcanic ash concentrations exceed the threshold value of 2 mg m^{-3}. Although the new maps generated by the London VAAC allowed higher and lower volcanic ash load in the atmosphere to be distinguished, they remained qualitative concerning the absolute mass concentration due to model and source parameter uncertainties. It still remains unclear if the new EU volcanic ash concentration threshold of 2 mg m^{-3}, above which flights would not be permitted, has sufficient technical and scientific justification. Furthermore, it is restricted to Europe.

This new approach stimulated the need to determine absolute mass concentrations of volcanic ash, but this still remains challenging with the currently available instruments, satellite retrievals and numerical models. Although substantial progress in remote sensing of volcanic ash by satellite has been achieved since the volcanic eruption of Eyjafjallajökull in Iceland in 2010, huge uncertainties remain. Prata and Prata (2012) estimated volume mass concentrations of volcanic ash for the 2010 Eyjafjallajökull eruption. Evaluation of volume mass concentrations of volcanic ash showed that it is difficult to determine those values with an uncertainty less than a factor of 2, which has important consequences for threshold values of volcanic ash concentration for aviation safety. Prata and Prata (2012) suggest a logarithmic scale for assigning ash concentration thresholds. In addition, they argue for an ash dosage for the transit through a volcanic ash cloud. Similar dosage thoughts are presented in Clarkson *et al.* (2016), who conducted research on ash concentration tolerances for jet turbines. These studies also emphasise to need to revise the current threshold values for aviation safety.

Although the 2010 Eyjafjallajökull eruption is characterised as an eruption with moderate intensity, it revealed the huge vulnerability of modern societies caused by a natural event (Gudmundsson *et al.*, 2012; Langmann *et al.*, 2012). Airspace over Europe and other countries was closed for six consecutive days in April 2010 and again during May 2010 (Wilson *et al.*, 2012). The disruption of air traffic and the economic consequences were the biggest in history. In April 2010, 108,000 flights were cancelled, 10.5 million passengers were affected and the cost for the airline industry was estimated to exceed $1.7 billion (Eurocontrol, 2010). When taking into account the insurance industry and industries dependent on access to the European markets, the costs are even higher.

According to Parker (2015), the 2010 Eyjafjallajökull eruption represents a cascading crisis, as it developed from an aviation safety crisis to a mobility crisis with huge economic consequences. Due to a lack of pre-existing procedures and planning, the authorities were caught by surprise. Decision making was characterised by high uncertainty, requiring flexibility and ingenuity (Alexander, 2013). As civil aviation is such an important transport mode, the 2010 Eyjafjallajökull ash cloud dramatically illustrated the vulnerability of modern society (Budd, 2011). However, it also revealed the catalytic power of crises as a starting point for necessary reforms (Parker, 2015).

One year later, the effects for aviation during the Grimsvötn volcanic eruption on Iceland in May 2011 were moderate compared to the 2010 Eyjafjallajökull eruption. One reason was the northward transport direction of the volcanic ash (Gudmundsson *et al.*, 2012), which left the European air space widely unaffected and allowed an easily organised redirection of air traffic over the North Atlantic. Another reason is the much better preparation of the authorities one year after the *Eyjafjallajökull crisis* (Parker, 2015). Since 2011, VAAC London has published many alerts for volcanic ash clouds originating mainly from potential Icelandic volcanic eruptions (http://www.metoffice.gov.uk/aviation/vaac/vaacuk_vag.html), but luckily no major volcanic ash event took place after May 2011 in the North Atlantic and European region.

12.6 Lessons Learned and Lessons to Work on

Although no aviation accident occurred during the 2010 Eyjafjallajökull eruption, a proper risk assessment providing suitable guidelines will be necessary (Alexander, 2013; Parker, 2015) to be better prepared for future volcanic ash events affecting Europe. On the policy side, as well as for the scientific understanding, a lot of progress has been achieved since the 2010 Eyjafjallajökull eruption, although further steps are required.

Scientifically, the 2010 Eyjafjallajökull eruption triggered research beyond pure volcanology, to mention in particular remote sensing of volcanic ash and atmospheric dispersion modelling. Numerous scientists, who were not involved in volcanic ash research before, became interested in the topic. Three special issues about the 2010 Eyjafjallajökull eruption were published in the scientific journals *Atmospheric Environment, Journal of Geophysical Research* and *Atmospheric Chemistry and Physics.* In web-of-science the search string 'aviation volcanic ash' returns 120 publications since 2010 out of 174 in total. Searching for Eyjafjallajökull in web-of-science gives in total 430 results of which 410 were published starting with the 2010 Eyjafjallajökull eruption.

The progress in satellite remote sensing is extensively documented in the literature. A lot of work was done to reduce false detection of volcanic ash by satellites, e.g. by taking into account a few additional retrieval wavelengths (Pavolonis *et al.*, 2006, 2013) or by using hyperspectral data in the thermal infrared wavelength region — TIR (Clarisse *et al.*, 2010; Gangale *et al.*, 2010). Further inter-comparisons of the different advanced retrieval algorithms developed in recent years (including retrievals for ground-based measurements) will be necessary to evaluate their strengths and weaknesses, in addition to the often shown comparison with the traditional approach of Wen and Rose (1994). On the other hand, a stronger focus on the volcanic ash cloud with sensors in the microwave region could offer possibilities for the detection of aggregation (clusters of ash/ice particles sticking together due to e.g. collision) (Durant *et al.*, 2009) thus allowing measurement of near-source quantities such as total erupted mass, which represents important input information for long-range transport modelling of volcanic ash and insights into the volcanic processes in the eruption plume.

Because the optical properties of ash are among the most critical parameters in the mass retrieval algorithms for satellite data, an improved knowledge of volcanic ash refractive indices in the TIR wavelength range for different chemical composition is crucially needed. Similar to the work of Sokolik and Toon (1999) on optical properties of mineral dust, ash refractive indices could be estimated based on the volcanic glass and mineral composition. However, besides the variability in composition, the mixing state (external versus internal mixture) and the occurrence of minerals (amorphous versus crystalline forms) may considerably modify optical properties. Nevertheless, a database of refractive indices in the TIR of major components of volcanic ash, would allow choosing different combinations. Thereby, it could replace the data of Volz (1973) and Pollack *et al.* (1973), reduce uncertainties in mass retrieval algorithms for volcanic ash and introduce a physico-chemical basis for the choice of volcanic ash optical properties.

Further work will be necessary on volcanic ash cloud height and thickness to reduce uncertainties in column and volume mass concentrations. Besides associated improvements in temperature and humidity profiles, more accurate retrievals for volume mass concentrations are necessary to contribute to a reliable determination of threshold values for aviation safety using satellite monitoring of volcanic ash.

Finally, an extension of the detection from single-layer volcanic ash clouds to multiple layer clouds (Poulsen *et al.*, 2012) could offer considerable more insight into the dispersion of volcanic ash clouds, in particular during long-lasting volcanic eruptions when aged air-masses loaded with volcanic ash meet fresh volcanic ash air-masses.

In addition to the advisory centre models (Witham *et al.*, 2007) various other dispersion models have been applied during the Eyjafjallajökull volcanic eruptions (Langmann *et al.*, 2012; Folch, 2012). The quality of model simulation results of airborne volcanic ash concentration depends on:

  (i) a realistic description of the source term (eruption rate, start and end of the eruption, column height, vertical distribution of mass, particle size distribution);

  (ii) the quality of meteorological data;

  (iii) atmospheric removal processes (gravitational settling, wet deposition, aggregation processes).

All models differ concerning the definition of the source term, atmospheric removal processes and driving meteorological data, typically provided by different numerical weather prediction models. Although all the models can reproduce volcanic ash dispersions qualitatively well, the quantitative prediction of volcanic ash concentration and thresholds for aviation remains challenging. Measured and modelled volcanic ash concentration may deviate by a factor of two and more (Folch *et al.*, 2012; Heinold *et al.*, 2012; Matthias *et al.*, 2012; O'Dowd *et al.*, 2012; Webster *et al.*, 2012). The reason for the uncertainties in the numerical modelling of atmospheric ash concentrations originates in the poorly constrained quantities and processes necessary for volcanic ash dispersion modelling, like source term variables and removal processes (see above). An expert consensual document has been put together on the 2nd IUGG-WMO Workshop on ash dispersal forecast and civil aviation in Geneva, Switzerland in November 2013 (Bonadonna *et al.*, 2014). Besides a number of various uncertainties mentioned in the consensual document, wet ash aggregation and wet deposition processes have been identified to be insufficiently handled in all ash dispersion models (Bonadonna *et al.*, 2014). The inclusion of robust aggregation models in volcanic ash forecast models is still in a preliminary stage and no operational model considers aggregation yet (Folch, 2012). This omission may lead to model results, which underestimate fine ash fallout and overestimate airborne ash concentrations during long-range transport (Rose and Durant, 2011).

From the policy point of view, evaluations of the risk for aviation by volcanic ash on international level still calls for standardised measures including thresholds of volcanic ash concentrations for safe aviation (Alexander, 2013). European wide the authorities agreed on rules for co-operation during a volcanic ash event (Parker, 2015), and even carried out common exercises, e.g. the April 2011 Grismvötn simulation exercise — shortly before Grismvötn really erupted in May 2011. However, Europe still lacks interlinked European transport system as alternative for air transportation during volcanic ash events (Parker, 2015). This is of particular importance during a long-lasting volcanic eruption affecting Europe to reduce the vulnerability of the mobile society.

References

Alexander, D. (2013). Volcanic ash in the atmosphere and risks for civil aviation: A Study in European crisis management, *International Journal Disaster Risk Science*, **4**, 9–19.

Ansmann, A., Tesche, M., Gross, S. *et al.* (2010). The 16 April 2010 major volcanic ash plume over central Europe: EARLINET lidar and AERONET photometer observations at Leipzig and Munich, Germany, *Geophysical Research Letter*, **37**, L13810.

Ayres, P. M. and Delmelle P. (2012). The immediate environmental effects of tephra emission, *Bulletin of Volcanology*, **74**, 1905–1936.

Baltensperger U., Bukowiecki N., Zieger P. *et al.* (2010). Physical and chemical properties of the volcanic ash aerosol from the Eyjafjoll volcano eruption, Book of Abstracts, *International Aerosol Conference*, August 30–3 September, 2010, Helsinki, Finland.

Bonadonna, C., Webley, P., Hort M., Folch, A., Loughlin, S. and Puempel, H. (2014). 2nd IUGG-WMO Workshop on Ash Dispersal Forecast and Civil Aviation, Geneva, Switzerland, November 2013, Consensual Document, available from: http://www.unige.ch/sciences/terre/mineral/CERG/Workshop2/results-2/2nd-IUGG-WMO-WORKSHOP-CONS-DOC2.pdf.

Brown, R. J., Bonadonna, C. and Durant, A. J. (2012). A review of volcanic ash aggregation, *Physics and Chemistry of the Earth*, **45–46**, 65–78.

Budd, L., Griggs, S., Howarth, D. and Ison, S. (2011). A Fiasco of Volcanic Proportions? Eyjafjallajokull and the Closure of European Airspace, *Mobilities*, **6**, 31–40.

Clarisse, L., Hurtmans, D., Prata, A. J., Karagulian, F., Clerbaux, C., De Mazière, M. and Coheur, P.-F. (2010). Retrieving radius, concentration, optical depth, and mass of different types of aerosols from high-resolution infrared nadir spectra, *Applied Optics*, **49**, 3713–3722.

Christoudias, T., Pozzer, A. and Lelieveld, J. (2012). Influence of the North Atlantic Oscillation on air pollution transport, *Atmospheric Chemistry Physics*, **12**, 869–877.

Clarkson, R. J., Majewicz, E. J. and Mack, P. (2016). A re-evaluation of the 2010 quantitative understanding of the effects volcanic ash has on gas turbine engines, Proceedings of the Institution of Mechanical Engineers, *Journal of Aerospace Engineering*, 0, 1–18, doi:10.1177/0954410015623372.

Dagsson-Waldhauserova, P., Arnalds, O. and Olafsson, H. (2014). Long-term variability of dust events in Iceland (1949–2011), *Atmospheric Chemistry and Physics*, **14**, 13411–13422.

Davies, S. M., Larsen, G., Wastegard, S., Turney, C. S. M., Hall, V. A., Coyle, L. and Thordarson, T. (2010). Widespread dispersal of Icelandic tephra: how does the Eyjafjoll eruption of 2010 compare to past Icelandic events?, *Journal of Quaternary Science*, **25**, 605–611.

Dunn, M. G., Baran, A. J. and Miatech, J. (1996). Operation of gas turbine engines in volcanic ash clouds, *Journal Engineering for Gas Turbines and Power*, **118**, 724–731.

Durant, A. J. and Rose, W. I. (2009). Sedimentological constraints on hydrometeor-enhanced particle deposition: 1992 Eruptions of Crater Peak, Alaska, *Journal of Volcanology and Geothermal Research,* **186**, 40–59.

Eurocontrol (2010). Available from https://www.eurocontrol.int/sites/default/files/content/documents/official-documents/facts-and-figures/statfor/ash-impact-air-traffic-2010.pdf.

EU Commission. (2010). Report on the actions undertaken in the context of the impact of the volcanic ash cloud crisis on the air transport industry, Brussels. Available from: http://ec.europa.eu/transport/doc/ash-cloud-crisis/2010_06_30_volcano-crisis-report.pdf.

Flentje, H., Claude, H., Elste, T., Gilge, S., Koehler, U., Plass-Duelmer, C., Steinbrecht, W., Thomas, W., Werner, A. and Fricke, W. (2010). The Eyjafjallajokull eruption in April 2010-detection of volcanic plume using in-situ measurements, ozone sondes and lidar-ceilometer profiles, *Atmospheric Chemistry and Physics,* **10**, 10085–10092.

Folch, A. (2012). A review of tephra transport and dispersal models: Evolution, current status, and future perspectives, *Journal of Volcanology and Geothermal Research,* **235–236**, 96–115.

Folch, A., Costa, A. and Basart, S. (2012). Validation of the FALL3D ash dispersion model using observations of the 2010 Eyjafjallajökull volcanic ash clouds, *Atmospheric Environment,* **48**, 165–183.

Francis, P. N., Cooke, M. C. and Saunders, R. W. (2012). Retrieval of physical properties of volcanic ash using Meteosat: A case study from the 2010 Eyjafjallajökull eruption, *Journal of Geophysical Research,* **117**, D00U09, doi:201210.1029/2011JD016788.

Gangale, G., Prata, A. J. and Clarisse, L. (2010). The infrared spectral signature of volcanic ash determined from high-spectral resolution satellite measurements, *Remote Sensing Environment,* **114**, 414–425, doi:10.1016/j.rse.2009.09.007.

Gasteiger, J., Gross, S., Freudenthaler, V. and Wiegner, M. (2011). Volcanic ash from Iceland over Munich: Mass concentration retrieved from ground-based remote sensing measurements, *Atmospheric Chemistry and Physics,* **11**, 2209–2223.

Gauthier, P.-J., Sigmarsson, O., Gouhier, M., Haddadi, B. and Moune, S. (2016). Elevated gas flux and trace metal degassing from the 2014–2015 fissure eruption at the Bárðarbunga volcanic system, Iceland, *Journal Geophysical Research,* 1610–1630, doi:10.1002/2015JB012111.

Gudmundsson, G. (2011). Respiratory health effects of volcanic ash with special reference to Iceland. A review, *Clinical Respiratory Journal,* **5**, 2–9.

Gudmundsson, M. T., Thordarson, T., Hoskuldsson, A. *et al.* (2012). Ash generation and distribution from the April-May 2010 eruption of Eyjafjallajokull, Iceland, *Scientific Reports,* **2**, 572, doi:10.1038/srep0057.

Guffanti, M., Casadevall T. J. and Budding, K. (2010). 1953–2009: Encounters of Aircraft with Volcanic Ash Clouds, A Compilation of Known Incidents, U.S. Geological Survey Data Series 545, 12 p., Plus 4 Appendixes Including the Compilation Database, http://pubs.usgs.gov/ds/545/.

Heinold, B., Tegen, I., Wolke R., Ansmann, A., Mattis, I., Minikin, A., Schumann, U. and Weinzierl, B. (2012). Simulations of the 2010 Eyjafjallajökull volcanic ash dispersal over Europe using COSMO-MUSCAT, *Atmospheric Environment,* **48**, 195–204.

Horwell, C. J. and Baxter, P. J. (2006). The respiratory health hazards of volcanic ash: a review for volcanic risk mitigation, *Bulletin Volcanology,* **69**, 1–24, doi:10.1007/s00445-006-0052-y.

Hunton, D. E., Viggiano, A. A., Miller, T. M. *et al.* (2005). In-situ aircraft observations of the 2000 Mt. Hekla volcanic cloud: Composition and chemical evolution in the Arctic lower stratosphere, *Journal Volcanology Geothermal Research,* **145**, 23–34.

Langmann, B., Folch, A., Hensch, M. and Matthias, V. (2012). Volcanic ash over Europe during the eruption of Eyjafjallajokull on Iceland, April-May 2010, *Atmospheric Environment,* **48**, 1–8.

Langmann, B. (2013). Volcanic ash versus mineral dust: Atmospheric processing and environmental and climate impacts, *ISRN Atmospheric Sciences,* 2013, doi:10.1155/2013/245076.

Leadbetter, S. J., Hort, M. C., von Lowis, S., Weber, K. and Witham, C. S. (2012). Modeling the resuspension of ash deposited during the eruption of Eyjafjallajokull in spring 2010, *Journal of Geophysical Research,* **117**, D00U10, doi:10.1029/2011JD016802.

Marshall, J., Johnson, H. and Goodman, J. (2001). A Study of the Interaction of the North Atlantic Oscillation with Ocean Circulation, *Journal of Climatology,* **14**, 1399–1421.

Martinsson, B. G., Brenninkmeijer, C. A. M., Carn, S. A., Hermann, M., Heue, K.-P., van Velthoven, P. F. J. and Zahn, A. (2009). Influence of the 2008 Kasatochi volcanic eruption on sulfurous and carbonaceous aerosol constituents in the lower stratosphere, *Geophysical Research Letter*, **36**, L12813, doi:10.1029/2009GL038735.

Marzano, F. S., Lamantea, M., Montopoli, M., Herzog, M., Graf, H. and Cimini, D. (2013). Microwave remote sensing of the 2011 Plinian eruption of the Grimsvotn Icelandic volcano, *Remote Sensing Environment*, **129**, 168–184.

Matthias, V., Aulinger, A., Bieser, J. *et al.* (2012). The ash dispersion over Europe during the Eyjafjallajökull eruption — comparison of CMAQ simulations to remote sensing and in-situ observations, *Atmospheric Environment*, **48**, 184–194.

Miller, T. P. and Casadevall, T. J. (2000). Volcanic ash hazards to aviation, in Sigurdsson, H. (ed.), *Encyclopedia of Volcanoes* (Academic Press: San Diego, CA), 915–930.

Newhall, C. and Self, S. (1982). The Volcanic Explosivity Index (vei) — an Estimate of Explosive Magnitude for Historical Volcanism, *Journal of Geophysical Research*, **87**, 1231–1238.

O'Dowd, C., Varghese, S., Martin D. *et al.* (2012). The Eyjafjallajökull Ash Plume — Part 2: Forecasting ash cloud dispersion, *Atmospheric Environment*, **48**, 143–151.

Parker, C. F. (2015). Complex Negative Events and the Diffusion of Crisis: Lessons from the 2010 and 2011 Icelandic Volcanic Ash Cloud Events, *Geografiska Annaler: Series A, Physical Geography*, **97**, 97–108.

Pavolonis, M. J., Feltz, W. F., Heidinger, A. K. and Gallina, G. M. (2006). A Daytime Complement to the Reverse Absorption Technique for Improved Automated Detection of Volcanic Ash, *Journal Atmospheric Oceanic Technology*, **23**, 1422–1444, doi:10.1175/JTECH1926.1.

Pavolonis, M. J., Heidinger, A. K. and Sieglaff, J. (2013). Automated retrievals of volcanic ash and dust cloud properties from upwelling infrared measurements, *Journal Geophysical Research Atmospheres*, **118**, 1436–1458, doi:10.1002/jgrd.50173.

Pollack, J. B., Toon, O. B. and Khare, B. N. (1973). Optical properties of some terrestrial rocks and glasses, *Icarus*, **19**, 372–389, doi:10.1016/0019-1035(73)90115-2.

Poulsen, C. A., Siddans, R., Thomas, G. E., Sayer, A. M., Grainger, R. G., Campmany, E., Dean, S. M., Arnold, C. and Watts, P. D. (2012). Cloud retrievals from satellite data using optimal estimation: Evaluation and application to ATSR, *Atmospheric Measurement Techniques*, **5**, 1889–1910, doi:10.5194/amt-5-1889-2012.

Prata, A. J. and Prata, A. T. (2012). Eyjafjallajökull volcanic ash concentrations determined using Spin Enhanced Visible and Infrared Imager measurements, *Journal of Geophysics Research*, **117**, D00U23, doi:10.1029/2011JD016800.

Robock, A. (2000). Volcanic eruptions and climate, *Reviews Geophysics*, **38**, 191–219.

Rose, W. I. and Durant, A. J. (2009). Fine ash content of explosive eruptions, *Journal of Volcanology Geothermal Research*, **186**, 32–39.

Rose, W. I. and Durant, A. J. (2011). Fate of volcanic ash: Aggregation and fallout, *Geology*, **39**, 895–896.

Schumann, U., Weinzierl, B., Reitebuch, O. *et al.* (2011). Airborne observations of the Eyjafjalla volcano ash cloud over Europe during air space closure in April and May 2010, *Atmospheric Chemistry Physics*, **11**, 2245–2279.

Simkin, T. and Siebert, I. (1994). *Volcanoes of the World* (Smithonia Institution, Misoula, Montana. Geoscience Press, USA).

Sokolik, I. N. and Toon, O. B. (1999). Incorporation of mineralogical composition into models of the radiative properties of mineral aerosol from UV to IR wavelengths, *Journal of Geophysical Research Atmospheres*, **104**, 9423–9444, doi:10.1029/1998JD200048.

Sparks, R. S. J., Bursik, M. I., Carey, S. N., Gilbert, J. S., Glaze, L. S., Sigurdsson, H. and Woods, A.W. (1997). *Volcanic Plumes* (John Wiley & Sons Ltd, Chichester, UK).

Thordarson, T. and Larsen, G. (2007). Volcanism in Iceland in historical time: Volcano types, eruption styles and eruptive history, *Journal of Geodynamics.*, 43, 118–152.

Thorsteinsson, T., Johannsson, T., Stohl, A. and Kristiansen, N. I. (2012). High levels of particulate matter in Iceland due to direct ash emissions by the Eyjafjallajokull eruption and resuspension of deposited ash, *Journal of Geophysical Research*, **117**, B00C05, doi:10.1029/2011JB008756.

Trickl, T., Giehl, H., Jäger, H. and Vogelmann, H. (2013). 35 yr of stratospheric aerosol measurements at Garmisch-Partenkirchen: from Fuego to Eyjafjallajökull, and beyond, *Atmospheric Chemistry Physics*, **13**, 5205–5225.

Volz, F. E. (1973). Infrared optical constants of ammonium sulphate, Sahara dust, volcanic pumice, and flyash, *Applied Optics*, **12**, 564–568, doi:10.1364/AO.12.000564.

Wastegard, S. and Davies, S. M. (2009). An overview of distal tephrochronology in northern Europe during the last 1000 years, *Journal of Quaternary Science*, **24**, 500–512.

Webley, P. W., Dean, K., Peterson, R., Steffke, A., Harrild, M. and Groves, J. (2012). Dispersion modeling of volcanic ash clouds: North Pacific eruptions, the past 40 years: 1970–2010, *Natural Hazards*, **61**, 661–671.

Webster, H. N., Thomson, D. J., Johnson, B. T. *et al.* (2012). Operational prediction of ash concentrations in the distal volcanic cloud from the 2010 Eyjafjallajökull eruption, *Journal of Geophysical Research*, **117**, doi:10.1029/2011JD016790.

Webster, H. N., Witham, C. S., Hort, M. C., Jones A. R. and Thomson D. J. (2013). NAME modelling of aircraft encounters with volcanic ash plumes from historic eruptions, Forecasting Research Technical Report No: 552, Met Office, UK.

Wen, S. and Rose, W. I. (1994). Retrieval of sizes and total masses of particles in volcanic clouds using AVHRR bands 4 and 5, *Journal Geophysical Research Atmospheres*, **99**, 5421–5431, doi:199410.1029/93JD03340.

Wilson, T. M., Stewart, C., Sword-Daniels, V., Leonard, G. S., Johnston, D. M., Cole, J. W., Wardman, J., Wilson, G. and Barnard, S. T. (2012). Volcanic ash impacts on critical infrastructure, *Physics and Chemistry of the Earth*, **45–46**, 5–23.

Witham, C. S., Hort, M. C., Potts, R., R. Servranckx, R., Husson, P. and Bonnardot, F. (2007). Comparison of VAAC atmospheric dispersion models using the 1 November 2004 Grimsvötn eruption, *Meteorological. Applications*, **14**, 27–38.

Zehner, C. (2010). Monitoring volcanic ash from space. in *Monitoring Volcanic Ash from Space*, vol. STM-280, (Frascati, Italy).

Zimanowski, B., Wohletz, K., Dellino, P. and Buttner, R. (2003). The volcanic ash problem, *Journal of Volcanology Geothermal Research*, **122**, 1–5.

Chapter 13

East Asia (2010): Continent Wide Dust

I-Chien Lai*, Peter Brimblecombe[†] and Chon-Lin Lee*

Marine Environment and Engineering,
University of Sun Yat-sen University, Taiwan
[†]*School of Energy and Environment, City University*
of Hong Kong, Hong Kong

13.1 Introduction

Asian dust storms are a form of natural disaster; the occurrence subject to synoptic weather conditions at certain times of the year, mainly in spring (March–May). Areas such as the Taklamakan Desert and the Gobi Desert in Mongolia and northern China occupy 3.57 million km^2 and have been recognised as part of the world's largest dust storm regions (Gao and Washington, 2009; Natsagdorj *et al.*, 2003; Qian *et al.*, 2002; Washington *et al.*, 2003). The annual emission rate of dust aerosols from East Asia is around 100–800 Tg, about 15–20% of global emission (Laurent *et al.*, 2006; Tanaka and Chiba, 2006; Zhang *et al.*, 2008). Dust storms are also related to human activities. Recent studies indicate increasing frequency and strength of Asian dust storms in the past two decades that are believed to be associated with inappropriate anthropogenic activities, such as over-exploitation for expanding the cropped area necessary to feed a growing population. It leads to rapid deforestation or land surface degradation on a large scale (Kurosaki *et al.*, 2011; Tan *et al.*, 2014; Wang *et al.*, 2014). Dust aerosols accompanied by anthropogenic pollutants and potential allergens can transport across to the North Pacific Ocean (Cottle *et al.*, 2013; Jaffe *et al.*, 1999; McKendry *et al.*, 2008; Uno *et al.*, 2001), even throughout the Northern Hemisphere (Biscaye *et al.*, 1997; Grousset *et al.*, 2003; Huang *et al.*, 2015; Uno *et al.*, 2009). The transport and loading processes induce various effects on atmospheric chemistry, land/ocean ecosystems, human health, climate and socio–economic activities. Increasing toxicity of Asian dust particles in terms of anthropogenic pollutants from urban or industrial areas carried with dust particles is another important issue of serious concern to the environment and human health over past decades. The assessment of annual economic loss from dust storms in China amounted to about USD 195 million in 2002 (Lu and Wu, 2002); however, the annual economic loss increased to USD 321 million between 2010 and 2013 (UNEP, 2016).

Over the past 20 years, the magnitude of dust storm intensity seems to have increased (Tan *et al.*, 2014; Tatarov *et al.*, 2012), though the occurrence frequency in the desert regions seems to decline (Qian *et al.*, 2002; Wang *et al.*, 2008, 2014). In spring 2010, several extraordinary dust storms originating from the Gobi desert with an unprecedented distribution of dust and in the unique meteorological conditions affected the North Hemisphere for over a month (Cottle *et al.*, 2013; Uno *et al.*, 2011). The most profound one occurred during the period 19th–23rd March; a large-scale dust storm event which covered and affected the whole East Asian continent, that included north and northeast China, Korea, Japan, Taiwan and Hong Kong and the western Pacific Ocean (Li *et al.*, 2011; Yeh *et al.*, 2015), even reaching the Arctic (Huang *et al.*, 2015). In China, 16 provinces were hit by the dust storm, affecting an area of about 2.82 million km^2 and with 270 million people (Wang *et al.*, 2014). The visibility in the dust source region was almost reduced to 200 m, with numerous livestock deaths and damage to buildings and agricultural ecosystems. Rapidly elevated particulate matter (PM) concentrations led to serious air quality deterioration, classified as *hazardous* in East Asia (Hsu *et al.*, 2013; Li *et al.*, 2011; Park and Cho, 2013; Wang *et al.*,

2013; Yeh *et al.*, 2015; Zhao *et al.*, 2011). Dust budget simulations including the dust emission, deposition and atmospheric loading for March were all larger than the previous strong dust storms that had occurred in 2002 and 2006 (Li *et al.*, 2011). The event of April 2010, which is not as important as the case in March 2010; nevertheless had a dust storm strength that allowed it to be as a *black sandstorm* (the top grade of dust storm definition in China), with visibility lower than 50 m and a maximum wind velocity of 28 m s^{-1}. It brought devastation to most of the dust source regions: the Hexi Corridor, Inner Mongolia and northwestern China (Li *et al.*, 2013; Qian *et al.*, 2011). The economic loss to Minqin County, Gansu Province was around 250 million Renminbi Yuan (USD\$30 million) and almost 10 times higher than the well-known historical case in May 1993. However, the total cost for recovering from damage to buildings and population health must be higher than these estimated values.

Apart from the impacts on human health and ecosystems, large quantities of dust aerosols conveyed across the whole North Hemisphere during these unique dust storm activities are associated with a thick dust aerosol belt occupying the North Pacific region (Liu *et al.*, 2013; Uno *et al.*, 2011) and may directly and indirectly affect the climate system, by inducing radiative forcing in the atmosphere.

13.2 Dust Storm Identification

Dust storm types are classified based on horizontal visibility reduction and wind velocity in source regions. The protocol of World Meteorological Organisation classifies dust events into four categories of:

(1) *Dust-in-suspension*: widespread dust in suspension, not raised at or near the station at the time of observation; visibility is usually not greater than 10 km;
(2) *Blowing Dust*: raised dust or sand at the time of observation, reducing visibility to 1 km to 10 km;
(3) *Dust Storm*: strong winds lift large quantities of dust particles, reducing visibility to between 200 m and 1,000 m; and

(4) *Severe Dust Storm*: very strong winds lift large quantities of dust particles, reducing visibility to less than 200 m.

The China Meteorological Administration defines dust storm strength into five types: floating dust, blowing dust, dust storm, strong dust storm and serious dust storm is also called a *black storm* in some regions. The definitions of floating dust and blowing dust are similar to the protocol of World Meteorological Organisation; the visibility of floating dust is less than 10 km, of blowing dust is between 1,000 m and 10 km and of dust storm is less than 1,000 m. Strong dust storm categorised as the visibility is between 50 and 200 m and of serious dust storm is less than 50 m and wind velocity 25 m s^{-1} (China Meteorological Administration, 2016). With regard to downwind areas, the definition of dust storm differs from country to country; the visibility reduction at different levels between source regions and downwind environments and physical–chemical processes of dust particles on transport should be taken into account. In Japan, the Japan Meteorological Agency defines a heavy Asian dust storm day using the criterion of visibility <10 km couple with a light detection and ranging (LIDAR) network, as well as particulate matter measurements from the Japanese Ministry of the Environment (Watanabe *et al.*, 2011). The Korea Meteorological Administration defines dust storm events when PM$_{10}$ concentration exceeds 190 μg m^{-3} for at least 2 h (Chung *et al.*, 2005). In Taiwan, the Environmental Protection Administration uses PM$_{10}$ concentration variation (dramatically elevated 100–200 μg m^{-3}) at background monitoring stations and in data from Moderate-Resolution Imaging Spectroradiometer (MODIS) to identify dust storm events (Taiwan, Environmental Protection Administration, 2016).

13.3 Background

13.3.1 *History*

The earliest written record related to dust storms in the world was found in 1150 BC in the Chinese historical book: *Zhu Shu Ji Nian: the Bamboo Annals* or chronicles recorded on bamboo slips (Liu *et al.*,

1981). In the ancient Chinese literature, dust fall was referred to as "dust rain", "dust fog" or "yellow fog". In the neighbouring regions, dust event was given similar meaning names: *Hwangsa* or *Woo-Tou* in Korea and *kosa* in Japan (Chun *et al.*, 2001). Phonetic of *Woo-Tou* represents "dust rain" in Chinese sound. *Woo-Tou* in Chinese characters was the most frequently used term in East Asian ancient literature to denote dust events. Dust events in the ancient Korea were considered as God's punishment or a warning to the ruler and the earliest record can be found in Korean literature is from 174 AD (Chun *et al.*, 2008). In these historical records, dust storms were most frequent in the spring months. Apart from the records in historical literature, scientific evidence from paleo-climatological data (i.e. dating ice cores, sediments and tree rings) can be used to reconstruct long-range transport evidence of Asian dust storm, dust storm frequencies and trends at different time scales. Biscaye *et al.* (1997) compared the clay mineralogy and metal isotope (i.e. Sr, Nd and Pb) compositions between ice dust in Greenland and coeval aeolian sediments revealing that the earliest dust transport from eastern Asian to Greenland can be traced back 20 ka. Other relevant records in the East Asia region have been found in sediments of Taiwan's mountain lake (the Great Ghost Lake) and of Cheju Island in Korea. These sedimentary records suggested dust fall fluxes associated with dust storm activities originating from northern China back to the Little Ice Age (LIA) (Chen *et al.*, 2001, 2004; Lim and Matsumoto, 2006; Lou *et al.*, 1997).

13.3.2 *Long-term trend*

Trend analysis through ancient Chinese records, historical climate information and weather reports suggest that two marked dust storm periods found in AD 1430–1570 and AD 1700–1890 both took place in LIA (Chen *et al.*, 2013; Yang *et al.*, 2007; Zhang, 1984). Reconstruction of paleo-climatological records (Chen *et al.*, 2013; He *et al.*, 2015; Yang *et al.*, 2007) proved that stronger dust storms occurring during the LIA were associated with a strengthening of the Siberian High (Fig. 13.1). Modern meteorological data in China, available since the 1950s indicates a declining trend in the frequency of occurrence of dust storms during the period of

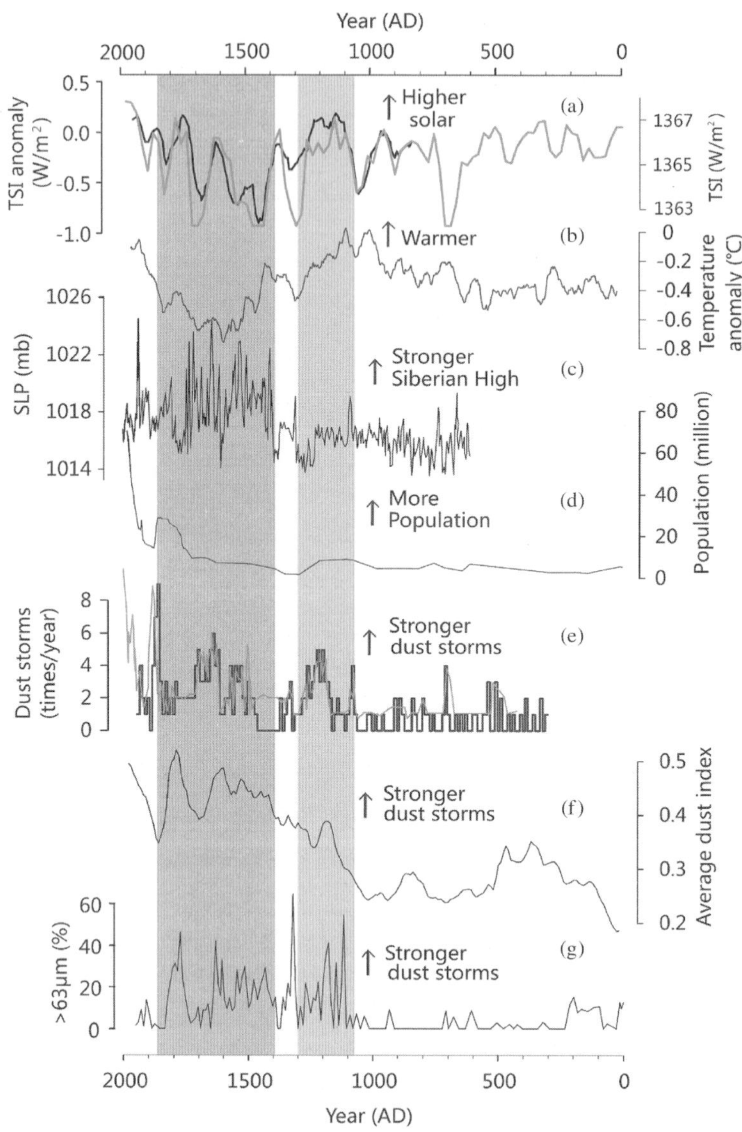

Fig. 13.1.   Variations in dust storm events in northern China compared with global and regional conditions. (a) Reconstructed total solar irradiance (TSI) records (the black line is from 1200 years data, whereas grey line from 9400 years data). (b) Northern Hemisphere mean temperature anomaly. (c) Reconstructed Siberian High strength. (d) Total population of five provinces in northwestern China (Shanxi, Gansu, Ningxia, Qinghai and Xinjiang). (e) Historical dust storms

1954–2000 (Natsagdorj *et al.*, 2003; Qian *et al.*, 2002; Wang *et al.*, 2004, 2008; Sun *et al.*, 2003). However, in different periods, there are spatial changes to the dust storms. For instance, dust storms appearing in northwestern China were frequent in the 1980s, but the highest frequency in northern China (the eastern part of Inner Mongolia) was apparent in the mid-1960s (Qian *et al.*, 2004; Wang *et al.*, 2005). The diminishing trend observed over the past 50 years correlates with temperature increases, which may be induced by anthropogenic activities (i.e. global warming effect on climate). These may weaken the strength of the westerlies and the temperature gradient between northern China and Mongolian regions (Qian *et al.*, 2002; Wang *et al.*, 2008; Zhu *et al.*, 2008).

By contrast, more dust outbreaks have been observed in downwind areas since 2000 (Kurosaki *et al.*, 20111; Kim, 2008), though a declining trend is observed in the desert regions (Qian *et al.*, 2002; Wang *et al.*, 2004, 2008). Surface vegetation loss and soil moisture reduction resulting from intensified drought conditions after the mid-1990s over Mongolia and Inner Mongolia could be responsible for the increasing trend in the downwind areas (Kurosaki *et al.*, 2011; Han *et al.*, 2008; Lee and Sohn, 2011). Tan *et al.* (2014) defined a new index for dust storm intensity consisting of a comprehensive set of parameters, such as the frequency, duration and visibility data to evaluate the trend of dust storms. The frequency and intensity after 2000 were underestimated in comparison with the studies only concerned the frequency of dust activities alone. A recent study, using the analysis of particle size, hydrological data and historical documents for the past 2000 years, substantiated the onset of frequent dust storm period to around AD 1100. This thus

---

Fig. 13.1. (*Continued*) frequency records from northern and eastern China. (f) The 50-year averaged synthesis dust storm record across the mid-latitude Asia. (g) Percentage of >63 μm particles from Lake Gahai. Grey shadings indicate the cool LIA period from AD 1400 to AD 1850 while light grey indicates the early onset of frequent dust storms at ~AD 1100 within the warm Medieval Warm. Modified from He *et al.* (2015).

occurred during the relatively warm period within the LIA and was superimposed on high total solar radiation or warm/dry weather conditions resulting in reduction coverage of vegetation and insufficient moisture in northern China as seen in Fig. 13.1 (He *et al.*, 2015). This finding may be able to explain the present situation with frequent dust storms occurring after 2000. Moreover, the expanded areas of desertification, as the consequence of the global warming and dust aerosol, itself could be increasing aridity and the availability of dust as a source in northern China (He *et al.*, 2015; Han *et al.*, 2008).

The findings from these studies categorise the factors influencing the frequency of dust storm and include wind energy (Yang *et al.*, 1998), temperature gradients and cyclone frequencies (Qian *et al.*, 2002), the geopotential height over the Mongolian Plateau and Siberia (Mao *et al.*, 2011; Wang *et al.*, 2008), human activity (Chen *et al.*, 2013; Wang *et al.*, 2004) and climate change (Yang *et al.*, (2007). With the exception of human activity, these factors suggest that the long-term trend of dust storms in Asia is mainly influenced by the variation of large-scale atmospheric circulation and climate. Nevertheless, anthropogenic effects, such as inappropriate land development and global warming, might take over the role of natural climate effects on Asian dust storm occurrence in the future.

13.4 Dust Storm Generation

13.4.1 *Generation mechanism*

Dust storm generation is a complex mechanism affected by synoptic circulation patterns on both regional and global scales and land-surface conditions (Zhang, 1984). Atmospheric dynamics, thermal conditions, dry climate and low vegetation coverage are the crucial to generating Asian dust storm (Chun *et al.*, 2005; Qian *et al.*, 2002; Yang *et al.*, 2007). Strong wind usually linked to a deep cold frontal system and a steep pressure gradient at the surface are the essential and pivotal dynamic conditions for *dust storm formation* (Chun *et al.*, 2001; Kok *et al.*, 2012; Qiang *et al.*, 2007). A dry climate and low vegetation coverage both contribute to soil exposure and the provision of transportable materials. Dust storm weather in Mongolia analysed for the period 1937–1999 indicates that dust storms generally took place with wind speeds from 6 m s^{-1} to 20 m s^{-1} and low relative humidity around 20–40% (Natsagdorj *et al.*, 2003); both wind velocity and relative humidity conditions are relevant to the background processes in dust storm generation.

Dust storm weather is mainly associated with cold air outbreaks. The outbreaks of cold air, in northern China, include the Mongolian cyclone, the northeast China low and the Yellow River cyclone (Qian *et al.*, 2002). Sun *et al.* (2001) classified outbreaks of cold air, according to air masses moving direction relative to dust source regions, into three routes: north, northwest and west directions. The northern route: the air masses originate from Lake Baikal and move southward across central Mongolia and China carrying dust aerosols from the southern Gobi Desert of Mongolia and from northern China. The dust source region for the northwest route of air masses is the Hexi Corridor and the Gobi Desert in northern China. The third route, from west direction, is related to the dust source regions in the Taklamakan Desert, the Hexi Corridor and the Gobi deserts. These routes occur with frequencies of 41%, 32% and 27% for the northwest route, the north route and the west route, respectively.

The interaction between the Siberian High, cyclones/cold fronts and topographical features of dust source regions affects the extent, occurrence frequency and transport pathway of dust storms. As mentioned previously, the high frequency of dust storm occurrence was associated with the intensity variation of Siberian High. Thus, the strength and duration of Siberian High are also decisive factors of dust storm generation. The Siberian High is the most dominant atmospheric circulation in Asian winter, usually accompany with cold and dry weather conditions. A strong Siberian High may cause frozen soil, reduce vegetation coverage and facilitate cyclone formation in the arid and semiarid regions in winter and early spring (Qian *et al.*, 2002; Yang *et al.*, 2007). Topographic characteristics, such as mountain terrains would physically alter the dust aerosol transport mechanism (Sun *et al.*, 2001). For instance, the

Taklamakan Desert is flanked by high mountain terrains over the southwest side with the average elevation of 4,000 m. In this dust region, the mountain blocking effect may facilitate vertical transported of aeolian dust up to 5,000 m, then horizontal transport far out of the source regions, even around the whole North Hemisphere by the jet stream zone in the middle troposphere (Uno *et al.*, 2009); or redeposition in the neighbouring areas under weak vertical transport (Qian *et al.*, 2004). With respect to the dust storm sources in northwest and northern China, the cyclone activities mostly occur at an altitude of 3,000 m and move southeastward and northeastward. The interaction between cyclone activity and topographical features presents a regional discrepancy in spatial-temporal characteristics of dust storm weather (Qian *et al.*, 2002, 2004). The synoptic patterns of long-range dust transport and the transport pathways are determined by the westerly jets, regional weather and topography (Alfaro *et al.*, 2003; Arimoto *et al.*, 2006; Huebert *et al.*, 2003; Jacob *et al.*, 2003).

13.4.2 *Generation mechanisms in 2010*

In 2010, there were two notable severe dust storms, one observed during 19th–23rd March and the other during 24th to 26th April. Extremely low temperatures and droughts in the preceding winter (WMO, 2010) provided a large number of transportable materials for dust storm formation in 2010. Research on atmospheric dynamic and thermal conditions and meteorological parameters of these severe dust storms found similar circulation patterns in the upper troposphere, associated with a baroclinic trough and strong frontal zones, but there were different surface circulations between these two dust storms (Guo *et al.*, 2011; Wang *et al.*, 2014).

For the event in March, an intense cyclone rapidly developed in southern Mongolia (48°N, 102°E) interacting with dynamic and thermal conditions in the upper troposphere. This was the dominant cause of this large-scale event (Di *et al.*, 2014; Wang *et al.*, 2014). On 18th March, a trough from the western Siberia moved southeastward with strong cold advection behind, a jet stream at 500 hPa moving southward induced cold air outbreaks, intensifying the Mongolian cyclone and quickly moving the trough in southerly and easterly directions to mid-latitudes. The interaction between intense thermal advection and vortex advection created a high wind velocity in the cyclone centre with upward flow from the surface and the horizontal airflow at high altitudes leading to dust generation and transport. The maximum wind velocity observed during dust storm was 24 m s^{-1} (Di *et al.*, 2014). The dust storm regions encompassed the Gobi desert, Inner Mongolia and northern China (Wang *et al.*, 2014).

By contrast, the major surface dynamic factor, for the case in April 2010, was governed by the development of surface thermal low pressure resulting in large temperature and pressure gradients and baroclinic instability circulation for convection development (Qian *et al.*, 2011). Additionally, a strong vertical circulation caused the ascending air flows from the surface to the upper troposphere. Such as a warm conveyor ahead of the cold front, the convergence activity of the subtropical jet and the second circulation from the polar jet all facilitated the generation of a *black storm* in the April case (Guo *et al.*, 2011; Li *et al.*, 2011; Qian *et al.*, 2011). The interaction among these dynamic and thermal conditions generated a regional *black storm* in the Hexi Corridor, Inner Mongolia and northwestern China with the maximum wind velocity of 28 m s^{-1} on 24th April at 19:00 (local time) and persisted for three hours, ended up at 22:00 (Qian *et al.*, 2011). With respect to the strength of the dust storms, the case in April was stronger than that of March for the dust source regions. However, the horizontal helicity value of April's event was −1028.3 m^2 s^{-2} lower than the event in March with −921.3 m^2 s^{-2} (Li *et al.*, 2013). The impacts of the April's case were more significant in mid-latitude zones, as well as in the North Pacific Ocean and North America; whereas the event in March covered the whole East Asian continent, southward to the tropics and northward to the Arctic (Huang *et al.*, 2015) and southward to South China Sea (Hsieh and Liao, 2013; Yeh *et al.*, 2015). The interaction of synoptic circulation between the surface and upper atmosphere may explain the differences in the extent of impacts and the transport pathway between these two events. The vertical structure and horizontal movement of synoptic circulation control the generation, duration and transport pathway, as well as the strength of dust storms. Thus, the understanding of the interaction among vertical and horizontal of synoptic circulation and meteorological parameters during dust storm occurrence is useful as background information for establishing dust storm forecast models. Moreover, these synoptic conditions can be directly related to the dust transport mechanism and further affect dust aerosol chemical composition and physical properties.

13.5 Spatiotemporal Variations of PM Concentrations and Transport

Large quantities of dust aerosol generated during the dust storm in March 2010 caused deteriorating air quality in both the source regions and downwind areas, as observed in air quality monitoring records and satellite data. The main dust transport pathways split into three branches: an eastern pathway, a southeastern pathway and a northern pathway. The dust storm originated in the Gobi Desert (at the mid-latitude 45°N) moved eastwards across eastern China, the Korean peninsula (Kwak *et al.*, 2012; Li *et al.*, 2011) and Japan, and then down to South China (Shanghai and Xiamen), Hong Kong and Taiwan (Hsy *et al.*, 2013; Yeh *et al.*, 2015). The northern path was split from the easterly path by cyclones and the enhanced East Asia Trough. Dust was transported from northwest China, crossing eastern China and Japan and then moved northward crossing Siberia to the Arctic within five days (Huang *et al.*, 2015). Satellite images collected from Moderate Resolution Imaging Spectroradiometer (MODIS) also showed the dust storm originated in Inner Mongolia and northern China on 19th March, and then rapidly moved southeastward carrying a large quantities dust particles passing through urban and industrial areas in eastern China (Fig. 13.2).

Daily average PM_{10} concentrations observed at air monitoring stations in the East Asian region varied along the dust storm path. Figure 13.3 displays the variations of daily average concentrations of PM_{10} (particles aerodynamic diameter are equal to 10 μm or less) in

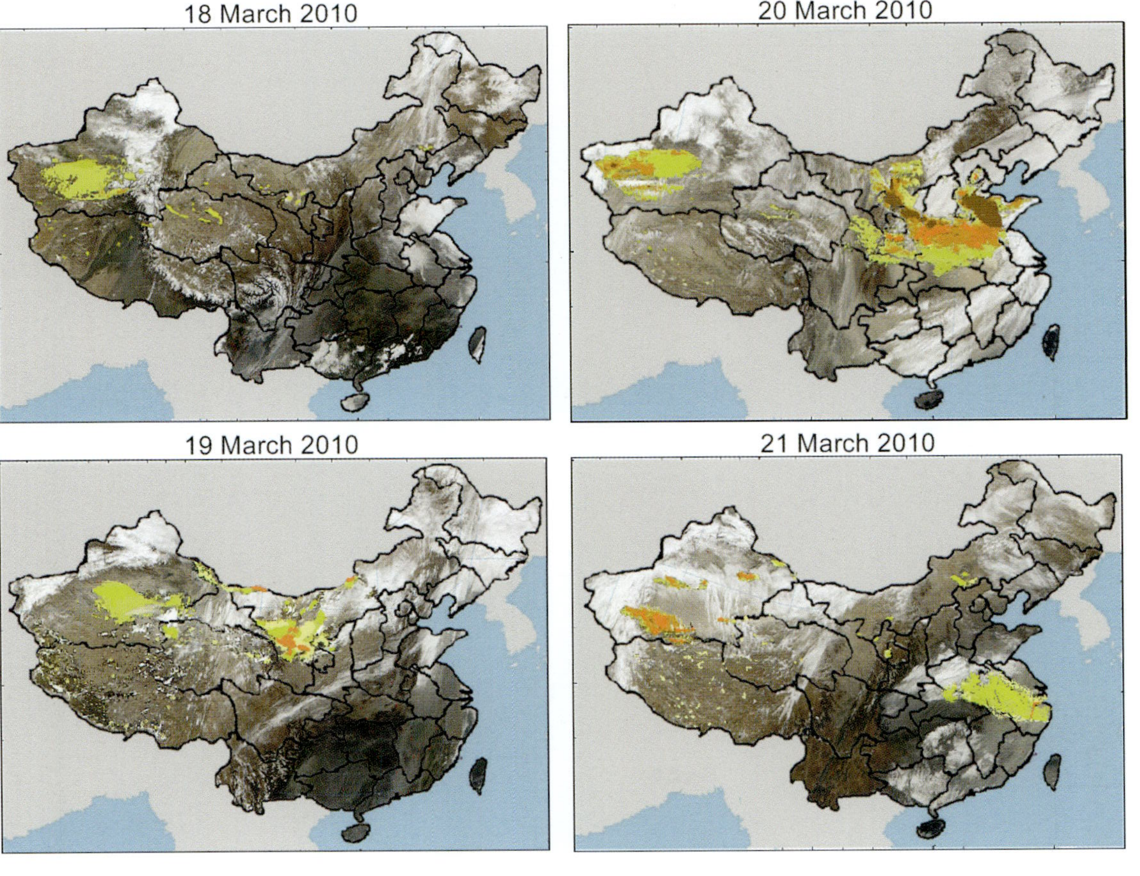

Fig. 13.2.   MODIS images of the dust storm observed on 18th–21st March 2010.

Source: Han *et al.,* (2012).

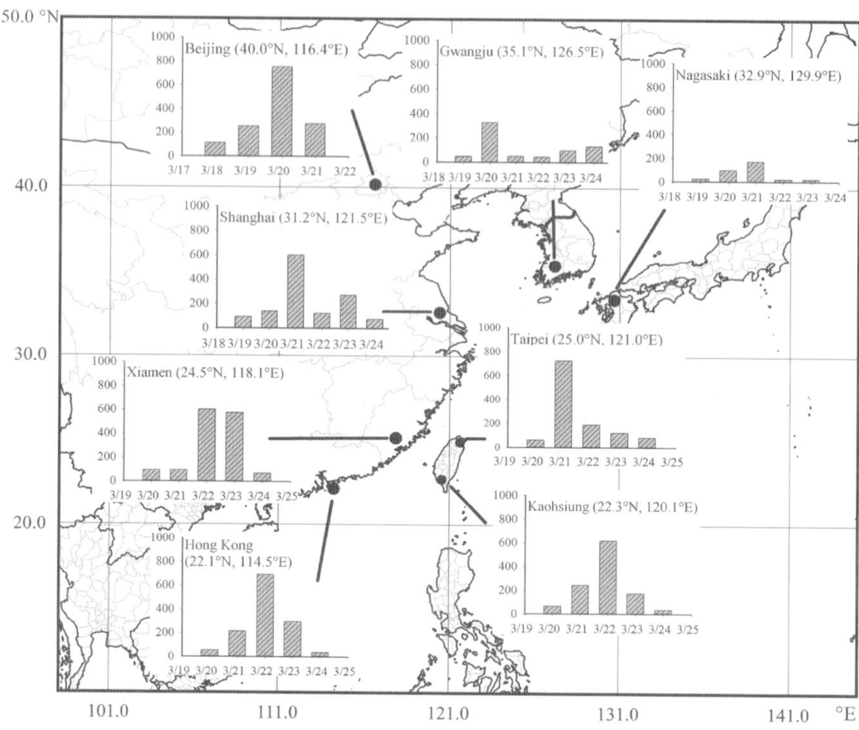

Fig. 13.3. Spatiotemporal variations of daily mean PM_{10} concentrations at air quality monitoring stations in East Asia. The solid circles represent monitoring stations.

Source: Yeh *et al.* (2015).

some major cities in downwind areas over the dust storm period. For the eastern pathway, the dust storm reached Gwangju (Korea) at 12:00 local time (LT) on 20th March with the highest hourly PM_{10} concentration of 1,864 μg m^{-3} (Park and Cho, 2013) and Nagasaki (Japan) at 21:00 (LT) on 20th March with the highest hourly PM_{10} of 727 μg m^{-3} (Air Environment Early Warning System, 2016). For the southerly route, the dust storm arrived in Beijing (Chen *et al.*, 2012; Li *et al.*, 2011) and Shanghai (Wang *et al.*, 2013) on 20th March and arrived in Xiamen (Zhao *et al.*, 2011), Hong Kong and Taiwan on 21st March and then in the South China Sea (Dongsa station) on 22nd March (Hsu *et al.*, 2013; Wang *et al.*, 2011).

The maximum hourly concentrations of PM_{10} were 5,080 μg m^{-3}, 727 μg m^{-3}, 1700 μg m^{-3}, 990 μg m^{-3}, 1,720 μg m^{-3}, 791 μg m^{-3} and 557 μg m^{-3} for Beijing, Shanghai, Xiamen, Hong Kong, Taipei, Kaohsiung and Dongsa, respectively. As with PM_{10}, the observations of $PM_{2.5}$ (i.e. particles less than or equal to aerodynamic diameter 2.5 μm), in the downwind cities also displayed a striking increasing of hourly peak concentrations (Japan Ministry of the Environment, Particulate Monitoring Data, 2016; Taiwan, Environmental Protection Administration, 2016; Zhao *et al.*, 2011). Both PM_{10} and $PM_{2.5}$ showed significant increases in concentration about two to three times that of non-dust periods. The maximum hourly concentrations and the daily average concentrations of PM which characterised the event in March exerted striking impacts on air quality in downwind areas, especially for East Asia (Li *et al.*, 2011; Wang *et al.*, 2013; Yeh *et al.*, 2015). There were thirteen cities in China which exceeded the 24-h standard of Chinese Ambient Air Quality Standards such that that are classified as hazardous in terms of air quality based risk category (Chen *et al.*, 2012; Li *et al.*, 2011).

13.6 Chemical Composition and Physical Properties

Chemical composition and physical properties of dust aerosols have been broadly studied over the past 30 years because of their importance for the climate system, global biogeochemistry and human health. Chemical composition and physical properties vary greatly with the source regions, transport pathways and pollutants encountered during the transport (Hsu *et al.*, 2013; Liu *et al.*, 2013; Yan *et al.*, 2015). Dust aerosols serve as carriers providing surfaces for many chemical and physical processes during transport (Onishi *et al.*, 2012; Zhao *et al.*, 2011). Of particular concern are the large amounts of anthropogenic particles (i.e. containing persistent organic compounds and contaminating metals) increasingly produced by the rapid urbanisation and industrialisation in Asia, which can modify the physico-chemical properties of dust particles, e.g. particle size and chemical compositions and enhance toxicity of dust particles (Arimato *et al.*, 2016; Guo *et al.*, 2004; Onishi *et al.*, 2012; Yan *et al.*, 2015).

13.6.1 *Particle size distribution*

Particles size is vital to understanding the transport distance of dust aerosols, as well as the effects on radiative balance through altering optical properties. Particle size can be modified during transport processes, the size of 100–200 μm in dust source regions sediment or break apart to less than 50 μm, which is the major particle size transported in the boundary layer (Mahowald *et al.*, 2014). Analyses of dust particle size distributions in downwind areas showed that coarse particles ($PM_{2.5-10}$) were the dominant particle size fraction during dust storm periods (Chen *et al.*, 2001; Kim *et al.*, 2007; Yeh *et al.*, 2015).

Size fraction ratios are useful in determining dust particle size for affected areas. Three field observations in south Taiwan (Kaohsiung and Pingtung) and in south China (Xiamen) compared particle fraction ratios of $PM_{2.5}$, $PM_{2.5-10}$, PM_{10} and TSP between dust storm events (ADS) and non-dust events (NDS) in March 2010. The results illustrated that the highest enhancements of particle mass were in $PM_{2.5-10}$ and PM_{10} (Tsai *et al.*, 2012; Yeh *et al.*, 2015; Zhao *et al.*, 2011). Particle mass ratios of ADS/NDS for $PM_{2.5}$, $PM_{2.5-10}$, PM_{10} and TSP were 1.3, 7.2, 3.0 and 3.5 times in the Kaohsiung observations (Yeh *et al.*, 2015). The investigation of particle size in urban areas neighbouring the dust source regions found that the dust aerosols from the Taklamakan desert affected particles in the range 1.0–2.5 μm, while that from the Gobi desert influenced on the particles larger than 2.5 μm (Zhao *et al.*, 2015). Particle size distribution might be indirect evidence that the source region of dust storms in March 2010 is the Gobi deserts.

13.6.2 *Metallic elements*

Heavy metals have been a concern as hazards to human health, though they being only a small fraction of the total mass in dust aerosol. Research on metallic elements of dust aerosols is useful to characterise chemical species and to identify the source of metallic particles on the long-range transport processes. During dust storms, crustal elements are usually observed in downwind areas; although the composition and concentration of crustal elements and anthropogenic elements depend on dust storm transport pathway and local pollution (Kwah *et al.*, 2012; Onishi *et al.*, 2012; Sun *et al.*, 2015; Yan *et al.*, 2015).

Several mineral element analyses for the events in 2010 showed that the concentrations of crustal (i.e. Al, Ca, Ti, Fe, Mg, Ba and K) and anthropogenic species (i.e. Zn, V, Ni, As, Cd and Pb) both significantly increased compared with non-dust storm periods. The concentrations of crustal elements were about 3–10 times higher than non-dust storm periods (Kwak *et al.*, 2012; Park and Cho, 2013; Tasi *et al.*, 2012; Yeh *et al.*, 2015; Yu *et al.*, 2013; Zhao *et al.*, 2011). The concentration comparison of the metallic elements among the downwind areas including Korea, southern China and southern Taiwan found the major elements for each area vary with dust transport pathways. For instance, the main crustal elements observed in Gwangju, Korea were Al, Ca and Fe; while Na and Mg were also major elements, with exception of Al, Ca and Fe, obtained from south China and south Taiwan (Park and Cho, 2013; Tasi *et al.*, 2012; Yeh *et al.*, 2015; Zhao *et al.*, 2011). Among the dominant crustal elements, Al concentration showed a dramatic increase over the downwind areas and can be used as an indicator to identify Asian dust storms (Hsu *et al.*, 2013) The variation of anthropogenic elements among these monitoring sites also depends on the features of polluted areas along the transport path, with local pollution characteristics also influencing the components of dust aerosols when they deposit at receptor sites. Differences in transport paths illustrate how dust aerosols transported from the dust source region to Xiamen, passed through a broad swath of polluted areas, such as northern China and the East coast of China (Fig. 13.2), resulting in the higher concentrations of anthropogenic elements observed in Xiamen than Kaohsiung (Yeh *et al.*, 2015; Zhao *et al.*, 2011).

13.6.3 *Water soluble ions*

Water soluble ions transported with dust aerosols can be important materials for ecosystems and affect climate. Additionally, secondary organic aerosols are a major source of water soluble ions (Wang *et al.*, 2005) e.g. Fe(II), K^+ and NO_3^- ions can be nutrients for biogenic productivity in marine ecosystems (Goudie and Middleton, 2001). Partially soluble ions, such as sulphate, could influence aerosol solubility as they can act as cloud condensation nuclei, which is critical to the atmospheric radiative budget (Gibson *et al.*, 2007).

Measurements of water soluble ions from areas downwind of dust storms, such as Shanghai, Xiamen, Korea and Taiwan, showed that SO_4^{2-}, NO_3^-, NH_4^+, Ca^+ and Na^+ were dominant; and the concentrations of these ions increased by factors of 2–3 between dust storm periods and non-dust storm periods in March and April 2010 (Li *et al.*, 2012; Park and Cho, 2013; Tasi *et al.*, 2012). Typically, high concentrations of crustal elements are observed, while the concentrations of SO_4^{2-}, NO_3^- and NH_4^+ decrease at receptor sites. The striking elevated concentrations of SO_4^{2-} and NO_3^- species implied these ions were affected by the long-range transport from polluted aerosol source regions, exemplifying chemical transformation of SO_2 emitted from polluted regions in northeastern China (Park and Cho, 2012; Sun *et al.*, 2015; Tsai *et al.*, 2012).

13.7 Potential Impacts of Asian Dust Storms

The massive quantities of dust aerosols are driven up into the atmospheric boundary layer and transported by dust storm activities have many environmental consequences. Dust storms not only affect human health and alter the radiation budget both in land surface and in the atmosphere, but also supply bioavailable nutrients (i.e. iron) to the marine ecosystem. The toxicity of dust aerosols varies dependent on the original source of elemental composition, chemical change along transport routes, particle sizes fractionation and dust loading. Recently, urban areas (i.e. northern China) are seen to enrich pollutants and have become additional second source regions of dust aerosol components especially for fine particles (Yan *et al.*, 2015). This may increase the number of anthropogenic particles (elevated in persistent organic compounds and contaminating metal) which mix with suspended desert dust and create an additional potential to affect human health, ecosystems and the climate.

13.7.1 *Human health*

The human health concern from dust storms is typically related to the impacts of abruptly elevated PM (i.e. PM_{10} and $PM_{2.5}$) concentrations causing air quality deterioration and visibility reduction for both dust source regions and downwind areas (Chen *et al.*, 2012; Jugder *et al.*, 2011; Liu *et al.*, 2006, 2003). High particulate loading for areas near the source and downwind can far exceed healthy levels and exacerbate symptoms of respiratory diseases, particularly for vulnerable groups (i.e. asthmatics, children and elderly). Numerous epidemiological studies with various study designs and statistical methods suggest that Asian dust storms may elevate hospital admissions and mortality for respiratory symptoms, cardiovascular diseases for both source regions and downwind areas (Chan *et al.*, 2008; Kanatani *et al.*, 2010; Lee *et al.*, 2013).

In the past decade, increasing clinical visits resulting from various allergic symptoms for respiratory illness, ocular effects and skin irritation have been found during Asian dust storm periods (Choi *et al.*, 2011; Otani *et al.*, 2011; Watanabe *et al.*, 2011; Yoo *et al.*, 2008). Scientists found dust particles were associated with lung function decrement among asthmatics, especially children (Otani *et al.*, 2011; Yoo *et al.*, 2008). Hsieh and Liao (2013) investigated the effects of dust particles on lung function via estimating forced expiratory volume in one second (FEV_1) for the dust storm periods in 2010 in Taiwan. They found there was 50% probability of decreasing percentage of FEV_1 about 16.9%, 18.9% and 7.1% in northern, centre and south Taiwan, respectively. In addition, contaminating chemicals (i.e. polycyclic aromatic hydrocarbons) or anthropogenic metals adhering to dust particles have been found to cause deleterious effects on the skin i.e. contact dermatitis (Hong *et al.*, 2010; Onishi *et al.*, 2015). Dust particles collected in Japan during 21st March, 2010 have been used to investigate the relationship between skin symptoms and Asian dust particles, the results suggest that dust particles-bound metal could be the factor causing skin symptoms among study subjects (Otani *et al.*, 2012).

13.7.2 *Impacts on ecosystems*

Desert-dust deposition on the Earth's surface and open oceans bring both positive and negative influences. On the one hand, massive dust aerosols loading on the land surface, which are significant negative effects for dust source regions, correlate with the destruction of agriculture and of natural vegetation. In dust source regions, damage of plants, water supply systems and nutrient deficiency caused by dust aerosol deposition and transport reduce crop productivity (Sivakumar, 2015). Additionally, air quality deterioration may enhance the menace to livestock (Mu *et al.*, 2013). Regular soil erosion also accelerates land degradation and desertification as a result of vegetation loss. Dust particles have potential influence on plants due to pathogens carried by dust aerosols, which might result in plant diseases in receptor areas (Brown *et al.*, 2002).

On the other hand, dust particles are able to neutralise acid rain and to export rich nutrients (i.e. iron and phosphorus) to the oceans and to terrestrial ecosystems in distant areas and are part of well-known positive feedback for ecosystems. Acid rain caused by using high-sulphur fuel, such as coal, for industrial or heating activities is a serious environment problem in East Asian continent, particularly for northern China. However, pH values observed from precipitation in northern China, between 6.0 and 7.2, were apparently higher than in southern China, between 3.5 and 5. This phenomenon is due to alkaline elements, such as calcium, in dust particles acting as buffers to neutralise acidity during atmospheric reactions, and (Larssen and Carmichael, 2000; Wang *et al.*, 2002). The acid neutralising capacity of dust particles hence reduces the effects of acid deposition on ecosystems.

The mineral aerosol is an important nutrient source for biological species, such as phytoplankton (Mahowald *et al.*, 2015, 2008). Chlorophyll concentration and primary production enhancement are found in the coastal seas of China and in the North Pacific, which are attributed to dust events from Asian continent (Bishop *et al.*, 2002; Tan *et al.*, 2011). With respect to the March case in 2010, Tsai *et al.* (2013) estimated the impact of soluble irons on the marginal sea of China by calculating dry deposition for the dust event. A total contribution of soluble iron from dust deposition was about 3,926 μg m^{-2}, which enhanced the concentrations of chlorophyll-*a* in the East China Sea by about 1.1 mg m^{-3} over the first two weeks after the dust passage. The daily maximum deposition flux of soluble iron increased by a factor of 40 compared to the daily average observed in the East China Sea in spring (Hsu *et al.*, 2010). A model simulation of deposited iron in the ocean region of East Asia suggested the total soluble iron deposition, including the dry and wet depositions, was ~327 tonnes; this contribution may supply primary productivity for the whole spring (Li *et al.*, 2012).

Iron-bearing dust is also a nutrient source for the diazotrophs of nitrogen fixation relating to primary productivity. Nitrogen input from air and iron from desert dust to the oceans enhancing phytoplankton growth-altering atmospheric temperature through increasing sequestration of atmospheric CO_2 and potentially has cooling effect on the global atmosphere (Mahowald, 2010). This biochemical interaction as a factor reflects a role of dust storms in climate. However, overloading nutrients in the oceans may produce toxic algal blooms in coastal environments and cause harmful impacts on marine ecosystems, as well as human health (Griffin and Kellogg, 2004; Tan *et al.*, 2011).

13.7.3 *Impacts on climate*

Dust aerosol transported from the soil surface to the atmosphere perturbs regional and global climate through direct, indirect and semi-direct effects. Unlike greenhouse gases, the effects of dust aerosol on climate usually show substantial spatial and temporal variation (Ramanathan *et al.*, 2001). Dust loadings may directly affect radiative energy balance through the absorption and scattering of solar radiation from the surface to the top atmosphere, i.e. radiative forcing. Direct radiative forcing (RF) is determined by the aerosol optical depth (AOD), the single-scattering albedo and heating rate. The estimation of the annual mean direct RF of mineral dust, on the global scale, is -0.1 ± 0.2 W m^{-2} (IPCC, 2013). With respect to RF of dust aerosols on different levels, the global annual mean RF values are -1.46 W, -0.4 W and 1.86 W m^{-2} at the surface, the top of atmosphere and within the atmosphere, respectively (Zhang *et al.*, 2010). Generally, dust aerosols elevate the temperature within the atmosphere, but with a cooling effect at the surface. Dust aerosols are well known as one of the most important ice nucleation agent for cloud droplet and ice crystal formation (Gibson *et al.*, 2007). Through modification of cloud macrophysical and microphysical properties, dust aerosols can have significant effects on cloud formation and then change radiative fluxes and hydrological cycle, which is called indirect radiative forcing.

In the present day, multi-satellite sensors can be useful techniques to track transport pathways and to characterise chemical compositions and physical properties of dust storm along their transport processes. Cao *et al.* (2014) analysed several ground based and satellite data, such as ground-based Aerosol robotic Network (AERONET), Moderate Resolution Imaging Spectroradiometer (MODIS), Atmospheric Infrared Sounder (AIRS) to investigate the effects of dust storms on aerosol optical properties in Beijing during 2005–2010. Table 13.1 lists the variation of aerosol optical depth (AOD) and Ångström Exponent (AE) during the dust storm event and the days prior to and after the dust storm. Near zero or slightly negative AE values represents coarse particles observed from dust storm activities (Eck *et al.*, 2005). The higher AOD value and lower value of AE during the dust storm compared with the days before and afterwards indicated an obvious increase of coarse particles.

Table 13.1.   AOD, AE and SSA from AERONET for the event on 22nd March 2010 and the days prior to and after the dust storm event.

Date	AOD (675 nm)	AE (440–870 nm)	SSA (440–1020 nm)
17 March 2010	0.190	1.316	0.791
22 March 2010	3.169	−0.119	0.900
25 March 2010	0.084	1.135	0.883

Source: Cao *et al.* (2014).

The average aerosol single scattering albedo (SSA: the ratio of the light scattered by the aerosol to total extinction) for the dust storm day in Beijing was higher than the non-dust storm days and increases at higher wavelength (Cao *et al.*, 2014) indicating a large number of scattering particles caused by the dust storm. Nevertheless, the average mid-visible (550 nm) SSA during the dust events was ~0.87 lower than the value (~0.89) averaged from long-term observation in Beijing. It suggested that pollutants mixed were well with dust aerosols as the dust storm passed through Beijing and may modify absorption of dust aerosols (Cao *et al.*, 2014; Eck *et al.*, 2005).

13.8 Conclusion

Dust storms occurred in spring 2010 were the most severe dust outbreaks over the past two decades and affected the North Hemisphere for over a month. Synoptic circulation was a pivotal factor responsible for these outbreaks. The two remarkable dust storms brought about extreme devastation to dust source regions and downwind areas. The most took place from 19th to 23rd March and impacted a vast area which including the whole East Asian continent, North Pacific Ocean and Arctic. The other observed in April was classified as a *black storm* in dust source regions. The generation mechanisms of both dust storm events indicated the discrepancy of thermal and dynamic conditions between surface and upper atmosphere creating strong surface wind to generate the storms. Additionally, strong frontal systems and a trough from western Siberia speedy extended southward and northeastward in the upper troposphere facilitating transport northward to Arctic and southward to tropical areas. The vertical structure and horizontal movement of synoptic circulation controlled the generation, duration, transport path and the strength of the storms. Additionally, inappropriate land-use resulting in deforestation and land surface degradation contributed to large quantities of dust particles as one of decisive factors in dust storm formation.

Several effects of the dust storm event in March 2010 have been investigated. There was deterioration in air quality in both dust source regions and downwind, particularly for the East Asian continent. The records of $PM_{2.5}$ and PM_{10} concentrations suggested that air quality in most dust particle receptor areas reached hazardous levels, which posed a serious risk of respiratory effects in the general population and aggravated symptoms in persons with cardiopulmonary disease. Moreover, some of anthropogenic metallic elements (i.e. Ni) may affect the skin. The dust may also have contributed nutrients to the ocean. Satellite observations and the analyses of dust suggested that the chemical composition and physical properties of dust particles were modified during transport and as the particles passed through polluted regions in northeastern China. This represented a second source region for the dust aerosol. The optical properties of dust aerosols suggested anthropogenic pollutants carried by dust particles may increase radiation absorption of dust particles, even during the dust storm period.

References

Air Environment Early Warning System (Nagasaki) (2016). PM_{10} Monitoring Information in Nagasaki http://nagasaki-taiki.aa0.netvolante.jp/graph/item, 21st June, 2016 and (in Japanese).

Alfaro, S. C., Gomes, L., Rajot, J. L., Lafon, S., Gaudichet, A., Chatenet, B., Maille, M., Cautenet, G., Lasserre, F., Cachier, H. and Zhang, X. Y. (2003). Chemical and optical characterization of aerosols measured in spring 2002 at the Ace-Asia supersite, Zhenbeitai, China, *Journal of Geophysical Research Atmosphere*, **108**, 8641, doi:10.1029/2002JD003214.

Arimoto, R., Kim, Y. J., Kim, Y. P., Quinn, P. K., Bates, T. S., Anderson, T. L., Gong, S., Uno, I., Chin, M., Huebert, B. J., Clarke, A. D., Shinozuka, Y., Weber, R. J., Anderson, J. R., Guazzotti, S. A., Sullivan, R. C., Sodeman, D. A., Prather, K. A. and Sokolik, I. N. (2006). Characterization of Asian Dust During Ace-Asia, *Global and Planetary Change*, **52**, 23–56.

Biscaye, P., Grousset, F., Revel, M., Van der Gaast, S., Zielinski, G., Vaars, A. and Kukla, G. (1997). Asian Provenance of glacial dust (Stage 2) in the Greenland Ice Sheet Project 2 Ice Core, Summit, Greenland, *Journal of Geophysical Research. (Oceans)*, **102**, 26765–26781.

Bishop, J. K. B., Davis, R. E. and Sherman, J. T. (2002). Robotic observations of dust storm enhancement of carbon biomass in the North Pacific, *Science*, **298**, 817–821.

Brown, J. K. M. and Hovmøller, M. S. (2002). Aerial dispersal of pathogens on the global and continental scales and its impact on plant disease, *Science*, **297**, 537–541.

Cao, C., Zheng, S. and Singh, R. P. (2014). Characteristics of aerosol optical properties and meteorological parameters during three major dust events (2005–2010) over Beijing, China, *Atmospheric Research*, **150**, 129–142.

Chan, C.-C., Chuang, K.-J., Chen, W.-J., Chang, W.-T., Lee, C.-T. and Peng, C.-M. (2008). Increasing cardiopulmonary emergency visits by long-range transported Asian dust storms in Taiwan, *Environmental Research*, **106**, 393–400.

Chen, C.-T A. and Wann, J.-K. (2004). Factors regulating the distribution of elements in the sediments of a seasonally anoxic lake in tropical Taiwan, *Terrestrial Atmospheres Oceanic Sciences*, **15**, 785–811.

Chen, C.-T A., Wann, J.-K. and Lou, J.-Y. (2001). Aeolian flux of metals in Taiwan in the Past 2600 Years, *Chemosphere*, **43**, 287–294.

Chen, F., Qiang, M., Zhou, A., Xiao, S., Chen, J. and Sun, D. (2013). A 2000 year dust storm record from lake sugan in the dust source area of arid China, *Journal of Geophysical Research and Atmospheres*, **118**, 2149–2160.

Chen, H., Zhao, L., Zhao, L., Tian, H., Wu, H. and Huan, N. (2012). Effects of sand dust weather on the air quality of Beijing, *Research of Environmental Sciences*, **25**, 609–614 (in Chinese).

China Meteorological Administration, Dust Storm Definition (2016). http://www.cma.gov.cn/kppd/kppdqxtb/201303/t20130315_207755.html, 1st May 2016 (in Chinese).

Choi, H., Shin, D. W., Kim, W., Doh, S.-J., Lee, S. H., Noh, M. (2011). Asian dust storm particles induce a broad toxicological transcriptional program in human epidermal keratinocytes, *Toxicology Letters*, **200**, 92–99.

Chun, Y., Boo, K. O., Kim, J., Park, S. U. and Lee, M. (2001). Synopsis, transport, and physical characteristics of Asian Dust in Korea, *Journal of Geophysical Researched Atmospheres*, **106**, 18461–18469.

Chun, Y., Cho, H.-K., Chung, H.-S. and Lee, M. (2008). Historical Records of Asian Dust Events (Hwangsa) in Korea, *Bulletin of the American Meteorology Society*, **89**, 823–827.

Chung, Y.-S., Kim, H.-S., Park, K.-H., Jugder, D. and Gao, T. (2005). Observations of dust-storms in China, Mongolia and associated dust falls in Korea in Spring 2003, *Water, Air, Soil Pollution Focus*, **5**, 15–35.

Cottle, P., Strawbridge, K., McKendry, I., O'Neill, N. and Saha, A. (2013). A Pervasive and persistent Asian Dust event over North America during spring 2010: Lidar and sunphotometer observations, *Atmospheric Chemistry and Physics*, **13**, 4515–4527.

Di, X., Zhang, X., Liu, X., Honge, S. (2014). Comparative analysis of two severe sandstorm events in Gansu, *Journal of Arid Meteorology*, **32**, 81–86.

Eck, T. F., Holben, B. N., Dubovik, O., Smirnov, A., Goloub, P., Chen, H. B., Chatenet, B., Gomes, L., Zhang, X. Y., Tsay, S. C., Ji, Q., Giles, D. and Slutsker, I. (2005). Columnar aerosol optical properties at aeronet sites in central Eastern Asia and aerosol transport to the tropical mid-pacific, *Journal of Geophysical Research-D, Atmosphere*, **110**, D06202, doi:10.1029/2004JD005274.

Gao, H. and Washington, R. (2009). The spatial and temporal characteristics of Toms Ai over the Tarim Basin, China, *Atmospheric Environment*, **43**, 1106–1115.

Gibson, E. R., Gierlus, K. M., Hudson, P. K. and Grassian, V. H. (2007). Generation of internally mixed insoluble and soluble aerosol particles to investigate the impact of atmospheric aging and heterogeneous processing on the ccn activity of mineral dust aerosol, *Aerosol Science Technology*, **41**, 914–924.

Goudie, A. S. and Middleton, N. J. (2001). Saharan dust storms: nature and consequences, *Earth-Science Reviews*, **56**, 179–204.

Griffin, D. W. and Kellogg, C. A. (2004). Dust storms and their impact on ocean and human health: Dust in Earth's atmosphere, *EcoHealth*, **1**, 284–295.

Grousset, F. E., Ginoux, P., Bory, A., Biscaye, P. E. (2003). Case study of a Chinese dust plume reaching the French Alps, *Geophysical Research Letters*, **30**, 1277, doi:1210.1029/2002GL016833.

Guo, J., Rahn, K. A. and Zhuang, G. (2004). A mechanism for the increase of pollution elements in dust storms in Beijing, *Atmospheric Environment*, **38**, 855–862.

Guo, P., Yin, X., Liu, S. and Liu, D. (2011). Analysis of a heavy dust storm occurred in middle part of Hexi corridor and forecast methods study for dust storm weather, *Journal of Arid Meteorology*, **29**, 110–115 (in Chinese).

Han, Y., Dai, X., Fang, X., Chen, Y. and Kang, F. (2008). Dust aerosols: A possible accelerant for an increasingly arid climate in North China, *Journal Arid Environment*, **72**, 1476–1489.

He, Y., Zhao, C., Song, M., Liu, W., Chen, F., Zhang, D. and Liu, Z. (2015). Onset of frequent dust storms in Northern China at ~AD 1100, *Science Reports*, **5**, doi: 10.1038/srep17111.

Hong, Y.-C., Pan, X.-C., Kim, S.-Y., Park, K., Park, E.-J., Jin, X., Yi, S.-M., Kim, Y.-H., Park, C.-H., Song, S. and Kim, H. (2010). Asian dust storm and pulmonary function of school children in seoul, *Science of the Total Environment*, **408**, 754–759.

Hsieh, N.-H. and Liao, C.-M. (2013). Assessing exposure risk for dust storm events-associated lung function decrement in asthmatics and implications for control, *Atmospheric Environment*, **68**, 256–264.

Hsu, S. C., Tsai, F., Lin, F. J., Chen, W. N., Shiah, F. K., Huang, J. C., Chan, C. Y., Chen, C. C., Liu, T. H., Chen, H. Y., Tseng, C. M., Hung, G. W., Huang, C. H., Lin, S. H. and Huang, Y. T. (2013). A super Asian dust storm over the East and South China seas: Disproportionate dust deposition, *Journal of Geophysical Research Atmospheres*, **118**, 7169–7181.

Hsu, S.-C., Wong, G. T. F., Gong, G.-C., Shiah, F.-K., Huang, Y.-T., Kao, S.-J., Tsai, F., Candice Lung, S.-C., Lin, F.-J., Lin, I. I., Hung, C.-C. and Tseng, C.-M. (2010). Sources, solubility, and dry deposition of aerosol trace elements over the East China Sea, *Marine Chemistry*, **120**, 116–127.

Huang, Z., Huang, J., Hayasaka, T., Wang, S., Zhou, T. and Jin, H. (2015). Short-cut transport path for Asian Dust directly to the Arctic: A Case Study, *Environmental Research Letters*, **10**, 114018.

Huebert, B. J., Bates, T., Russell, P. B., Shi, G., Kim, Y. J., Kawamura, K., Carmichael, G. and Nakajima, T. (2003). An Overview of Ace — Asia: Strategies for quantifying the relationships between Asian Aerosols and their climatic impacts, *Journal of Geophysical Research Atmosphere*, **108**, 8633, doi:8610.1029/2003JD003550.

Intergovernmental Panel on Climate Change (IPCC) (2013). The physical science basis, in Contribution of Working Group I to the Fifth

Assessment Report of the Intergovernmental Panel on Climate Change.

Jacob, D. J., Crawford, J. H., Kleb, M. M., Connors, V. S., Bendura, R. J., Raper, J. L., Sachse, G. W., Gille, J. C., Emmons, L. and Heald, C. L. (2003). Transport and chemical evolution over the Pacific (Trace-P) aircraft mission: Design, execution, and first results, *Journal of Geophysical Research Atmosphere*, **108**, 9000, doi:10.1029/2002JD003276.

Jaffe, D., Anderson, T., Covert, D., Kotchenruther, R., Trost, B., Danielson, J., Simpson, W., Berntsen, T., Karlsdottir, S., Blake, D., Harris, J., Carmichael, G. and Uno, I. (1999). Transport of Asian air pollution to North America, *Geophysical Research Letters*, **26**, 711–714.

Japan Ministry of the Environment, Particulate Matter Monitoring Data (2016). http://www.env.go.jp/air/osen/pm/monitoring.html, 17th July 2016 (in Japanese).

Jugder, D., Shinoda, M., Sugimoto, N., Matsui, I., Nishikawa, M., Park, S.-U., Chun, Y.-S. and Park, M.-S. (2011). Spatial and temporal variations of dust concentrations in the Gobi Desert of Mongolia, *Global Planet Change*, **78**, 14–22.

Kanatani, K. T., Ito, I., Al-Delaimy, W. K., Adachi, Y., Mathews, W. C., Ramsdell, J. W., Toyama Asian Desert, D. and Asthma Study, T. (2010). Desert dust exposure is associated with increased risk of Asthma hospitalization in children, *American Journal of Respiratory Critical Care Medicine*, **182**, 1475–1481.

Kim, J. (2008). Transport routes and source regions of Asian dust observed in Korea during the past 40 years (1965–2004), *Atmospheric Environment*, **42**, 4778–4789.

Kim, J., Jung, C. H., Choi, B.-C., Oh, S.-N., Brechtel, F. J., Yoon, S.-C. and Kim, S.-W. (2007). Number size distribution of atmospheric aerosols during Ace-Asia dust and precipitation events, *Atmospheric Environment*, **41**, 4841–4855.

Kok, J. F., Parteli, E. J., Michaels, T. I. and Karam, D. B. (2012). The physics of wind-blown sand and dust, *Report on Progress Physics*, **75**, 106901, 4.

Kurosaki, Y., Shinoda, M. and Mikami, M. (2011). What caused a recent increase in dust outbreaks over East Asia?, *Geophysical Research Letters*, **38**, L11702, doi:11710.11029/12011GL047494.

Kwak, J.-H., Kim, G., Kim, Y.-J. and Park, K. (2012). Determination of heavy metal distribution in PM_{10} during asian dust and local pollution events using laser induced breakdown spectroscopy (Libs), *Aerosol Science Technology*, **46**, 1079–1089.

Larssen, T. and Carmichael, G. R. (2000). Acid rain and acidification in China: The importance of base cation deposition, *Environmental Pollution*, **110**, 89–102.

Laurent, B., Marticorena, B., Bergametti, G. and Mei, F. (2006). Modeling mineral dust emissions from Chinese and Mongolian Deserts, *Global Planetary Change*, **52**, 121–141.

Lee, E.-H. and Sohn, B.-J. (2011). Recent Increasing trend in dust frequency over Mongolia and Inner Mongolia regions and its association with climate and surface condition change, *Atmospheric Environment*, **45**, 4611–4616.

Lee, H., Kim, H., Honda, Y., Lim, Y.-H. and Yi, S. (2013). Effect of Asian dust storms on daily mortality in seven metropolitan cities of Korea, *Atmospheric Environment*, **79**, 510–517.

Li, J., Han, Z. and Zhang, R. (2011). Model study of atmospheric particulates during dust storm period in March 2010 over East Asia, *Atmospheric Environment*, **45**, 3954–3964.

Li, J., Wang, Z., Zhuang, G., Luo, G., Sun, Y. and Wang, Q. (2012). Mixing of Asian mineral dust with anthropogenic pollutants over East Asia: A Model case study of a super-duststorm in March 2010, *Atmospheric Chemistry Physics*, **12**, 7591–7607.

Li, Y.-Y., Xu, D.-B. and Chen, Y. (2013). The turning causes of trough to ridge in a typical black sandstrom process occurred in North-West China, *Journal Desert Research*, **23**, 187–194 (in Chinese).

Lim, J. and Matsumoto, E. (2006). Bimodal grain-size distribution of Aeolian Quartz in a Maar of Cheju Island, Korea, During the last 6500 years: Its flux variation and controlling factor, *Geophysical Research Letters*, **33**, L21816, doi:21810.21029/22006GL027432.

Liu, C.-M., Young, C.-Y. and Lee, Y.-C. (2006). Influence of Asian dust storms on air quality in Taiwan, *Science & Total Environment*, **368**, 884–897.

Liu, T.-S., Gu, S.-F., An, Z.-S. and Fan, Y.-X. (1981). The dust fall in Beijing, China, on 18 April 1980., Desert Dust, in Pewe, T. L. (ed.), *Geological Society of America Special Paper*, **186**, 149–157.

Liu, Z., Fairlie, T. D., Uno, I., Huang, J., Wu, D., Omar, A., Kar, J., Vaughan, M., Rogers, R., Winker, D., Trepte, C., Hu, Y., Sun, W., Lin, B. and

Cheng, A. (2013). Transpacific transport and evolution of the optical properties of Asian Dust, *Journal of Quantitative Spectroscopy Radiative Transfer*, **116**, 24–33.

Lou, J.-Y., Chen, C.-T. A. and Wann, J.-K. (1997). Paleoclimatological records of the Great Ghost Lake in Taiwan, *Science in China Series D Earth Sciences*, **40**, 284–292.

Lu, Q. and Wu, B. (2002). Disaster assessment and economic loss budget of desertification in China, *China Population, Research and Environment*, **2**, 011 (in Chinese).

Mahowald, N. M., Baker, A. R., Bergametti, G., Brooks, N., Duce, R. A., Jickells, T. D., Kubilay, N., Prospero, J. M. and Tegen, I. (2005). Atmospheric global dust cycle and iron inputs to the ocean, *Global Biogeochemical Cycles*, **19**, GB4025, doi:10.1029/2004GB002402.

Mahowald, N. M., Engelstaedter, S., Luo, C., Sealy, A., Artaxo, P., Benitez-Nelson, C., Bonnet, S., Chen, Y., Chuang, P. Y., Cohen, D. D., Dulac, F., Herut, B., Johansen, A. M., Kubilay, N., Losno, R., Maenhaut, W., Paytan, A., Prospero, J. M., Shank, L. M. and Siefert, R. L. (2008). Atmospheric iron deposition: Global distribution, variability, and human perturbations, *Annual Review of Marine Science*, **1**, 245–278.

Mahowald, N. M., Kloster, S., Engelstaedter, S., Moore, J. K., Mukhopadhyay, S., McConnell, J. R., Albani, S., Doney, S. C., Bhattacharya, A., Curran, M. A. J., Flanner, M. G., Hoffman, F. M., Lawrence, D. M., Lindsay, K., Mayewski, P. A., Neff, J., Rothenberg, D., Thomas, E., Thornton, P. E. and Zender, C. S. (2010). Observed 20th century desert dust variability: Impact on climate and biogeochemistry, *Atmospheric Chemistry and Physics*, **10**, 10875–10893.

Mahowald, N., Albani, S., Kok, J. F., Engelstaeder, S., Scanza, R., Ward, D. S. and Flanner, M. G. (2014). The size distribution of desert dust aerosols and its impact on the Earth system, *Aeolian Research*, **15**, 53–71.

Mao, R., Ho, C.-H., Shao, Y., Gong, D.-Y. and Kim, J. (2011). Influence of Arctic oscillation on dust activity over Northeast Asia, *Atmospheric Environment*, **45**, 326–337.

McKendry, I. G., Macdonald, A. M., Leaitch, W. R., van Donkelaar, A., Zhang, Q., Duck, T. and Martin, R. V. (2008). Trans-pacific dust events observed at whistler, British Columbia during Intex-B, *Atmospheric Chemistry and Physics*, **8**, 6297–6307.

Meng, Z. and Lu, B. (2007). Dust events as a risk factor for daily hospitalization for respiratory and cardiovascular diseases in Minqin, China, *Atmospheric Environment*, **41**, 7048–7058.

Mu, H., Otani, S., Shinoda, M., Yokoyama, Y., Onishi, K., Hosoda, T., Okamoto, M. and Kurozawa, Y. (2013). Long-term effects of livestock loss caused by dust storm on mongolian inhabitants: A survey 1 year after the dust storm, *Yonago Acta Med.*, **56**, 39–42.

Natsagdorj, L., Jugder, D. and Chung, Y. S. (2003). Analysis of dust storms observed in Mongolia during 1937–1999, *Atmospheric Environment*, **37**, 1401–1411.

Onishi, K., Kurosaki, Y., Otani, S., Yoshida, A., Sugimoto, N. and Kurozawa, Y. (2012). Atmospheric transport route determines components of Asian dust and health effects in Japan, *Atmospheric Environment*, **49**, 94–102.

Onishi, K., Otani, S., Yoshida, A., Mu, H. and Kurozawa, Y. (2015). Adverse health effects of asian dust particles and heavy metals in Japan, *Asia-Pacific Journal of Public Health*, **27**, NP1719–NP1726.

Otani, S., Onishi, K., Mu, H. and Kurozawa, Y. (2011). The effect of Asian dust events on the daily symptoms in Yonago, Japan: A pilot study on healthy subjects, *Archives of Environmental Occupational Hazard*, **66**, 43–46.

Otani, S., Onishi, K., Mu, H., Yokoyama, Y., Hosoda, T., Okamoto, M. and Kurozawa, Y. (2012). The relationship between skin symptoms and allergic reactions to Asian dust, *International Journal of Environmental Research Public Health*, **9**, 4606–4614.

Park, S. S. and Cho, S. Y. (2013). Characterization of organic aerosol particles observed during Asian dust events in spring 2010, *Aerosol Air Quality Research*, **13**, 1019–1033.

Qian, L., Yang, Y. and Wang, R. (2011). Analysis of black strom cause in Hexi Corridor on 24 April 2010, *Plateau Meteorology*, **30**, 1653–1660 (in Chinese).

Qian, W., Quan, L. and Shi, S. (2002). Variations of the dust storm in China and its climatic control, *Journal of Climatology*, **15**, 1216–1229.

Qian, W., Tang, X. and Quan, L. (2004). Regional characteristics of dust storms in China, *Atmospheric Environment*, **38**, 4895–4907.

Qiang, M., Chen, F., Zhou, A., Xiao, S., Zhang, J. and Wang, Z. (2007). Impacts of wind velocity on sand and dust deposition during dust storm as inferred from a series of observations in the Northeastern Qinghai–Tibetan Plateau, China, *Powder Technology*, **175**, 82–89.

Ramanathan, V., Crutzen, P. J., Kiehl, J. T. and Rosenfeld, D. (2001). Aerosols, climate, and the hydrological cycle, *Science*, **294**, 2119–2124.

Sivakumar, M. V. K. (2005). Impacts of sand/dust storms on agriculture, In: Sivakumar, M. V. K., Motha, R. P. and Das, H. P. (eds.), *Natural Disasters and Extreme Events in Agriculture* (Springer, New York), pp. 159–177.

Sun, J., Zhang, M. and Liu, T. (2001). Spatial and temporal characteristics of dust storms in China and its surrounding Regions, 1960–1999: Relations to Source Area and Climate, *Journal of Geophysical Research Atmosphere*, **106**, 10325–10333.

Sun, L., Zhou, X., Lu, J., Kim, Y.-P. and Chung, Y.-S. (2003). Climatology, trend analysis and prediction of sandstorms and their associated dust-fall in China, *Water, Air, Soil Pollution Focus*, **3**, 41–50.

Sun, Y., Zhuang, G., Wang, Y., Zhao, X., Li, J., Wang, Z. and An, Z. (2005). Chemical composition of dust storms in Beijing and implications for the mixing of mineral aerosol with pollution aerosol on the pathway, *Journal of Geophysical Research Atmosphere*, **110**, D24209, doi:10.1029/2005JD006054.

Taiwan, Environmental Protection Administration, Asia Dust Storm Monitoring Network (2016). http://dust.epa.gov.tw/dust/tw/b0307.aspx, 15th June 2016 (in Chinese).

Tan, M., Li, X. and Xin, L. (2014). Intensity of dust storms in China from 1980 to 2007: A new definition, *Atmospheric Environment*, **85**, 215–222.

Tan, S. C., Shi, G. Y., Shi, J. H., Gao, H. W. and Yao, X. (2011). Correlation of Asian Dust with chlorophyll and primary productivity in the coastal seas of china during the period from 1998 to 2008, *Journal of Geophysical Research (Biogeosci.)*, **116**, G02029, doi:10.1029/2010JG001456.

Tanaka, T. Y. and Chiba, M. (2006). A numerical study of the contributions of dust source regions to the global dust budget, *Global Planetary Change*, **52**, 88–104.

Tatarov, B., Müller, D., Noh, Y. M., Lee, K. H., Shin, D. H., Shin, S. K., Sugimoto, N., Seifert, P. and Kim, Y. J. (2012). Record heavy mineral dust outbreaks over Korea in 2010: Two cases observed with multiwavelength aerosol/depolarization/Raman-Quartz Lidar, *Geophysical Research Letters*, **39**, L14801, doi:14810.11029/12012GL051972.

Tsai, F.-J., Fang, Y.-S. and Huang, S.-J. (2013). Case study of Asian dust event on 19–25 March 2010 and Its impact on the marginal sea of China, *Journal of Marine Science Technology*, **21**, 353–360.

Tsai, J.-H., Huang, K.-L., Lin, N.-H., Chen, S.-J., Lin, T.-C., Chen, S.-C., Lin, C.-C., Hsu, S.-C. and Lin, W.-Y. (2012). Influence of an Asian dust storm and Southeast Asian biomass burning on the characteristics of seashore atmospheric aerosols in Southern Taiwan, *Aerosol Air Quality Research*, **12**, 1105–1115.

United Nations Environment Programme, (UNEP) (2016). Global assessment of sand and dust storms, 55.

Uno, I., Amano, H., Emori, S., Kinoshita, K., Matsui, I. and Sugimoto, N. (2001). Trans-pacific yellow sand transport observed in April 1998: A numerical simulation, *Journal of Geophysical Research Atmospheres*, **106**, 18331–18344.

Uno, I., Eguchi, K., Yumimoto, K., Liu, Z., Hara, Y., Sugimoto, N., Shimizu, A. and Takemura, T. (2011). Large Asian dust layers continuously reached North America in April 2010, *Atmospheric Chemistry Physical*, **11**, 7333–7341.

Uno, I., Eguchi, K., Yumimoto, K., Takemura, T., Shimizu, A., Uematsu, M., Liu, Z., Wang, Z., Hara, Y. and Sugimoto, N. (2009). Asian dust transported one full circuit around the Globe, *Nature Geoscience*, **2**, 557–560.

Wang, L., Du, H., Chen, J., Zhang, M., Huang, X., Tan, H., Kong, L. and Geng, F. (2013). Consecutive transport of anthropogenic air masses and dust storm plume: Two case events at Shanghai, China, *Atmospheric Research*, **127**, 22–33.

Wang, R. F., Feng, Q. and Shang, K. Z. (2014). A severe sand-dust storm over China in the Spring of 2010, *Arid Land Geographic*, **37**, 31–44 (in Chinese).

Wang, S., Wang, J., Zhou, Z. and Shang, K. (2005). Regional characteristics of three kinds of dust storm events in China, *Atmospheric Environment*, **39**, 509–520.

Wang, S.-H., Tsay, S.-C., Lin, N.-H., Hsu, N. C., Bell, S. W., Li, C., Ji, Q., Jeong, M.-J., Hansell, R. A., Welton, E. J., Holben, B. N., Sheu, G.-R., Chu, Y.-C., Chang, S.-C., Liu, J.-J. and Chiang, W.-L. (2011). First detailed observations of long-range transported dust over the Northern South China Sea, *Atmospheric Environment*, **45**, 4804–4808.

Wang, X., Dong, Z., Zhang, J. and Liu, L. (2004). Modern dust storms in China: An Overview, *Journal Arid Environment,* **58**, 559–574.

Wang, X., Huang, J., Ji, M. and Higuchi, K. (2008). Variability of East Asia dust events and their long-term trend, *Atmospheric Environment,* **42**, 3156–3165.

Wang, Z., Akimoto, H. and Uno, I. (2002). Neutralization of soil aerosol and its impact on the distribution of acid rain over east Asia: Observations and model results, *Journal of Geophysical Research Atmosphere,* **107**, ACH 6.1–6.12.

Washington, R., Todd, M., Middleton, N. J. and Goudie, A. S. (2003). Dust-storm source areas determined by the total ozone monitoring Spectrometer and surface observations, *Annals of the Association of American Geographers,* **93**, 297–313.

Watanabe, M., Noma, H., Kurai, J., Sano, H., Mikami, M., Yamamoto, H., Ueda, Y., Touge, H., Fujii, Y., Ikeda, T., Tokuyasu, H., Konishi, T., Yamasaki, A., Igishi, T. and Shimizu, E. (2016). Effect of Asian dust on pulmonary function in adult asthma patients in Western Japan: A panel study, *Allergology International,* **65**, 147–152.

Watanabe, M., Yamasaki, A., Burioka, N., Kurai, J., Yoneda, K., Yoshida, A., Igishi, T., Fukuoka, Y., Nakamoto, M., Takeuchi, H., Suyama, H., Tatsukawa, T., Chikumi, H., Matsumoto, S., Sako, T., Hasegawa, Y., Okazaki, R., Horasaki, K. and Shimizu, E. (2011). Correlation between Asian dust storms worsening asthma in Western Japan, *Allergology International,* **60**, 267–275.

World Meteorological Organization (WMO) (2010). A snapshot of some extreme events over the past decade, 5.

Yan, Y., Sun, Y. B., Weiss, D., Liang, L. J. and Chen, H. Y. (2015). Polluted dust derived from long-range transport as a major end member of urban aerosols and its implication of non-point pollution in Northern China, *Science Total Environment,* 506–507, 538–545.

Yang, B., Bräuning, A., Zhang, Z. Dong, Z. and Esper, J. (2007). Dust storm frequency and its relation to climate changes in Northern China during the past 1000 years, *Atmospherical Environment,* **41**, 9288–9299.

Yang, X., Fang, D. and Li, X. (1998). Analysis on the variation trend of sandstorm in Northern China, *Quarterly Journal of Applied Meteorology,* **3**, 352–358 (in Chinese).

Yeh, C.-F., Lee, C.-L., Brimblecombe, P. and Lai, I. C. (2015). Markers of East Asian dust storms in March 2010, *Atmospheric Environment*, **118**, 219–226.

Yoo, Y., Choung, J. T., Yu, J., Kim, D. K. and Koh, Y. Y. (2008). Acute effects of asian dust events on respiratory symptoms and peak expiratory flow in children with mild asthma, *Journal of Korean Medicine Science*, **23**, 66–71.

Yu, L., Wang, G., Zhang, R., Zhang, L., Song, Y., Wu, B., Li, X., An, K. and Chu, J. (2013). Characterization and source apportionment of $Pm_{2.5}$ in an urban environment in Beijing, *Aerosol Air Quality Research*, **13**, 574–583.

Zhang, B., Tsunekawa, A. and Tsubo, M. (2008). Contributions of sandy lands and stony deserts to long-distance dust emission in China and Mongolia during 2000–2006, *Global Planetary Change*, **60**, 487–504.

Zhang, D. (1984). Synoptic-climatic studies of dust fall in China since historic times, *Science China Chemistry*, **27**, 825–836.

Zhang, R., Shen, Z., Cheng, T., Zhang, M. and Liu, Y. (2010). The elemental composition of atmospheric particles at Beijing during asian dust events in spring 2004, *Aerosol Air Quality Research*, **10**, 67–75.

Zhao, J., Zhang, F., Xu, Y., Chen, J., Yin, L., Shang, X. and Xu, L. (2011). Chemical characteristics of particulate matter during a heavy dust episode in a coastal city, Xiamen, 2010, *Aerosol Air Quality Research*, **11**, 299–308.

Zhao, S., Yu, Y., Xia, D., Yin, D., He, J., Liu, N. and Li, F. (2015). Urban particle size distributions during two contrasting dust events originating from Taklamakan and Gobi Deserts, *Environmental Pollution*, **207**, 107–122.

Zhu, C., Wang, B. and Qian, W. (2008). Why do dust storms decrease in Northern China concurrently with the recent global warming?, *Geophysical Research Letters*, **35**, L18702, doi:18710.11029/12008GL034886.

Chapter 14

Beijing (2013): A Driver of Change

Zhi Ning*, Junke Zhang[†] and Jian Gao[‡]

*School of Energy and Environment, City University of
Hong Kong, Hong Kong
[†]Faculty of Geosciences and Environmental Engineering,
Southwest Jiaotong University, China
[‡]Chinese Research Academy of Environmental Sciences, China

14.1 Background

Beijing (39°56′N, 116°20′E), the political, economic, cultural and international exchange centre and the capital of China is one of the largest megacities in the world with more than 21.5 million residents and 5.6 million vehicles in operation by the end of 2014. The energy consumption was 68.3 million tonnes of standard coal in 2014 (Beijing Municipal Bureau of Statistics, 2015).

In the west, north and northeast, Beijing is surrounded by the Taihang and Yanshan mountains (Fig. 14.1) at approximately 1,000–1,500 m, a.s.l. (Chen *et al.*, 2015). This typical U-shape terrain surrounding the city provides a possible trap for air pollutant from the south and southeast in the city and can result in poor air quality under stagnant conditions with weak winds (Zhang *et al.*, 2012). According to the Beijing Environmental Statement 2013 and 2014, the $PM_{2.5}$ (Particulate Matter with aerodynamic diameter less than 2.5 μm) concentrations reached an annual mean of 85.9 μg m^{-3} in

345

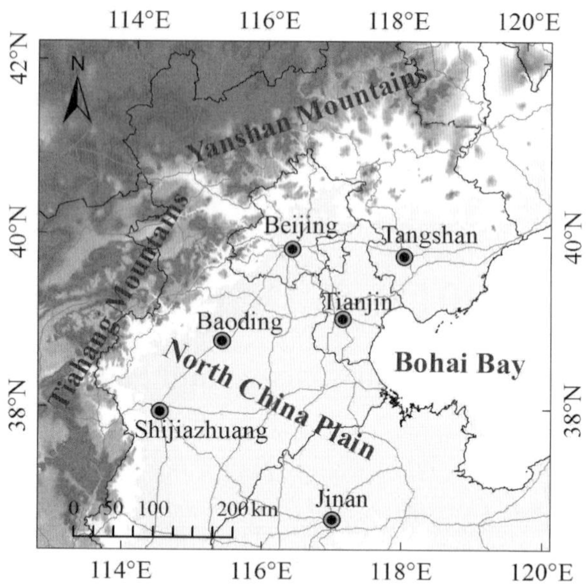

Fig. 14.1.   Satellite map of Beijing and the major cities in surrounding area.

2014, which is 2.5 times higher than the new national ambient air quality standards (NAAQS) (35 μg m^{-3}) and 9 times higher than the air quality guideline (10 μg m^{-3}) recommended by the World Health Organization (WHO). This value was only 3.6 μg m^{-3} lower than the value of 89.5 μg m^{-3} in 2013 (http://www. bjepb.gov.cn/), indicating that fine particulate matter (PM) pollution remains a very serious issue in Beijing (Zhang *et al.*, 2016).

The weather in Beijing is a typical north temperate semi-humid continental monsoon climate. Generally, the four seasons of Beijing are characterised by variable meteorological conditions: spring features high wind speeds and low precipitation, summer high temperature and frequent rain, autumn sunny days and northwest winds, and winter cold and dry air (Zhang *et al.*, 2013). The seasonal distribution of precipitation is very uneven. For example, in 2014, the annual precipitation was 461.5 mm, mainly occurring in summer (262 mm), but only 4.8 mm in winter.

As the capital city of China, the air quality of Beijing has been a spotlight of national and international attention since 2008 Olympic

Games. The local authorities spent tremendous efforts in tackling its air pollution problem. With the fast economic development and expansive urbanisation, the characteristics and sources of air pollution in the region also changed. These were mainly suspended dust and coal combustion in the later 1990s, while their contributions now have decreased. On the other hand, the vehicle emissions have increased considerably in the inventory, together with the secondary sources of PM pollution (BEPB, 2013). The fast shift of emission inventory and pollution source profiles witnessed the development of the city but also posed challenges in the development and implementation of pollution control measures.

Air pollution, especially $PM_{2.5}$, was not the centre of media attention in China a decade ago. The 2008 Olympic Games in Beijing raised considerable public interest in air pollution both from a scientific (Streets *et al.*, 2007) and media perspective (Min and Zhen, 2010). Along with the economic development, public awareness and attention to environmental issues also increased, and the air quality management authorities were urged to take immediate and effective actions in order to meet public expectation. Implementation of various emission control policies resulted in significant improvement in Beijing air quality during the games (Rich *et al.*, 2012). In the years since the Olympics, however, air quality in Beijing has not witnessed a continued and significant improvement. According to the long-term air monitoring record in Beijing, although the annual $PM_{2.5}$ concentrations have shown a general declining trend over the years, their levels in winter season, especially during the heating period, clearly increased. The contradiction indicates both the success in pollution control, and also the complexity of urban pollution characteristics and sources.

14.2 General Description

14.2.1 *Description of episode*

The winter of 2013 in Beijing experienced more pollution episodes than earlier years (Wang *et al.*, 2014). One of the monumental events of the pollution episodes occurred in January 2013, marked with extremely severe and persistent haze in Beijing and surrounding regions. This month has been reported as the haziest month in the past 60 years in Beijing (Sun *et al.*, 2014) and the highest hourly $PM_{2.5}$ concentration reached 1,000 μg m^{-3} in some heavily polluted areas of Beijing. Up to 0.8 billion people were affected covering a large area in Northern and Eastern China (Zhang *et al.*, 2014). During the episode period, 70% of the daily average $PM_{2.5}$ mass concentrations exceeded the Grade II standard (GB3095-2012) while only less than 10% of the days met the Grade I standard. The worst visibility was recorded on 29 January (1.8 km), similar to average visibility (1.9 km) of serious fog days (see Fig. 14.2) when relative humidity approaches 100% (Ding and Liu, 2014), although much greater visibility reduction was experienced in the London episode of 1952 (Brimblecombe, 2017). There was much research investigation on the formation reasons of the severe pollution episode. Stagnant meteorological conditions, coal combustion, secondary production and regional transport are regarded as the four major factors driving the formation and evolution of the haze (Ji *et al.*, 2014; Sun *et al.*, 2014; Zheng *et al.*, 2015).

According to the analysis of Zheng *et al.* (2015), there were no significant change in the emission inventory between the winters of 2013 and earlier year of 2012. Zheng *et al.* (2016) also estimated that the emissions were relatively stable on the seasonal scale (except when intensive biomass-burning activities occur; the fireworks during the Spring Festival were not observed in that study because of instrument breakdown), whereas meteorology varies on an hourly basis. Thus, they argued the unfavourable synoptic conditions were a key factor for the severe pollution of January 2013. Simulation results further confirmed the point: Zheng *et al.* (2015) found that under the same emission level, changing the meteorological conditions from 2012 to 2013 resulted in a monthly average $PM_{2.5}$ increase of 10–40 μg m^{-3} in the Beijing area, and up to 120 μg m^{-3} over the whole north China plain. This suggests that the severe haze episodes in January 2013 were most likely due to unfavourable meteorology, rather than an abrupt increase in emissions.

The dominant PM chemical components during the episode were sulphate, nitrate, ammonium, organic materials (OM) and elemental

Fig. 14.2.   Comparative view from the Institute of Atmospheric Physics (IAP), Chinese Academy of Sciences (CAS) in the smog on 13 January 2013; and on 24 January 2013 (photo by Dr. Junke Zhang).

carbon (EC) (Wang *et al.*, 2013; Zhang *et al.*, 2013). Depending on the different phases of episode, they have shown different relative abundancy and characteristics. For example, the organic mass fraction dominated in the clean period (74–77%) and decreased slightly during the transition (48–49%) and polluted (35–42%) periods (Guo *et al.*, 2014). Zheng *et al.* (2015) found that with increasing pollution level, the EC fraction decreased slightly, OC fraction decreased significantly, while SO_4^{2-} and NO_3^- contributions increased sharply. Secondary inorganic aerosol species were suggested to be the major contributor to severe haze, based on offline $PM_{2.5}$ analysis (Quan *et al.*, 2014). In addition, the contribution of secondary organic aerosol (SOA) also increased with the aggravation of pollution (Zhang *et al.*, 2014). The diurnal variation of the main $PM_{2.5}$ components, such as OM, SO_4^{2-}, NO_3^-, NH_4^+ exhibited similar patterns during the episode, with high concentrations at night and low during the day (Zhang *et al.*, 2014, Han *et al.*, 2016). There was also obvious shift of peak particle size in different phases of episode, i.e. fine mode maximum shifted from 0.43–0.65 μm on clear days ($PM_{-2.5} < 75\ \mu g\ m^{-3}$) to 0.65–1.1 μm on lightly polluted days ($75 < PM_{2.5} < 150\ \mu g\ m^3$) and to 1.1–2.1 μm on heavily polluted days with a minor coarse mode component from 4.7 μm to 5.8 μm (Tian *et al.*, 2014).

14.2.2 *The post-episode impact*

The January 2013 episode had profound impact in many aspects. It provided a pivot for the new initiatives in environmental policy making, technology development and scientific investigations. For example, starting from 2013, MEP officially initiated the enactment of *The Emergency Management of China's Heavily Polluted Weather* and followed by issuing related guidelines. The authorities also paid more attention to the science driven air quality management and started nationwide $PM_{2.5}$ source apportionment. Over 40 cities have completed their individual source appointment investigation up to 2015. Monitoring practices and research have started to focus on high time resolution mass spectrometry and isotopic composition data for source apportionment to establish the sources of PM during heavy pollution episodes. This can be further extended to link with air quality data, meteorological data, satellite data for comprehensive research. A large number of significant findings support the importance of targeting fundamental sciences in understanding air pollution episodes. These research initiatives greatly boosted the development of atmospheric sciences and technology, and improved the science-based policy making process in China.

14.3 Formation Mechanisms

14.3.1 *Role of meteorology*

14.3.1.1 *Meteorological conditions at the surface*

During the episode, the average temperature on unpolluted and polluted days was –4.3°C and –2.2°C, respectively, and the average humidity was 25.5% and 61.6%, respectively (Zhang *et al.*, 2014). The higher humidity on polluted days resulted in higher moisture content in the particles and more weight of certain chemical components, such as secondary organic matter, sulphate, nitrate, and ammonium (Quan *et al.*, 2014; Zhang *et al.*, 2015). In addition, the wind speed is also lower in haze episode than in no-haze episode (Zhang *et al.*, 2014), which is not conducive to the dispersion of pollutants. The wind direction on polluted days was dominantly from the south, allowing the transport of pollutants from the adjacent areas south of Beijing. However, the wind direction on unpolluted days was variable, including northwest and north winds that passed over some clean areas, and serving the role of pollution dilution and removal.

Zhang *et al.* (2014) analysed long-term meteorological data from National Centers for Environmental Prediction (NCEP) and surface meteorological observations from January 1981 to 2013. They found an abnormally weak East Asian winter monsoon over eastern China during January 2013, compared to the climatological average. This weak winter monsoon allowed more frequent penetration of humid air from the south and led to calm and stable conditions. Surface weather maps from polluted periods were generally characterised by a weak high-pressure (1,034–1,037 hPa) centre northeast of Beijing, which could result in low surface wind speed and prevent the influx of northwest clean air. In contrast, weather patterns for the clean hours were characterised by strong high-pressure (up to 1,046 hPa) centres northwest of Beijing (Zheng *et al.*, 2015). Similar effects resulted from the change in climate and synoptic systems was previously reported in Qu *et al.* (2013), who found that intensification and extension of western Pacific subtropical high caused more stable and humid weather and contributed to the deterioration of summer visibility over eastern China.

14.3.1.2 *Atmospheric boundary layer*

There is probably a feedback between the planetary boundary layer (PBL) height and aerosol loading (Quan *et al.*, 2013). The enhancement of aerosols tends to depress the development of PBL by decreasing solar radiation, while the repressed structure of PBL, will in turn, weaken the diffusion of pollutants, leading to the heavy pollution (Quan *et al.*, 2014). During the pollution episodic days, the PBL height decreased significantly. For example, on the heavily polluted days of 12th and 28th of January, the mixed layer heights were approximately 200–300 m only one-tenth of that on a clear day (2,000–3,000 m) (Wang *et al.*, 2013, 2014), which prevented the vertical dispersion of pollutants and led to an increase in their surface concentrations (Quan *et al.*, 2014). Zhang *et al.* (2015) simulated the extreme case on the 12 January 2013 and compared the result with three typical winter episodes during January 2010–2014. They found a stronger temperature inversion and a much compressed PBL during 2013 episode, to as low as 40 m. In a larger scale, the height of the PBL across China in 2013 was about 200 m lower than that in 2012, a clear evidence of PBL's role in the haze formation (Zheng *et al.*, 2015).

14.3.2 *Contribution of secondary pollutants*

A large number of studies have shown that secondary inorganic and organic components increased significantly during the severe pollution periods (Huang *et al.*, 2014; Sun *et al.*, 2014; Zhang *et al.*, 2014; Zheng *et al.*, 2015; Ho *et al.*, 2016). Heterogeneous chemistry was considered as a critical pathway for the secondary PM formation, especially for secondary inorganic species (Quan *et al.*, 2015; Zheng *et al.*, 2015). During the episode, $PM_{2.5}$ mass concentration and RH were both high, resulting in low visibility. Meanwhile, the O_3 concentrations and solar radiations were low, which limited photochemical conversion from gas to particles. However, at such high humidity (80%), the radius of aerosol particles could be doubled by water coating the particle surface and providing an excellent site for heterogeneous aqueous reactions (Quan *et al.*, 2015). The main reaction mechanism can be summarised as: (1) for inorganic species, under very high RH conditions, water-rich particles and fog droplets are able to capture more soluble gases such as ammonia, nitric acid, NO_x and SO_2, leading to the more efficient production of ammonium nitrate and sulphate (Ge *et al.*, 2012). Sun *et al.* (2013) also found that at elevated RH levels, aqueous-phase processing plays a more significant role and particularly affects sulphate. They estimated that the fog processing of SO_2 was responsible for approximately 70% of sulphate production during wintertime in Beijing (Sun *et al.*, 2013); (2) for organic species, the dissolution of water-soluble volatile organic compounds (VOCs) may be enhanced in fog drops and deliquesced particles and the aqueous-phase reactions of these compounds may generate low-volatility species that remain in the particle phases after water evaporates. The aqueous-phase SOA production was observed based on single particle mass spectrometry in a fog event in London (Dall'Osto *et al.*, 2009).

The mass concentrations of NO_3^- and SO_4^{2-} were higher in the lower visibility condition than in the higher visibility condition. When visibility is >5 km, the percentage of NO_3^- and SO_4^{2-} were 14% and 13%, while 17% and 25% with visibility <2 km condition (Quan *et al.*, (2014), with NH_4^+ a further water-soluble ion contributing to haze formation (Yang *et al.*, (2015). Some online research results, which focused on non-refractory submicron particles ($NR\text{-}PM_1$), also emphasised the important role of secondary pollutants during severe pollution events ($PM_{2.5}$ >75 μg m^{-3}) in this month.

For example, Zhang *et al.* (2014) found that between unpolluted days and polluted days, the contributions of three major secondary inorganic species (SO_4^{2-}, NO_3^- and NH_4^+) increased from 42% to 59%. For the organic species, the contribution of secondary species increased from 35% to 63%. Although, the total contribution of secondary species increased significantly, their respective effects are different. Taking three inorganic species (SO_4^{2-}, NO_3^- and NH_4^+) as an example, Han *et al.* (2016) suggested the contributions of SO_4^{2-} and NH_4^+ were higher than NO_3^-. Meanwhile, with the aggravation of air quality, the $[NO_3^-]/[SO_4^{2-}]$ mass ratio decreased, indicating that the SO_2 emitters contributed more to form $PM_{2.5}$.

In addition, the important role of secondary pollutants during the severe pollution process almost happened in other cities in China. For example, Huang *et al.* (2014) investigated the chemical nature and sources of particulate matter at urban locations in Beijing and other mega cities in China during January 2013. They find that the severe haze pollution event was driven to a large extent by secondary aerosol formation, which contributed 30–77% and 44–71% (average for all cities) of $PM_{2.5}$ and of organic aerosol, respectively.

The rapid transformation of primary gaseous pollutants to secondary aerosols seems important in haze formation and contributes to the "explosive" and "sustained growth" of $PM_{2.5}$ (Wang *et al.*, 2013). Gaseous pollutants are transformed into secondary aerosols through heterogeneous reactions on the surface of fine particles, altering particle size and chemical composition. Zheng *et al.* (2015) incorporated several heterogeneous reactions on aerosol surface into the CMAQ model and reached more realistic predictions of sulphate and nitrate, underscoring the importance of heterogeneous chemistry during the 2013 episode. Such reactions mean that the proportion of secondary inorganic ions, such as sulphate and nitrate, gradually increased in $PM_{2.5}$, which enhances particle hygroscopicity and thereby accelerating the formation of haze pollution (also (Quan *et al.* 2014; Zheng *et al.* 2015).

14.3.3 *Regional transport*

Regional transport has been identified as an important contribution to the haze pollution in Beijing, especially with air masses originating from the east and south of Beijing. A large number of polluting industries (such as power plants, steel mills, cement mills and petrochemical industry) and several heavily polluted cities (such as Langfang, Tianjin, Baoding, Shijiazhuang and Jinan) are located in the east and south of Beijing (Zhang *et al.*, 2016). Based on the characteristics of saw-tooth cycles, Jia *et al.* (2008) estimated that regional transport on average contributed approximately 50% of PM (up to 70%) during the southerly flow. Based on an aerodyne aerosol chemical speciation monitor (ACSM) observation, Sun *et al.* (2014) evaluated the regional contributions during the peak of episode on 12 January and found that regional transport on average contributed 66% of PM_1 during the pollution process and the contributions varied by different chemical species, among which sulphate and coal combustion induced organic aerosol presented the largest regional contributions, accounting for 75% and 84%, respectively.

The Nested Air Quality Prediction Model System (NAQPMS) shows that the cross-city clusters transport outside Hebei-Beijing-Tianjin (HBT) and transport among cities inside HBT contributed 20–35% and 26–35% of $PM_{2.5}$ as compared with local emission, in HBT, respectively (Wang *et al.* 2014). Regional transport is important in haze formation and it is relevant the proportion of southerly winds increased from 58% in January 2012 to 63% in January 2013 (Yang *et al.* 2015), bring air from industrial areas to the south of Beijing. Zheng *et al.* (2015) found that high aerosol concentration was a regional phenomenon. The accumulation process of aerosol particles occurred successively from cities southeast of Beijing. The apparent sharp increase in $PM_{2.5}$ concentrations (up to several hundred μg m^{-3} h^{-1}) recorded in Beijing represented rapid recovery from an interruption to the continuous pollution accumulation over the region, rather than pure local chemical production.

14.3.4 *Primary source contribution*

14.3.4.1 *Heat plant and domestic coal burning*

Coal has been the primary energy source in China, accounting for nearly 70% of the total energy consumption in 2005. The use of coal in China ranges from power and heat plants, industries to domestic household (Chen and Xu, 2010). Winter time central heating in northern Chinese cities consumes large quantities of coal (Zhang *et al.*, 2016) and their emissions contributed to the seasonal occurrences of pollution episodes in winter and early spring (Ma *et al.*, 2011). Tao *et al.* (2014) estimated more than a quarter of PM pollutant comes from coal combustion in winter in Beijing. Using aerosol mass spectrometry method, Zhang *et al.* (2016) found the coal combustion OA (CCOA) was the largest primary source, on average accounting for 20–32% of OA during January 2013 pollution episode in Beijing.

During severe pollution episode of December 2015, a "red alert" was in Beijing issued for the first time and strict traffic control was implemented which allowed only cars to drive on alternate days under odd-even license policy. The policy was first implemented in the 2008 Beijing Olympic Games and later extended during "red alert" pollution days since 2015. Source appointment revealed that over 60% of PM came from primary pollutants emitted during coal combustion or secondary pollutants derived from the coal burning related emission of precursors. This is a clear evidence of the role of coal burning as primary emission. During the episode period in December 2015, $PM_{2.5}$ nitrate concentration dropped significantly while coal combustion related sulphate surged confirming the important role of domestic coal combustion.

Domestic household coal burning may also play a role in severe episodes during winter heating seasons in northern Chinese cities, but there is no direct evidence to support their causal relation due to their nature of scattering point source in households. In September 2014, the Chinese Research Academy of Environmental Science (CRAES) initiated a survey in rural area of Baoding, Hebei province, near Beijing and collected the energy-use data from 543 families in five villages. The results show coal is still the dominant source of energy in rural areas accounting for over 80% of the total consumption, followed by 10% as electricity and 5% as liquefied petroleum gas, while biomass combustion by firewood and straw usage only amounts to 5%. Estimated PM and SO_2 emissions from rural bulk coal combustion were 5.4×10^4 and 11.2×10^4 tonnes, respectively, exceeding the combined emissions of industrial and township domestic emissions in the emission inventory (MEP, 2013), clearly indicating the rural bulk coal combustion needs special attention in pollution control to improve air quality.

14.3.4.2 *Vehicle emissions*

Source apportionment studies also suggested vehicle emissions as possible local sources during severe episodes. However, there has been great inconsistency of the results reported by different groups, especially for the last few heavy pollution processes in October 2010, January 2013 and February 2014. For example, Zhang *et al.* (2013) estimated only 4% of the primary PM in Beijing was attributed to the motor vehicle emissions by PMF method, while many other studies [literature] indicated that the contribution may be between 5% and 25%. Official reports summarising real-time aerosol mass spectrometer data and chemical mass balance results suggest about 30% of local $PM_{2.5}$ concentration was attributed to the motor vehicle emissions. The controversial literature reports well represent the complexity of the PM pollution in Beijing especially during episodic periods in terms of its temporospatial variation and source contributions.

14.3.4.3 *Dust superposition*

The surface monitoring data indicated that PM concentrations were very high on 12 and 13 January over the Jing-Jin-Ji area. Satellite data analysis reveals that dust and polluted layers existed above the boundary layer over the Jing-Jin-Ji area. Meanwhile, Tao *et al.* (2014) found that the northwest winds in high altitudes with dust particles on the top boundary layer met southerly airflows, which not only transport industry pollutants to Beijing but also cause widespread haze pollution when moist air masses exist. Moreover, polluted continental aerosol, polluted dust, and smoke existed in the boundary layer on 12th January. Back trajectories revealed west upper air flow and slight south wind in the lower atmosphere. The results of the surface network monitoring indicated the hourly average values of the $PM_{2.5}$ concentration reached its maximums of 250 μg m^{-3} and 260 μg m^{-3} at Dunhuang and Shapotou, at 23:00 LT 11 January and 05:00 LT 12 January, respectively. Thus, the dust particles were carried into the Jing-Jin-Ji area by a northwest air flow, crossed the Taihang Mountains, and mixed with the local urban pollutants, which contributed to rare maximums of the hourly average values of the $PM_{2.5}$ concentrations, especially for Beijing, Shijiazhuang and Tianjin. Therefore, dust superposition may as well be an important factor for PM episode.

14.4 Public Health Implications

Air pollution has been long demonstrated to be closely linked with adverse health effects. Since Beijing made available the official daily monitoring data in the late 1990s, many studies have investigated the impact of air pollution on public health in China (Zhang *et al.*, 2010; Cheng *et al.*, 2013). Prior to March 2012, $PM_{2.5}$ was not included in Chinese air quality standards (Dominici and Mittleman, 2012), so many short-term health studies have primarily focused on the effect of PM_{10}. Wide-spread respiratory irritation symptoms associated with the high $PM_{2.5}$ levels in pollution episode (Zhang *et al.*, 2015), especially in winter time. Air pollution-induced respiratory distress became even commonly recognised by the public that a phrase *Beijing cough* was coined by Beijing-based expats as early as 1990s, to describe the "sporadic, dry cough or tickle in the throat that usually lasts from December through April", and it has now become a popular buzz word among local residents and visitors in Beijing (Ouyang, 2013).

During the January 2013 pollution episode, a statistically significant increase in hospital visits, including emergency room visits, outpatient visits and hospital admissions was observed (Chen *et al.*, 2013). The surge of the daily hospital visits waned down after the heavy smog, consistent with the trend of daily mortality during London's 1952 Great Smog (Brimblecombe, 2011). In addition, estimated "each 10 μg m^{-3} increase of PM$_{10}$ was associated with a 1.0% increase in emergency room visits, a 0.7% increase in outpatient visits, and a 3.9% increase in hospital admissions in Beijing" (Chen *et al.*, 2013). Xu *et al.* (2016) investigated a longer time series of PM$_{2.5}$ data and emergency room visits to ten general hospitals in six urban districts in Beijing from 1 January to 31 December, 2013 and found strong evidence of the association of PM$_{2.5}$ with the total and cause-specific respiratory diseases, including upper respiratory tract infection (URTI), lower respiratory tract infection (LRTI) and acute exacerbation of chronic obstructive pulmonary disease (AECOPD). Expanding the study area to a larger region including Beijing and adjacent Tianjin and Hebei Province, Zhou *et al.* (2015) found smog episodes were consistently and statistically significantly associated with higher total mortality and mortality from cardiovascular/respiratory diseases. The associations are significant among rural residents who live close to big and polluted cities, while less significant for urban residents.

Studies have also shown the biological fraction of atmospheric particles, although only a small fraction in PM mass, could potentially cause significant health effects beyond the chemical constituents induced particle toxicity (Boreson *et al.*, 2004). Particle-bound bacteria and fungi may have direct effects on the human health or trigger allergic respiratory symptoms such as asthma (Vermani *et al.*, 2010; Haas *et al.*, 2013). By comparing typical haze and non-haze days in Beijing, Gao *et al.* (2015) found there was negative association between the viable bacterial and fungi concentration with the pollution levels. They attributed this to the increased particle bound toxic and hazardous constituents reducing the viability of biotic components during haze days. Metagenomic analysis of microbial composition of the PM during the January episode suggests that the inhalable microorganisms were found to be soil-associated and non-pathogenic to human (Cao *et al.*, 2014).

14.5 Policy Implications

The occurrence of the January 2013 Beijing episode received wide attention in national and international scale. It provided a unique opportunity to understand the pollution sources, formation mechanisms, role of meteorology and impact on public health. More importantly, the various research investigations that followed the episode also had great implications on policy making on emission control strategies including changing the energy source structure, reducing local dust emissions, controlling vehicle emissions, and relocating major industrial emitters. The contribution of regional transport to the pollution episode indicates the importance of establishing a regional joint framework of policymaking and action system to improve regional air quality more effectively. Wang *et al.* (2013) pointed out potential problems and difficulties in the existing frame work of regional jointly policy-making (e.g. lack of provincial level coordination for regional pollution control efforts), control strategies, and effective enforcement (e.g. lack of a mandatory implementation action system). The large fraction of secondary components, i.e. sulphate, nitrate and organic materials in PM, implies an urgent policy need for the source control of precursors to secondary aerosol formation, which has not been the main focus in earlier policy targets which addressed mostly control of primary emissions. The underestimated role of domestic coal burning in severe winter pollution episodes requires a better understanding of their emission inventory and advancement of comprehensive control technologies. In March 2016, the MEP of China issued consultative documents on the guideline of establishing domestic coal emission inventory and coal emission control technologies, implying the policy to tackle the pollution from the coal combustion is likely to be imminent.

Nevertheless, the occurrence of severe pollution episodes led the government to invest more efforts and revamp the policy on the emergency prevention and control work on the heavy pollution process. The Ministry of Environmental Protection of China (MEP) issued a guideline in May 2013, just months after the episode, for the compilation of urban air pollution emergency plan, which is used to, firstly, promote the local governments' capability of forecasting, in order to give early warnings and develop emergency responses to severe air pollution. The guideline also provides a reference for local government to compile corresponding contingency plans. On 16th September 2013, an action plan for the control of air pollution was released by the State Council of China as the guideline for air pollution prevention and control on a national scale. The action plan set three ambitious objectives for the next five years including national scale overall air quality improvement, significant reduction of pollution episodes and significant air quality improvement in Jing-jin-ji region (Beijing, Tianjin and Hebei province), Yangtze River Delta and Pearl River Delta areas. The objectives have specific targets: (1) PM_{10} concentrations in prefecture-level cities or above to be reduced by more than 10% in 2017 compared to 2012; (2) continued increase of number of blue sky days; (3) $PM_{2.5}$ concentrations in Jing-jin-ji region, Yangtze River Delta and Pearl River Delta to be reduced by 25%, 20% and 15%, respectively, and the annual $PM_{2.5}$ concentrations in Beijing to be reduced below 60 μg m^3.

The enactment of action plans paved the way for rigorous air pollution control measures. On 29 August, 2015, the National People's Congress released the updated version of the 15-year-old *Law on Air Pollution Prevention and Control*, which dedicates a specific chapter on the response to air pollution episodes, including the establishment of the monitoring and forecasting system, and the compiling of the emergency plan. Since 2013, more than 30 cities in China have promulgated *Air Pollution Emergency Contingency Actions* to protect the public health from pollution episodes.

14.6 Conclusions

This event was monumental, because in its wake, the government saw the importance of a tighter and more science-driven approach towards air quality management. In parallel, there was great public awareness and a desire for a more transparent approach and access to real-time monitoring data.

References

BEPB. (2013). *A Review of Air Pollution Control in Beijing* (China: Beijing Environmental Protection Bureau).

Boreson, J., Dillner A. M. and Peccia, J. (2004). Correlating bioaerosol load with $PM_{2.5}$ and PM10cf concentrations: A comparison between natural desert and urban-fringe aerosols, *Atmospheric Environment*, **38**(35), 6029–6041.

Brimblecombe, P. (2011). *The Big Smoke: A History of Air Pollution in London since Medieval Times* (Routledge, London).

Brimblecombe, P. (ed.) (2017). London 1952: An enduring legacy, *Air Pollution Episodes* (World Scientific, Singapore).

Cao, C., Jiang, W. J., Wang, B. Y., Fang, J. H., Lang, J. D., Tian, G., Jiang, J. K. and Zhu, T. F. (2014). Inhalable microorganisms in Beijing's $PM_{2.5}$ and PM_{10} pollutants during a severe smog event, *Environmental Science & Technology*, **48**(3), 1499–1507.

Chen, C., Sun, Y. L., Xu, W. Q., Du, W., Zhou, L. B., Han, T. T., Wang, Q. Q., Fu, P. Q., Wang, Z. F., Gao, Z. Q., Zhang, Q. and Worsnop, D. R. (2015). Characteristics and sources of submicron aerosols above the urban canopy (260 m) in Beijing, China, during the 2014 APEC summit, *Atmospheric Chemistry and Physics*, **15**(22), 12879–12895.

Chen, R., Zhao, Z. and Kan, H. (2013). Heavy smog and hospital visits in Beijing, China, *American Journal of Respiratory and Critical Care Medicine*, **188**(9), 1170–1171.

Chen, W. and Xu, R. (2010). Clean coal technology development in China, *Energy Policy*, **38**(5), 2123–2130.

Cheng, Z., Jiang, J. K., Fajardo, O., Wang, S. X. and Hao, J. M. (2013). Characteristics and health impacts of particulate matter pollution in China (2001–2011), *Atmospheric Environment*, **65**, 186–194.

Dall'Osto, M., Harrison, R. M., Coe, H. and Williams, P. (2009). Real-time secondary aerosol formation during a fog event in London, *Atmospheric Chemistry and Physics*, **9**(7), 2459–2469.

Ding, Y. and Liu, Y. (2014). Analysis of long-term variations of fog and haze in China in recent 50 years and their relations with atmospheric humidity, *Science China Earth Sciences*, **57**(1), 36–46.

Dominici, F. and Mittleman, M. A. (2012). China's air quality dilemma: Reconciling economic growth with environmental protection, *Journal of the American Medical Association*, **307**(19), 2100–2102.

Gao, M., Qiu, T., Jia, R., Han, M., Song, Y. and Wang, X. (2015). Concentration and size distribution of viable bioaerosols during non-haze and haze days in Beijing, *Environmental Science and Pollution Research*, **22**(6), 4359–4368.

Ge, X. L., Zhang, Q., Sun, Y. L., Ruehl, C. R. and Setyan, A. (2012). Effect of aqueous-phase processing on aerosol chemistry and size distributions in Fresno, California, during wintertime, *Environmental Chemistry*, **9**(3), 221–235.

Guo, S., Hu, M., Zamora, M. L., Peng, J. F., Shang, D. J., Zheng, J., Du, Z. F., Wu, Z., Shao, M., Zeng, L. M., Molina, M. J. and Zhang, R. Y. (2014). Elucidating severe urban haze formation in China, *Proc Natl Acad Sci USA*, **111**, 17373–17378.

Haas, D., Galler, H., Luxner, J., Zarfel, G., Buzina, W., Friedl, H., Marth, E., Habib, J. and Reinthaler, F. F. (2013). The concentrations of culturable microorganisms in relation to particulate matter in urban air, *Atmospheric Environment*, **65**, 215–222.

Han, B., Zhang, R., Yang, W., Bai, Z., Ma, Z. and Zhang, W. (2016). Heavy haze episodes in Beijing during January 2013: Inorganic ion chemistry and source analysis using highly time-resolved measurements from an urban site, *Science Total Environment*, **544**, 319–329.

Ho, K.-F., Ho, S. S. H., Huang, R.-J., Chuang, H.-C., Cao, J.-J., Han, Y., Lui, K.-H., Ning, Z., Chuang, K.-J., Cheng, T.-J., Lee, S.-C., Hu, D., Wang, B. and Zhang, R. (2016). Chemical composition and bioreactivity of $PM_{2.5}$ during 2013 haze events in China, *Atmospheric Environment*, **126**, 162–170.

Huang, R. J., Zhang, Y., Bozzetti, C., Ho, K. F., Cao, J. J., Han, Y., Daellenbach, K. R., Slowik, J. G., Platt, S. M., Canonaco, F., Zotter, P., Wolf, R., Pieber, S. M., Bruns, E. A., Crippa, M., Ciarelli, G., Piazzalunga, A., Schwikowski, M., Abbaszade, G., Schnelle-Kreis J., Zimmermann, R., An, Z., Szidat, S., Baltensperger, U., El Haddad, I. and Prevot, A. S.,

(2014). High secondary aerosol contribution to particulate pollution during haze events in China, *Nature*, **514**(7521), 218–222.

Ji, D., Li, L., Wang, Y., Zhang, J., Cheng, M., Sun, Y., Liu, Z., Wang, L., Tang, G., Hu, B., Chao, N., Wen, T. and Miao, H. (2014). The heaviest particulate air-pollution episodes occurred in northern China in January, 2013: Insights gained from observation, *Atmospheric Environment*, **92**, 546–556.

Jia, Y., Rahn, K. A., He, K., Wen, T. and Wang, Y. (2008). A novel technique for quantifying the regional component of urban aerosol solely from its sawtooth cycles, *Journal of Geophysical Research*, **113**(D21), D21309.

Ma, N., Zhao, C. S., Nowak, A., Müller, T., Pfeifer, S., Cheng, Y. F., Deng, Z. Z., Liu, P. F., Xu, W. Y., Ran, L., Yan, P., Göbel, T., Hallbauer, E., Mildenberger, K., Henning, S., Yu, J., Chen, L. L., Zhou, X. J., Stratmann, F. and Wiedensohler, A. (2011). Aerosol optical properties in the North China Plain during HaChi campaign: an in-situ optical closure study, *Atmospheric Chemistry and Physics*, **11**(12), 5959–5973.

MEP. (2013). China Environmental Statistics Yearbook, *Ministry of Environmental Protection of the People's Republic of China*.

Min, W. and Zhen, X. (2010). Mirroring the Olympic Games — The Beijing 2008 Olympic Games in the American Media, *The International Journal of the History of Sport*, **27**(9–10), 1794–1808.

Ouyang, Y. (2013). China wakes up to the crisis of air pollution, *The Lancet Respiratory Medicine*, **1**(1), 12.

Qu, W., Wang, J., Gao, S. and Wu, T. (2013). Effect of the strengthened western Pacific subtropical high on summer visibility decrease over eastern China since 1973, *Journal of Geophysical Research: Atmospheres*, **118**(13), 7142–7156.

Quan, J., Gao, Y., Zhang, Q., Tie, X., Cao, J., Han, S., Meng, J., Chen, P. and Zhao, D. (2013). Evolution of planetary boundary layer under different weather conditions, and its impact on aerosol concentrations, *Particuology*, **11**(1), 34–40.

Quan, J., Liu, Q., Li, X., Gao, Y., Jia, X., Sheng, J. and Liu, Y. (2015). Effect of heterogeneous aqueous reactions on the secondary formation of inorganic aerosols during haze events, *Atmospheric Environment*, **122**, 306–312.

Quan, J., Tie, X., Zhang, Q., Liu, Q., Li, X., Gao, Y. and Zhao, D. (2014). Characteristics of heavy aerosol pollution during the 2012–2013 winter in Beijing, China, *Atmospheric Environment,* **88**, 83–89.

Rich, D. Q., Kipen, H. M., Huang, W. *et al.* (2012). Association between changes in air pollution levels during the beijing olympics and biomarkers of inflammation and thrombosis in healthy young adults, *Journal of the American Medical Association,* **307**(19), 2068–2078.

Streets, D. G., Fu, J. S., Jang, C. J., Hao, J., He, K., Tang, X., Zhang, Y., Wang, Z., Li, Z., Zhang, Q., Wang, L., Wang, B. and Yu, C. (2007). Air quality during the 2008 Beijing Olympic Games, *Atmospheric Environment,* **41**(3), 480–492.

Sun, Y. L., Jiang, Q., Wang, Z. F., Fu, P. Q., Li, J., Yang, T. and Yin, Y. (2014). Investigation of the sources and evolution processes of severe haze pollution in Beijing in January 2013, *Journal of Geophysical Research: Atmospheres,* **119**, 4380–4398.

Sun, Y. L., Wang, Z. F., Fu, P. Q., Jiang, Q., Yang, T., Li, J. and Ge, X. L. (2013). The impact of relative humidity on aerosol composition and evolution processes during wintertime in Beijing, China, *Atmospheric Environment,* **77**, 927–934.

Tao, M., Chen, L., Xiong, X., Zhang, M., Ma, P., Tao, J. and Wang, Z. (2014). Formation process of the widespread extreme haze pollution over northern China in January 2013: Implications for regional air quality and climate, *Atmospheric Environment,* **98**, 417–425.

Tian, S., Pan, Y., Liu, Z., Wen, T. and Wang, Y. (2014). Size-resolved aerosol chemical analysis of extreme haze pollution events during early 2013 in urban Beijing, China, *Journal of Hazardous Materials,* **279**, 452–460.

Vermani, M., Vijayan, V. K., Kausar, M. A. and Agarwal, M. K. (2010). Quantification of Airborne Aspergillus Allergens: Redefining The Approach, *Journal of Asthma,* **47**(7), 754–761.

Wang, H., Xu, J., Zhang, M., Yang, Y., Shen, X., Wang, Y., Chen, D. and Guo, J. (2014). A study of the meteorological causes of a prolonged and severe haze episode in January 2013 over central-eastern China. *Atmospheric Environment,* **98**, 146–157.

Wang, L. T., Wei, Z., Yang, J., Zhang, Y., Zhang, F. F., Su, J., Meng, C. C. and Zhang, Q. (2013a). The 2013 severe haze over the southern

Hebei, China: Model evaluation, source apportionment, and policy implications, *Atmospheric Chemistry and Physics Discussions*, **13**(11), 28395–28451.

Wang, Q., Cao, J., Shen, Z., Tao, J., Xiao, S., Luo, L., He, Q. and Tang, X. (2013b). Chemical characteristics of PM2.5 during dust storms and air pollution events in Chengdu, China, *Particuology*, **11**(1), 70–77.

Wang, Y. S., Yao, L., Wang, L. L., Liu, Z. R., Ji, D. S., Tang, G. Q., Zhang, J. K., Sun, Y., Hu, B. and Xin, J. Y. (2013). Mechanism for the formation of the January 2013 heavy haze pollution episode over central and eastern China, *Science China Earth Sciences*, **57**(1), 14–25.

Xu, Q., Li, X., Wang, S., Wang, C., Huang, F., Gao, Q., Wu, L., Tao, L., Guo, J., Wang, W. and Guo, X. (2016). Fine particulate air pollution and hospital emergency room visits for respiratory disease in urban areas in Beijing, China, in 2013, *PLoS ONE*, **11**(4), e0153099.

Yang, Y., Liu, X., Qu, Y., Wang, J., An, J., Zhang, Y. and Zhang, F. (2015). Formation mechanism of continuous extreme haze episodes in the megacity Beijing, China, in January 2013, *Atmospheric Research*, **155**, 192–203.

Zhang, F., Xu, L., Chen, J., Chen, X., Niu, Z., Lei, T., Li, C. and Zhao, J. (2013). Chemical characteristics of $PM_{2.5}$ during haze episodes in the urban of Fuzhou, China, *Particuology*, **11**(3), 264–272.

Zhang, J. F., Mauzerall, D. L., Zhu, T., Liang, S., Ezzati, M. and Remais, J. V. (2010). Environmental health in China: Progress towards clean air and safe water, *Lancet*, **375**(9720), 1110–1119.

Zhang, J. K., Cheng, M. T., Ji, D. S., Liu, Z. R., Hu, B., Sun, Y. and Wang, Y. S. (2016). Characterization of submicron particles during biomass burning and coal combustion periods in Beijing, China, *Science of the Total Environment*, **562**, 812–821.

Zhang, J. K., Sun, Y., Liu, Z. R., Ji, D. S., Hu, B., Liu, Q. and Wang, Y. S. (2014). Characterization of submicron aerosols during a month of serious pollution in Beijing, 2013, *Atmospheric Chemistry and Physics*, **14**(6), 2887–2903.

Zhang, J. P., Zhu, T., Zhang, Q. H., Li, C. C., Shu, H. L., Ying, Y., Dai, Z. P., Wang, X., Liu, X. Y., Liang, A. M., Shen, H. X. and Yi, B. Q. (2012). The impact of circulation patterns on regional transport pathways and air quality over Beijing and its surroundings, *Atmospheric Chemistry and Physics*, **12**(11), 5031–5053.

Zhang, L., Wang, T., Lv, M. and Zhang, Q. (2015a). On the severe haze in Beijing during January 2013: Unraveling the effects of meteorological anomalies with WRF-Chem, *Atmospheric Environment*, **104**, 11–21.

Zhang, Q., Qiu, M., Lai, K. and Zhong, N. (2015b). Cough and environmental air pollution in China, *Pulmonary Pharmacology & Therapeutics*, **35**, 132–136.

Zhang, R., Jing, J., Tao, J., Hsu, S. C., Wang, G., Cao, J., Lee, C., S. L., Zhu, L., Chen, Z., Zhao, Y. and Shen, Z. (2013). Chemical characterization and source apportionment of $PM_{2.5}$ in Beijing: seasonal perspective, *Atmospheric Chemistry and Physics*, **13**(14), 7053–7074.

Zhang, R. H., Li, Q. and Zhang, R. N. (2014). Meteorological conditions for the persistent severe fog and haze event over eastern China in January 2013, *Science China Earth Sciences*, **57**(1), 26–35.

Zheng, B., Zhang, Q., Zhang, Y., He, K. B., Wang, K., Zheng, G. J., Duan, F. K., Ma, Y. L. and Kimoto, T. (2015). Heterogeneous chemistry: A mechanism missing in current models to explain secondary inorganic aerosol formation during the January 2013 haze episode in North China, *Atmospheric Chemistry and Physics*, **15**(4): 2031–2049.

Zheng, G., Duan, F., Ma, Y., Zhang, Q., Huang, T., Kimoto, T., Cheng, Y., Su, H. and He, K. (2016). Episode-based evolution pattern analysis of haze pollution: method development and results from Beijing, China, *Environmental Science and Technology*, **50**(9), 4632–4641.

Zheng, G. J., Duan, F. K., Su, H., Ma, Y. L., Cheng, Y., Zheng, B., Zhang, Q., Huang, T., Kimoto, T., Chang, D., Pöschl, U., Cheng, Y. F. and He, K. B. (2015). Exploring the severe winter haze in Beijing: The impact of synoptic weather, regional transport and heterogeneous reactions, *Atmospheric Chemistry and Physics*, **15**(6), 2969–2983.

Zhou, M., He, G., Fan, M., Wang, Z., Liu, Y., Ma, J., Ma, Z., Liu, J., Liu, Y., Wang, L. and Liu, Y. (2015). Smog episodes, fine particulate pollution and mortality in China, *Environmental Research*, **136**, 396–404.